SUBSURFACE MIGRATION
OF
HAZARDOUS WASTES

SUBSURFACE MIGRATION
OF
HAZARDOUS WASTES

Joseph S. Devinny, Ph.D.
University of Southern California, Los Angeles

Lorne G. Everett, Ph.D.
Metcalf & Eddy, Santa Barbara, California

James C. S. Lu, Ph.D.
Calscience Engineering and Laboratories, Cypress, California

Robert L. Stollar
R. L. Stollar and Associates, Inc., Santa Ana, California

Environmental Engineering Series

VNR VAN NOSTRAND REINHOLD
New York

Copyright © 1990 by Van Nostrand Reinhold
Library of Congress Catalog Card Number 89-16670
ISBN 0-442-21868-0

Printed in the United States of America

Van Nostrand Reinhold
115 Fifth Avenue
New York, New York 10003

Van Nostrand Reinhold International Company Limited
11 New Fetter Lane
London EC4P 4EE, England

Van Nostrand Reinhold
480 La Trobe Street
Melbourne, Victoria 3000, Australia

Nelson Canada
1120 Birchmount Road
Scarborough, Ontario M1K 5G4, Canada

Library of Congress Cataloging-in-Publication Data

Devinny, Joseph S.
    Subsurface migration of hazardous wastes / Joseph S. Devinny, Lorne G.
  Everett, James C.S. Lu, Robert L. Stollar.
      p.   cm.—(Environmental engineering series)
    Includes bibliographical references.
    ISBN 0-442-21868-0
    1. Hazardous wastes—Environmental aspects.   2. Water,
  Underground—Pollution.   I. Everett, Lorne G.   II. Lu, James C.S.   III. Stollar, Robert L.
  IV. Title.   V. Series: Environmental engineering series (New York, N.Y.)

    TD427.H3D48   1989
    628.1'68—dc20
                                                                            89-16670
                                                                            CIP

—VNR ENVIRONMENTAL ENGINEERING SERIES—

Nelson L. Nemerow, Consulting Editor

Water Pollution Control

HANDBOOK OF NONPOINT POLLUTION by Vladimir Novotny

CONTROL AND TREATMENT OF COMBINED-SEWER OVERFLOWS, edited by Peter E. Moffa

WATER AND WASTEWATER TREATMENT FOR SMALL COMMUNITIES by Martin and Martin (Spring 1990)

INDUSTRIAL AND HAZARDOUS WASTE TREATMENT by Nelson Nemerow and Avijit Dasgupta (Fall 1990)

Water Resources Development

CURRENT TRENDS IN WATER-SUPPLY PLANNING by David W. Prasifka

HANDBOOK OF PUBLIC WATER SYSTEMS by Culp, Wesner, and Culp

HANDBOOK OF CHLORINATION, 2nd Ed., by Geo. Clifford White

ANALYSIS OF WATER DISTRIBUTION SYSTEMS by Thomas M. Walski

WATER CLARIFICATION PROCESSES by Herbert E. Hudson, Jr.

DESIGN AND CONSTRUCTION OF WATER WELLS by the National Water Well Association

CORROSION MANAGEMENT IN WATER SUPPLY SYSTEMS by W. Harry Smith

DISINFECTION ALTERNATIVES FOR SAFE DRINKING WATER by Ted Bryant and George Fulton (Fall 1990)

Solid Waste Management

SMALL-SCALE MUNICIPAL SOLID WASTE ENERGY RECOVERY SYSTEMS by Gershman, Brickner and Bratton, Inc.

Hazardous Waste Treatment

GROUNDWATER TREATMENT TECHNOLOGY by Evan Nyer

HAZARDOUS WASTE SITE REMEDIATION by O'Brien and Gere, Inc.

SUBSURFACE MIGRATION OF HAZARDOUS WASTES by Joseph S. Divinny, Lorne G. Everett, James C.S. Lu and Robert L. Stollar

General Environmental

PROJECT PLANNING AND MANAGEMENT: AN INTEGRATED SYSTEM FOR IMPROVING PRODUCTIVITY by Louis J. Goodman

COMPUTER MODELS IN ENVIRONMENTAL PLANNING by Stephen I. Gordon

ENVIRONMENTAL PERMITS by Donna C. Rona

BEYOND GLOBAL WARMING by A. John Appleby (Spring 1990)

Preface

Groundwater contamination has caught the nation unprepared. We have been surprised by its frequency, severity, and extent. Water supply experts, hydrogeologists, and environmental engineers have been overwhelmed by the size of a problem that was virtually unrecognized ten years ago. Estimates of cleanup costs have climbed astonishingly.

We failed to anticipate the seriousness of the problems for many reasons. Casual or accidental dumping of toxic substances had not been considered important, so no effort was made to record occurrences, and there was no general awareness of its ubiquity. People often failed to recognize the toxicity of the chemicals released.

But the central oversight was the failure to perceive the importance of the mobility of chemicals in soils and groundwater. Substances dumped in one place often appear in another. A leaking tank at a gas station can contaminate not only the water below, but that under the house across the street. A small industrial site can foul the water supply of a whole town.

Part of this failure to foresee comes from the interdisciplinary nature of the problem. Anticipating the subsurface migration of hazardous wastes requires knowledge from hydrology, chemistry, microbiology, soil science, and environmental engineering. There were very few individuals whose backgrounds spanned these fields. The hydrologists knew how water moved, but were little acquainted with chemistry and microbiology. Indeed, they may have drawn false comfort from the general impression that passage through soil tends to purify water. The microbiologists and chemists knew something of each other's craft, but were only vaguely aware of groundwater flow. Soil scientists understood the chemistry and biology of soil microbial ecosystems, but had little interest in the engineering of waste disposal and water supplies.

As the urgency of groundwater contamination problems has been recognized, all of these workers have seen the need to learn from other fields of expertise.

The interdisciplinary nature of work on the subsurface migration of hazardous wastes is reflected in the collection of authors for this book. We are: an environmental engineer specializing in chemistry and microbiology, an expert in vadose zone phenomena, a groundwater hydrologist, and a practitioner of hazardous waste management.

It is our hope that, by contributing our special skills and learning from each other, we have created a book that covers the breadth of knowledge pertinent to soil and groundwater contamination.

ACKNOWLEDGMENTS

I would like to give special appreciation to Jennifer Gregory who researched, compiled, and composed groundwater monitoring data and information for Chapter 6. Also, appreciation is given to Berge Basmajian for the preparation of the graphic material for the groundwater chapters. I would like to thank Susan Garcia for her help in researching material for Chapter 6 and special thanks to Mary Hume Jolly for reviewing and editing Chapters 3 and 6. Also, I would like to thank Linda Wade and Peggy Phillips for their typing assistance, as well as Catherine Laub for her overall assistance in this project.

Robert L. Stollar

The author wishes to acknowledge the financial and technical assistance provided by the Cooperative Agreement between the U.S. Environmental Protection Agency, Las Vegas Laboratory, and the University of California, Santa Barbara, Institute for Crustal Studies in supporting the preparation of Chapters 7 and 8 (ICS-0037-07HW and ICS-0038-08HW). Special recognition is extended to Mr. Larry A. Eccles, EPA project officer, for his insightful technical review.

Lorne G. Everett, Ph. D.

CONTENTS

Preface   vii

Acknowledgments   ix

CHAPTER 1   Introduction   1
            Joseph S. Devinny and *James C. S. Lu*

The Nature of Subsurface Pollution   1
    The Results of Subsurface Pollution   3
    The Response to Subsurface Pollution   4
Sources of Subsurface Hazardous Waste   4
    Surface Impoundments   7
    Landfills   8
    Spills   9
    Tanks and Pipelines   9
    Septic Tanks   10
    Agriculture   10
    Urban Runoff   10
    Deep Well Injection   11
    Illegal Dumping   11
    Liquids from Solids   11
    Site Assessment   12
References   13

CHAPTER 2   The Composition of Hazardous Wastes   15
            Joseph S. Devinny

Phase   15
    Solids   15
    Liquids   17
    Gases   17
Chemical Composition   18
    Speciation   19
    Trace Metals   21
    Organic Chemicals   22
    Gross Parameters   26

Extractability	29
Sample Fractionation	30
Microbiological Contents	33
Is It Hazardous?	33
Classification	34
Extractable Toxicity	34
Leachate Composition	35
References	39

CHAPTER 3   Principles of Groundwater Flow   40
Robert L. Stollar

Introduction	40
Groundwater Resources	41
Occurrence of Groundwater	42
Aquifers	44
The Hydrologic Cycle	47
Aquifer Geology	49
Groundwater Storage	51
Porosity	52
Specific Yield	52
Specific Storage	55
Groundwater Flow	57
Heads and Gradients	57
Hydraulic Conductivity	63
Darcy's Law and Groundwater Velocity	66
Recharge and Discharge	70
Effects of Pumping	72
Groundwater Contamination	77
Site Investigations	92
Groundwater Models	100
Case Study	105
References	114

CHAPTER 4   Microbiology of Subsurface Wastes   116
Joseph S. Devinny

Soil Microbiology	116
Microbial Metabolism	117
Growth Kinetics	126
Growth Limitation	127
Processes in Soils	132
Microbiology and Subsurface Migration	134
Processes in Landfills	134
Decomposition of Toxic Substances	135

Products of Decomposition   136
Production of Liquids, Solids, and Gases   136
Promotion of Biodegradation   137
References   141

**CHAPTER 5   Chemical and Physical Alteration of Wastes   142
              and Leachates**
              Joseph S. Devinny

Dilution   142
Mixing Before Disposal   142
Mixing During Disposal   143
Mixing in Groundwater   144
Waste Interactions   145
Dissolution and Precipitation   146
Control of Dissolution   146
Precipitation   149
Effects on Groundwater Quality   150
Effects on Soils   150
Filtration and Adsorption   150
Filtration   150
Adsorption   151
Principles of Adsorption   151
Adsorption of Organic Chemicals on Soils   161
Adsorption of Trace Metals on Soils   163
Chemical Reactions   163
Acid-Base Reactions   163
Hydrolysis   163
Redox Reactions   164
Complexation   164
Photochemical Degradation   165
Volatilization   165
References   166

CHAPTER 6   Groundwater Monitoring   169
              Robert L. Stollar

Introduction   169
Designing a Groundwater Monitoring Program   169
Well Drilling and Soil Sampling   174
Drilling Methods for Monitoring Wells   174
Soil Sampling Techniques   187
Monitoring Well Completion   194
Material Selection for Groundwater Monitoring Well Casings
    and Screens   194
Well Screen Types   198

Pipe and Screen Casing Fitting Types   203
Filter Packs   204
Grouting Materials for Monitoring Wells   205
Well Completion   207
Well Development Methods   210
Waste and Wastewater Disposal Procedures   215
Water Quality Sampling Protocol   215
Field Preparation   216
Groundwater Quality Sampling Methods and Devices   226
Sampling Procedures   248
Analytical Process   262
Summary   264
References   264

CHAPTER 7   Soil Core Monitoring   267
              Lorne G. Everett

Vadose Zone Description   267
Soil Zone   267
Intermediate Unsaturated Zone   268
Capillary Fringe   269
Flow Regimes   270
General Equipment Classification   272
Sampling with Multipurpose Drill Rigs   272
Hand-Operated Drilling and Sampling Devices   278
Hand-Held Power Augers   283
Criteria for Selecting Soil Samplers   283
Capability for Obtaining Various Sample Types   284
Sampling Various Soil Types   286
Site Accessibility and Trafficability   286
Relative Sample Size   286
Personnel Requirements   287
Compositing Samples   287
Preliminary Activities   288
Sample Collection with Multipurpose Drill Rigs   289
Sample Collection with Hand-Operated Equipment   295
Miscellaneous Tools   300
Decontamination   300
Laboratory Cleanup of Sample Containers   300
Field Decontamination   300
Safety Precautions   301
References   302

CHAPTER 8   Soil Pore-Liquid Monitoring   306
              Lorne G. Everett

Soil Moisture/Tension Relationships   307
Pore-Liquid Sampling Equipment   308

Ceramic-Type Samplers   308
Cellulose-Acetate Hollow Fiber Samplers   313
Membrane Filter Samplers   314
Pan Lysimeters   315
Criteria for Selecting Soil Pore-Liquid Samplers   316
Preparation of the Samplers   317
Surveying in the Locations of Sites and Site Designations   317
Installation Procedures for Vacuum-Pressure Pore-Liquid Samplers   318
Installing Access Lines   318
Step-by-Step Procedures for Installing Vacuum-Pressure Pore-Liquid
   Samplers   318
Operation of Vacuum-Pressure Sampling Units   321
Porous Segments in Lysimeters   324
Dead Space in Lysimeters   326
Special Problems and Safety Precautions   328
Hydraulic Factors   328
Physical Properties: Soil Texture and Soil Structure   328
Cup-Wastewater Interactions   329
Climatic Factors   330
Safety Precautions   330
Lysimeter Failure Confirmation   330
Pan Lysimeter Installation and Operation   331
Trench Lysimeters   332
Free-Drainage Glass-Block Samplers   333
Pan Lysimeter Limitations   334
References   334

CHAPTER 9   Control of Subsurface Migration   337
         James C. S. Lu

Introduction   337
Overview   337
Selection of Control Alternatives   337
Contaminant Detection and Monitoring   340
Regulatory Control   340
Waste Modification   340
Pretreatment   341
In Situ Treatment   343
Sorbent Addition   344
Fixation   344
Excavation   345
Pre-Excavation Activities   345
Excavation and Disposal Plan   346
Air Monitoring Plan   347
Health and Safety Plan   347
Other Mitigation Procedures   347
Contaminated Site Control   347
Infiltration Control   348

Drainage and Water Diversion   348
Surface Sealing   353
Migration Barriers   355
Leachate Containment   355
Slurry Walls   356
Grout Curtains   358
Sheet Piling Cutoff Wall   361
Groundwater Diversion   361
Subsurface Drains   361
Drainage Ditches   362
Removal of Contaminated Groundwater   363
Leachate Removal   363
Well Point Systems   363
Deep Well Systems   364
Treatment of Contaminated Groundwater   364
Groundwater Treatment   364
In Situ Treatment   365
On-Site or Off-Site Treatment   368
Treatment Methods   369
Biological Treatment of Leachate   369
Physical and Chemical Treatment of Leachate   370
Leachate Treatment Systems   371
References   371

INDEX   375

SUBSURFACE MIGRATION
OF
HAZARDOUS WASTES

1

INTRODUCTION

Joseph S. Devinny
Associate Professor of Civil and Environmental Engineering
University of Southern California
and

James C. S. Lu
Calscience Engineering and Laboratories, Inc.
Cypress, California

THE NATURE OF SUBSURFACE POLLUTION

We have discarded harmful materials since humans first gathered to form societies. For most of our history, the bulk of hazardous waste was ordinary sewage, potent in its capability for the spread of deadly contagious disease. Then, as now, discharge to the soil or the nearest river put the problem out of sight and mind most conveniently. Humanity has learned to deal so effectively with the hazards of sewage (at least where there is money to do so), that it is no longer thought of as hazardous. But at the same time we have begun to generate an astonishing variety of chemicals with equally various potentials for harm. Particularly since the end of World War II, the production and disposal of such materials has risen tremendously. We have learned about the health effects such chemicals may have, and how to detect their presence at very low concentrations.

The technology of safe disposal has not kept pace with the technology of chemical synthesis and production. Too often we have used the ancient practice of dumping the waste on the nearest piece of land or river. The results are familiar to anyone who reads a newspaper or watches the 11 o'clock news: hazardous wastes have become a major technical, social, and political problem.

This book examines a particular aspect of this problem. Hazardous wastes disposed on land can move downward and away through the soil, spreading to contaminate a region extending far beyond the disposal site. Moving groundwater will transport the waste even further. Where the groundwater is a resource, it carries the toxins to the population, threatening public health. Discovery of the contamination ends use of the groundwater, and a valuable resource is lost.

Ultimately, humanity will learn to prevent contamination of the soil and groundwater. Chemicals will be watched and controlled, waste disposal will become organized, reasonably economical treatment methods will be developed, and industrial processes will be modified to minimize the generation of hazardous wastes. Substantial progress has been made towards these ends already. But for the immediate future, careless disposal will occur, and there is a large accumulation of contaminated sites which must be discovered and neutralized.

Because it is not easily examined, the environment beneath the surface of the soil is much less known than that around us on the surface. The public is almost entirely unaware of the phenomena which determine the fate of pollutants below ground. Even the most knowledgeable scientists struggle with the inevitable difficulties of collecting data by expensive remote methods which disturb and modify the samples they collect.

Except in swamps, the region immediately below the surface of the soil is the vadose zone, where most of the soil pores are filled with air. Water and its included pollutants pass down through this zone to the saturated zone, where the soil pores are filled with water. The upper boundary of the saturated zone is the "water table." A hole drilled deeper than the water table will fill with water and can function as a well. Water in the saturated zone is rarely still. It moves from source areas to discharges such as springs or rivers, carrying pollutants with it at rates of up to tens of feet per day.

Somewhere, at the base of the saturated zone, the water is bounded by an impermeable layer which limits the dynamic system. This bedrock may be under a few inches of soil on a mountainside, or many thousands of feet deep under thick layers of clay and sand deposited by ancient rivers.

This picture is an oversimplification. The soil and rock which make up the system are commonly nonuniform. Sand permits the easy passage of large volumes of water, while irregular layers and lenses of clay retard it. Layers of easily permeated material separated by an impermeable layer may split the flow into separate currents of different velocities (Sutton and Barker, 1985). Adjacent layers of differing permeabilities exchange contaminants, causing the contaminants to move slower than the water in one layer and faster than the water in the other. Holes left by the decay of tree roots or the burrows of small animals allow the water to flow rapidly away from the source. Breaks and fissures may penetrate the bedrock. Fractured rock or even fractured clay beds produce a groundwater flow system which diverts water into pathways difficult to understand from the data available at the surface. Horizontal fractures can be closed by the weight of the strata while vertical fractures remain open. Further, the vertical fractures will not have random orientation, but may all trend in one direction (Spayd, 1985). Such a system will divert water from paths predicted by the unsuspecting investigator.

The pollutants may not move passively with the water. They are often retarded by adsorption on the soils or degraded by microorganisms. Contaminants evaporate into the atmosphere of the vadose zone and redissolve in groundwater some distance away. Pollutants which do not dissolve in water will float in layers on top of the saturated zone or sink to its bottom, and travel in different directions and at different speeds than the water. The result is a complex dynamic system for which descriptive data are very difficult to obtain (Figure 1-1). In some cases, it is not possible to determine even the source of pollution. Contamination of the groundwater in the

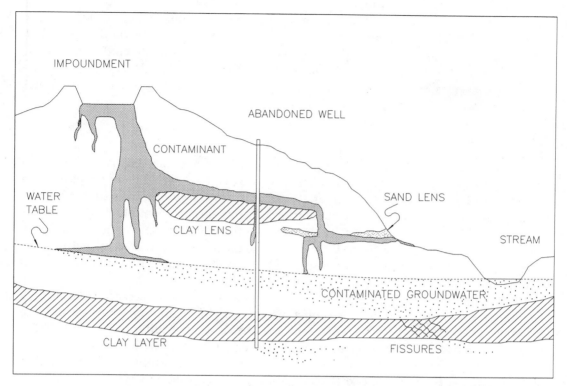

Figure 1-1. Contamination of soil and groundwater creates complex situations, often difficult to evaluate from data available at the surface.

San Gabriel and San Fernando Valleys in California by trichloroethylene has been thoroughly studied, but cannot be assigned to a specific source (Sjoberg, 1985).

The Results of Subsurface Pollution

Pollution of the subsurface has created immense problems for humanity. Public health has been damaged. Surveys have shown probable effects, such as prematurity and birth defects, near the worst sites (Goldman et al., 1985). Many of the chemicals involved have caused cancers at the higher concentrations which occur in occupational exposures, and are therefore suspect at the lower concentrations near polluted sites. In some cases, contaminated drinking water supplies have been used for some time before the contamination was discovered. The extent of harm to the public health, however, is very difficult to assess. For the most part, exposure has been to very low concentrations of chemicals over long periods of time. Science has not yet learned how to measure the health effects of low concentration exposures accurately.

The financial cost is more easily measured. Where a water supply is contaminated, it must be treated or abandoned. Either alternative is expensive. Many millions of dollars have gone into the importation of new water supplies or the purchase and operation of elaborate treatment plants.

Resolution of a contamination incident requires cleanup or containment measures. Again, the costs are tremendous. The Congressional Office of Technology Assessment (1985) has suggested that the cost of cleaning up superfund sites will be $100 billion over the next two decades. This estimate does not include the cost of the more numerous sites which will never be on the superfund list. Further, experience has shown that such long-range cost estimates are revised upwards several times before the work is complete.

The Response to Subsurface Pollution

The challenge of providing adequate hazardous waste management is generally being poorly met. The regulations governing waste disposal are changing rapidly, and are enforced in an irregular manner. In some cases, large amounts of money are being spent on trivial problems, while more serious threats to public health are not addressed. The regulatory agencies are too often staffed by new college graduates seeking a few year's experience before moving on to lucrative private sector jobs. The media treat the subject with sensationalism, and an undereducated public overreacts.

The majority of waste generators who genuinely wish to act as responsible citizens encounter many difficulties. The constant flux in the regulations means that the generators cannot stay abreast of the rules. Generators, consultants, and regulators cannot accumulate the long-term experience which gives the participants the knowledge of what is expected of them. Rules are still being reinterpreted by the regulators and the courts. Everyone has difficulty finding trained and experienced personnel.

The considerable technical uncertainty in hazardous waste issues makes some of this fumbling inevitable. The public cannot use its votes to tell the politicians what to do if the scientists and engineers do not provide clear and definite answers to the important questions. In many cases, those clear and definite answers do not yet exist. But a lack of political leadership and public responsibility has certainly made the problems worse. Any effort to solve hazardous waste problems requires the development of transportation systems and treatment facilities. But as an EPA official recently observed, the public reacts to solutions in exactly the same way that it reacts to problems. The citizens say "not in my back yard," the politicians follow with "not in my district," and no progress is made.

SOURCES OF SUBSURFACE HAZARDOUS WASTE

Problems of groundwater contamination are complicated by the many and various sources of contaminants. The most obvious and sensational case, in which wastes have been carelessly poured on the ground, is only one of many pathways by which chemicals enter the soil.

Contamination sources are in three major categories: facilities used for the direct disposal of hazardous materials, sites of indirect disposal of hazardous materials, and non-point sources (Figure 1–2). The number of direct sources of subsurface contamination in this country is not known with any certainty. One study made for the EPA's Office of Solid Waste indicated that there may be as many as 32,000 hazardous waste dump sites throughout the country. Others estimated the total number at 50,000 (USEPA, 1980). EPA's final groundwater protection strategy (USEPA, 1984(a)) includes an estimate that there are 93,500 landfills in the United

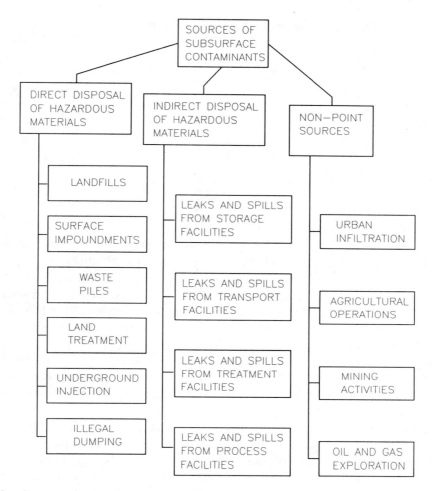

Figure 1-2. Sources of subsurface contaminants.

States, including 75,000 on-facility industrial sites and 18,500 municipal sites. A surface impoundment assessment report (USEPA, 1984(b)) released by the EPA identified 80,263 sites containing 181,973 surface impoundments. Nearly 50 percent of these are located over thin or permeable soils. About 70, 78, and 84 percent of the industrial, municipal, and agricultural impoundments, respectively, are unlined.

Other direct disposal facilities, such as waste piles, land treatment systems, and industrial extraction or injection wells, have not been carefully counted, and the total number could be millions. Septic tanks, usually neglected by investigations of hazardous subsurface contaminants, could be as numerous as 16.6 million (Langerman, 1983). Contamination from the indirect disposal of hazardous material and non-point sources is also undescribed.

There may be 4 million underground storage tanks (Table 1-1). The EPA estimated that 100,000 of these are presently leaking and another 350,000 will leak in the next five years (Lehmen, 1984). According to the American Petroleum Institute, 40 to 75 percent of the total are leaking now or will be shortly (Jafek, 1985). Some estimates and assumptions (Table 1-2) allow the calculation that several thousand billion gallons of contaminated water are generated each year. Septic tanks are pro-

Table 1–1. Number of Underground Storage Tanks.

ESTIMATED NUMBER OF UNDERGROUND STORAGE TANKS (IN MILLIONS)	REFERENCE
1.4 (Gasoline)	Water Information Center, Inc. (1983)
1.5 to 2.0	Aspen Systems Corp. (1984)
	Tejeda (1984)
	Lehman (1984)
>3.5	Dowd 91984)
>3.5	Jafek (1985)
4.0 (Petroleum)	Dwyer (1985)
0.2 to 0.4 (Chemicals)	
2.345	Hansen (1985)

ducing a greater quantity than other point sources of contamination. However, the contaminant release rates for leaking underground storage tanks are greater.

Under appropriate conditions, water will dissolve contaminants from the solids it contacts. This process is called leaching, and produces contaminated water called leachate. The availability of water is an important factor controlling leachate generation. For landfills, waste piles, poorly operated land treatment systems, and open dumps, the water may come from precipitation, surface run-on, groundwater intrusion, irrigation, waste decomposition, and liquid waste or sludge codisposal (Lu et al., 1985). Of these sources, the primary contributor is direct precipitation. For surface impoundments, the important leaching solution is stored liquids, which may be supplemented by precipitation and surface run-on.

Subsurface contamination by chemicals or wastes may come from storage, transport, treatment, and process facilities (Figure 1–2), generated by leaks or spills. Precipitation and surface run-on may carry releases even further from the contamination sources. Some of the important subsurface contamination sources can be depicted (Figure 1–3), and descriptions follow.

Table 1–2. Potential Leachate Generation from Major Subsurface Contamination Sources.

SOURCES	POLLUTED GROUNDWATER* (BILLION GALLONS/YEAR)
Landfills and dumps	450
Surface impoundments	300
Septic tanks	1,300
Leaking underground tanks	0.5 (2000)**
Others	Unknown

*Estimated based on the following assumptions:
1. 93,500 landfills and dumps, 17 acres, in average size, with an annual infiltration rate of 10 inches of water. 20 percent of landfills and dumps are situated 2 feet into the groundwater table with a subsurface soil permeability of 10^{-4} cm/sec.
2. 70 percent of 182,000 impoundments, 1000 ft² average size, are leaking at 3 feet of hydraulic head and 10^{-4} cm/sec subsurface soil permeability.
3. 25 percent of the total U.S. population are using septic tanks generating 70 gallons per day per person of leachate.
4. 400,000 underground tanks are leaking at an average rate of 3.5 million gallons per day.
**About 4000 gallons of groundwater are needed to dilute 1 gallon of pure chemical to the concentration of 1000 ppm.

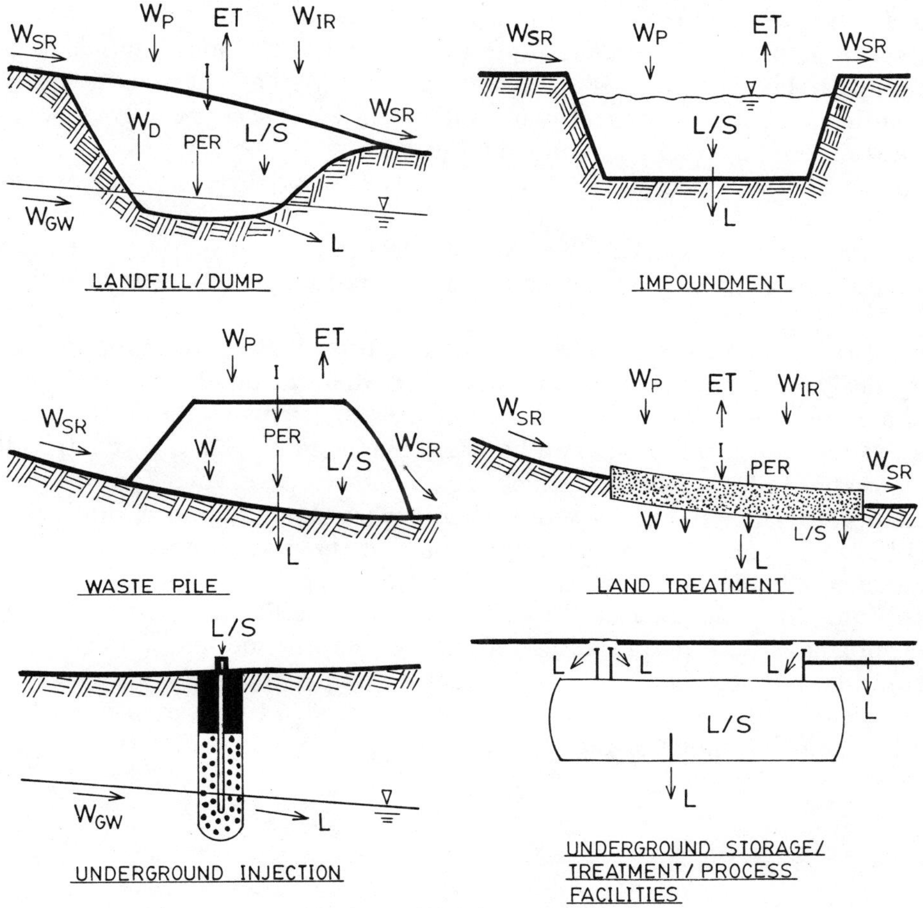

Figure 1-3. Generation of subsurface contaminants. W_{SR} = surface runon, W_p = precipitation, E_T = evapotranspiration, W_{IR} = irrigation water, W_{SR} = surface runoff, W_D = water from disposed materials, PER = percolation, L/S = liquid and/or sludge, W_{GW} = groundwater, L = leachate.

Surface Impoundments

Generators of hazardous wastes have commonly placed them in surface impoundments. For the disposer, these have had the utility that the liquids "disappeared" if enough time was allowed. One mechanism was evaporation, which may have contributed to air pollution. A second was percolation to the subsurface. Correspondingly, surface impoundments have been recognized as a major source of groundwater contamination.

Owners have attempted to improve the performance of impoundments. Most lined the ponds with some impermeable material, such as plastic sheets or layers of clay. Unfortunately, many of these carefully constructed impoundments have been little more successful at retaining wastes than their unlined predecessors, and some have become major sources of groundwater pollution (Ghassemi and Haro, 1985).

The clay liners have been penetrated by various solvents, which cause the clay to shrink so that cracks form. The permeabilities of clays may increase by as much as a thousand times when exposed to organic solvents (Brown et al., 1986).

The clay is excavated from natural deposits, so its quality varies. It may contain lumps which do not compact well, or pockets of sand or organic materials with high permeabilities (Peirce et al., 1986). In some cases, clays have been used which contain significant quantities of carbonate minerals. These were readily dissolved by acid wastes, creating cracks, cavities, and leaks.

Laboratory tests of permeability are relatively poor predictors of performance in the field.

Often, impoundments have failed simply because of poor quality control during construction. In many cases, for example, adequate compaction testing has not been done.

The plastic liners are synthetic and therefore uniform. However, they are delicate, and are easily punctured during construction or when the pond is in use. They must be installed in strips, with seams made between the strips as the installation progresses. Often, the seaming method has been inadequate, or quality control during installation has failed. A method which is easy and effective when done by engineers in the laboratory may be much less workable when done by poorly motivated personnel in a dusty or rainy environment. Some of the plastics or seam cements are also attacked by solvents in the pond contents.

Both kinds of pond liners have suffered from installer inexperience. Even the most senior have been doing the jobs for only a short time, and a long history of encountering and solving problems has not accumulated. Investigations of failed systems are difficult, and few have been documented. Moreover, the recent heavy demand means that much of the work is being done by contractors with no experience at all.

Landfills

Both municipal and private facilities used for burying wastes have proved potent sources of groundwater contamination. Many received materials in the past which are now recognized as hazardous. Construction and operation of these facilities failed to provide even minimal precaution for the prevention of migration of the contents.

A striking example of this past lack of concern is the old town gas sites (Salveson, 1984). "Town gas" was manufactured by heating coal. It was used at the turn of the century as natural gas is used today. Coal tar, a mixture of many petroleum hydrocarbons and their breakdown products, was a residue by-product. In many cases, the coal tar was just dumped "out back," creating massive deposits of carcinogenic material. Today, these haphazard landfills are resurfacing as a result of seepage or construction activity, and the cleanup costs will be very high.

Ordinary municipal garbage has been generally exempted from regulation as hazardous waste, but this was primarily an economic and political decision. Municipal solid waste inevitably contains significant amounts of hazardous materials discarded by householders and industries. Paints, solvents, pesticides, cleaners, and other household products are hazardous and constitute a threat of subsurface pollution when large amounts are collected in a sanitary landfill.

Liquids may be disposed of directly to the landfill, or may arise from chemical or biological transformations of landfill material. More often, the leachate from the landfill is rainfall or melting snow which has percolated through the waste mass, picking up hazardous constituents along the way. Copper wastes, for example, can dissolve to produce a leachate with high concentrations of toxic copper ion.

Some of the most notorious hazardous waste sites were created by early attempts to segregate and control hazardous materials. The infamous Stringfellow Acid Pits in Southern California, for example, were developed for controlled legal disposal of toxic wastes. The owner specialized in this business, and operated with the consent of local officials. It was only after a substantial period of operation that it was realized that the barriers to subsurface migration were grossly inadequate. Instead of a safe depository for wastes, the site became a notorious and horrendously expensive example of groundwater contamination.

Spills

Accidental releases of chemicals have created major problems. With the current attention being given to hazardous materials handling, spills from overturned trucks or ruptured drums are now cleaned up carefully. In the past, however, response was often limited to getting the truck back on the road. The chemicals spilled were left on the soil. The problems of such spills are particularly acute at large, old facilities such as refineries. Releases have occurred many times over the course of decades, generating large volumes of soil contaminated by a variety of chemicals in a complicated pattern.

Tanks and Pipelines

A particularly insidious kind of accidental spill occurs when an underground tank or pipeline begins to leak. If the leak is not large compared to the amount of material contained in the system, the leak may not be detected. Releases may occur for many years. Unfortunately, an amount of toxic chemical which is small in comparison to the amount of product used may be large in terms of the groundwater pollution problem created.

Because they are buried in damp soil, or in contact with the groundwater, metal tanks often corrode rapidly. The corrosion is highly irregular, even on a very small scale. It proceeds rapidly along microscopic cracks or other irregularities in the metal, so that pinhole leaks will occur even when the tank as a whole is intact.

Fiberglass tanks have been used to avoid this problem, but they also leak. Cracks resulting from poor manufacture, careless installation, or ground settling often occur. The surprising discovery has been made that many leaks in such tanks result from operators testing the level of product. When the measuring stick is dropped into the tank, it may hit the bottom hard enough so that repeated blows eventually rupture the tank.

Very often the leaks associated with tanks are in the pipes and fittings which serve the tank. These are installed piece by piece, and leaky joints are more probable than for a welded seam in the tank itself.

Chronic small spills further complicate the problems of underground tanks. While the tank is being filled from a truck, small amounts of product are commonly spilled. Over a period of time, these releases can accumulate to become a significant contamination problem. Often the remedial effort is extended in the frustrating effort to find a leak when the tank and piping are sound.

Septic Tanks

Many households and industrial facilities have used septic tanks for on-site disposal of waste from wash areas and bathrooms. These systems hold the waste in an underground tank where sedimentation and anaerobic microbial activity remove some of the contaminants. The partially treated water is released through perforated pipes to a subsurface "leach field." The soils in the leach field filter the waste, and provide an environment appropriate for further microbial degradation. The system is specifically intended to promote infiltration of water, which is the bulk of the waste.

Septic tanks are successful and appropriate systems where population density is low, soil conditions are good, and the waste is water with biodegradable contaminants. But public health problems occur where septic tanks are too numerous, the water table is too high, or subsurface channels carry leachate directly to a water supply aquifer. On many occasions, the transport of infectious organisms has caused epidemics, making septic tanks the number one source of waterborne disease outbreaks in the U.S. Nitrate concentrations may also be high, threatening the health of babies using water from the contaminated aquifer.

Septic tanks become a hazardous waste problem when materials other than ordinary sewage are put down the drains. Some household products are toxic. Tanks at industrial facilities may receive the usual spectrum of chemicals. Among these are toxins which are not quickly biodegraded. Some may also be relatively mobile in soil, not subject to the processes which attenuate the concentrations of sewage organics. Because the septic tank system is specifically designed to promote subsurface migration of the water, the included hazardous materials may become a serious problem for groundwater quality.

Agriculture

The pesticides and fertilizers used in farming have contaminated many water supplies. Ethylene dibromide, aldicarb, and others have become common contaminants in farming areas, and many wells have been contaminated with nitrates intended for plants. Fertilizers and pesticides are spread widely on the soil in large amounts. Irrigation water dissolves these and carries them downward as it infiltrates.

Manufacturing, loading, and mixing sites serving agriculture have produced some highly localized point sources of contamination. Loading of planes for aerial application of pesticides, for example, has polluted some airport areas.

Urban Runoff

Modern cities are built on the assumption that separate systems for wastewater collection are best: sewage from homes and industries is collected in an enclosed system and treated, while the much larger amount of clean rain water is collected in an open system and channeled to rivers, lakes, or the ocean. Recent studies are casting some doubt on this assumption. Runoff from cities contains significant quantities of contaminants. Lead, discharged in automobile exhausts and collecting in dust on surfaces, may appear in high concentrations in the first stormwater of the rainy season. Petroleum from spills and drippings in the street may also be present. Runoff water will stand on the soil for some period, and may infiltrate to contaminate groundwater.

Deep Well Injection

Hazardous waste is sometimes pumped into the ground. The wells used for this purpose have been many thousands of feet deep, and some effort has been made to establish the fact that there is no connection between the strata receiving the chemicals and shallower aquifers being used for water supply.

In some cases the connections have in fact been present, hidden by the complexity of the subsurface geology and the difficulty of making observations. The wastes may return to the shallow water supplies through fissures, through abandoned wells with corroded casings, or sometimes through the annular space surrounding a poorly installed injection well.

A recent survey, however, indicates that modern, well-operated systems have not contaminated useful aquifers (Anonymous, 1987).

Illegal Dumping

Probably the most common form of illicit disposal is the inclusion of hazardous materials in municipal garbage. If the amounts are small, it is relatively easy to conceal containers of chemicals in the solid waste going to the dump. Disposal of this kind is a significant factor in the production of hazardous leachates by municipal landfills.

Illegal dumping often produces large acute releases. Unethical individuals may wait for dark to deposit barrels of material in the countryside. These people rarely take care to see that the barrels are intact and closed so they can be picked up; more often they are tipped over or leaking. In one case, barrels of oil with PCBs were left in the desert and used for target practice. An expensive cleanup effort was required.

Smaller-scale, chronic dumping occurs. Householders or businesses may carelessly discard used motor oil or contaminated gasoline. One official has suggested that "every company has its back 40" where materials were dumped in the past.

Liquids from Solids

The hazardous wastes which migrate in the subsurface are almost entirely gases or liquids. Only the smallest particles of solid material in the most permeable soil move significant distances. Indeed, one way to make wastes less hazardous is to solidify them so that migration will not occur. This does not imply, however, that solids have no role in groundwater contamination. Rather, solids which generate hazardous liquids have created many of the most serious pollution episodes.

The solids may be municipal waste, sewage sludges, industrial sludges, process wastes, excess chemicals, and others. There are many mechanisms by which the conversion from immobile solid to mobile liquid can occur. Most are described later in this text, and they include simple dissolution, desorption, microbial degradation, and chemical reaction.

Site Assessment

The Environmental Protection Agency has prepared a standard procedure for ranking uncontrolled hazardous waste sites according to the risk they present to public health and environmental quality (USEPA, 1985). The method considers the possibilities of migration of toxic materials by groundwater, surface water, and air to "targets" of various sensitivities (Table 1–3).

The ranking techniques for the groundwater route provide a valuable overview of the site characteristics which contribute to the hazardousness of an uncontrolled site.

The "route characteristics" for groundwater include depth to the water-bearing strata, amounts of precipitation, soil permeability, and the physical characteristics of the waste. (This book treats vadose zone flow in Chapters 7 and 8, groundwater movement in Chapter 3, and the physical nature of the waste in Chapter 2.)

Waste characteristics considered include toxicity, persistence, and quantity (Chapters 4 and 5).

Groundwater uses, distance to the nearest well, and the number of people served by the well are evaluated to rate target vulnerability.

The nature of the containment itself, treated in Chapter 9, is also rated. Assessment is made of the soundness of the barriers to migration, and the likelihood of precipitation or run-on entering the waste mass (Table 1–4).

Table 1–3. Comprehensive List of Rating Factors.

| | HAZARD MODE: | | |
FACTOR CATEGORY	GROUND WATER ROUTE	SURFACE WATER ROUTE	AIR ROUTE
Route characteristics	Depth to aquifer of concern Net precipitation Permeability of unsaturated zone Physical state	Facility slope and intervening terrain One-year 24-hour rainfall Physical state	
Containment	Containment	Containment	
Waste	Toxicity/persistance Waste quantity	Toxicity/persistance Waste quantity	Reactivity/compatibility Waste quantity
Targets	Ground water use Distance to nearest well/population	Surface water use Distance to sensitive environment Population served/distance to water intake downstream	Land use Population within four-mile radius Distance to sensitive environment

Table 1-4. Containment Values for the Groundwater Route.*

A. Surface Impoundment	Assigned Value
Sound run-on diversion structure, essentially nonpermeable liner (natural or artificial) compatible with the waste, and adequate leachate collection system	0
Essentially nonpermeable compatible liner with no leachate collection system, or inadequate freeboard	1
Potentially unsound run-on diversion structure, or moderately permeable compatible liner	2
Unsound run-on diversion structure, no liner, or incompatible liner	3

B. Containers	Assigned Value
Containers sealed and in sound condition, adequate liner, and adequate leachate collection system	0
Containers sealed and in sound condition, no liner or moderately permeable liner	1
Containers leaking, moderately permeable liner	2
Containers leaking and no liner or incompatible liner	3

C. Piles	Assigned Value
Piles uncovered and waste stabilized; or piles covered, waste unstabilized, and essentially nonpermeable liner	0
Piles uncovered, waste unstabilized, moderately permeable liner, and leachate collection system	1
Piles uncovered, waste unstabilized, moderately permeable liner, and no leachate collection system	2
Piles uncovered, waste unstabilized, and no liner	3

D. Landfill	Assigned Value
Essentially nonpermeable liner, liner compatible with waste, and adequate leachate collection system	0
Essentially nonpermeable compatible liner, no leachate collection system, and landfill surface precludes ponding	1
Moderately permeable, compatible liner, and landfill surface precludes ponding	2
No liner or incompatible liner, moderately permeable compatible liner, landfill surface encourages ponding, no run-on control	3

Source: USEPA (1985)

*Assign containment a value of 0 if: (1) All of the hazardous substances at the facility are underlain by an essentially nonpermeable surface (natural or artificial) and adequate leachate collection systems and diversion systems are present: or (2) there is no ground water in the vicinity. The value ''0'' does not indicate no risk. Rather, it indicates a significantly lower relative risk when compared with more serious sites on a national level. Otherwise, evaluate the containment for each of the different means of storage or disposal at the facility using the following guidance.

Of course, all of this careful analysis is unnecessary if the subsurface migration of hazardous wastes to a water supply has been observed. Then the opportunity for prevention has passed, and cleanup must begin (Chapter 9).

REFERENCES

Anonymous, 1987. Have deep wells been shafted? *Civil Engineering,* 57(2):21.

Aspen Systems Corporation, 1984. The move to regulate storage tanks. *Hazardous Waste Report,* 5(11):2.

Brown, K. W., Thomas, J. C., and Green, J. W., 1986. Field cell verification of the effects of concentrated organic solvents on the conductivity of compacted soils. *Hazardous Waste and Hazardous Materials,* 3(1):1-19.

Dowd, Richard M., 1985. Leaking underground storage tanks. *Environ. Sci. Technol.,* 18(10):309A.

Dwyer, Paula, 1985. Coming up: Regulation of underground tanks. *Chemical Engineering,* April: 17-18.

Ghassemi, M., and Haro, M., 1985. Hazardous waste surface impoundment technology. *J. Env. Eng.,* 111(5):602-617.

Goldman, L. R., Paigen, B., Magnant, M. M., and Highland, J. H., 1985. Low birth weight, prematurity and birth defects in children living near the hazardous waste site, Love Canal. *Hazardous Waste and Hazardous Materials,* **2**(2):209–223.

Hansen, Penelope, 1985. LUST: Leaking underground storage tanks. In *Proceedings of the Third Annual Hazardous Materials Management Conference.* Tower Conference Management Co., Philadelphia, PA, June 4–6.

Jafek, Bev, 1985. Underground storage tanks: Key environmental issue. *Hazardous Materials and Waste Management Magazine,* March/April: 30–34.

Langerman, Paul T., 1983. *Groundwater Policy: The State Solution.* Washington, D.C.: The Heritage Foundation.

Lehman, John P., 1984. *Leaking Underground Storage Tanks.* The First Public Briefing on the 1984 Amendments to the Resource Conservation and Recovery Act.

Lu, J. C. S., Eichenberger, B., and Stearns, R. J., 1985. *Leachate from Municipal Landfills, Production and Management.* Park Ridge, NJ: Noyes Publications, 453 pp.

Office of Technology Assessment, 1985. *Superfund Strategy.* Washington, D.C., OTA-ITE-252, 282 pp.

Peirce, J. J., Sallfors, G., and Peterson, E., 1986. Clay liner construction and quality control, *J. Env. Eng.,* **112**(1):13–24.

Salveson, R. H., 1984. Downtown carcinogens—A gaslight legacy. In *Management of Uncontrolled Hazardous Waste Sites, Proceedings of the Fifth National Conference,* pp. 11–15. Hazardous Materials Control Research Institute.

Sjoberg, C. W., 1985. Underground storage tank program in Los Angeles County. In *Proceedings of the Hazardous Materials Management Conference/West,* pp. 39–42. Wheaton, Illinois: Tower Conference Management Company.

Spayd, S. E., 1985. Movement of volatile organics through a fractured rock aquifer. *Ground Water,* **23**(4):496–502.

Sutton, P. A., and Barker, J. F., 1985. Migration and attenuation of selected organics in a sandy aquifer—A natural gradient experiment. *Ground Water,* **23**(1):10–16.

Tejeda, Susan, 1984. Underground tanks contaminate groundwater. *EPA Journal,* January/February: 20–22.

United States Environmental Protection Agency. May, 1980. Controlling Hazardous Wastes. EPA-600/8-8—017.

United States Environmental Protection Agency. January, 1984(a). *Surface Impoundment Assessment.*

United States Environmental Protection Agency. August, 1984(b). *Final Groundwater Protection Strategy.*

United States Environmental Protection Agency. 1985. Uncontrolled hazardous waste site ranking system: A users' manual. Appendix A in 40 CFR Part 300, National Oil and Hazardous Substances Pollution Contingency Plan; Proposed Rule. *Federal Register* **50**(29).

Water Information Center, Inc., 1983. EPA to launch $1 million study aimed at underground storage tanks. *The Groundwater Newsletter,* **12**(22):1.

2

The Composition
of Hazardous Wastes

Joseph S. Devinny
Associate Professor of Civil and Environmental Engineering
University of Southern California

To understand the fates and effects of hazardous wastes, we must know their composition. It is a subject which takes in much of the science of chemistry. Hazardous wastes are extremely varied, including elements, ions, compounds, and complexes, as gases, solids, liquids, solutes, suspended particles, and adsorbed species. The fundamental chemistry remains an active area for research, and no doubt will be so for a long time.

PHASE

Among the most obvious characteristics of matter is its physical phase. Solids retain their shape and volume and will not flow. Liquids retain their volume but not their shape, and so may flow through soils. Gases retain neither volume nor shape, and flow readily. These characteristics are fundamental in determining the fate of pollutants which enter the soil.

Solids may be large objects, such as those often disposed of in municipal landfills. They may also be finely divided particulate material like soils. Liquids may be solvents or petroleum products, or more commonly, water carrying dissolved pollutants. Gases are sometimes released directly to the soil, but will more often be involved in subsurface migration of hazardous wastes because they are generated by evaporation, microbiological activity, or chemical reactions.

Solids

Particle Size and Shape.   Solids have the unique characteristic that only their surfaces will participate in chemical reactions or support microbiological growth. The surface-to-volume ratio therefore limits the activity of a solid. This ratio is in turn controlled by the geometry of the solid.

Small particles will have a far larger total surface area than the same mass of large particles. Surface area is inversely proportional to particle diameter, and so varies over many orders of magnitude for the materials commonly encountered in waste disposal and soils contamination.

Spherical particles have the minimum possible surface area for a given mass, while irregular particles display more. The shape factor becomes important in the case of porous particles. The difference in surface area for spherical and ellipsoidal particles is small, but porous particles may have surface areas a hundred times greater.

Seemingly solid rocks may have large effective surface areas in the form of pores. Some loosely consolidated sandstones transmit water easily, and form strata which are productive aquifers. Like soils, such rocks present large surfaces to the water passing through them, and may support high rates of precipitation, adsorption, or ion exchange reactions which influence the fate of subsurface contaminants.

Porosity.   The sizes and shapes of the particles will be important factors influencing the amount of space between them. The porosity of a soil or waste is the fraction of the volume not occupied by solid. In saturated soils, the space is entirely occupied by water. In dry soils, it is almost entirely air. The porosity has obvious effects on the density of the material: porous materials contain more air and less solid per unit volume.

Porosity also influences the water-holding capacity of wastes, and the ease with which they transmit water. Details of these relationships are presented in Chapter 3.

Adsorbed Substances.   Solids, either in a buried waste mass or in natural soils, strongly influence the composition of water passing through their pores. The solid itself may dissolve. Where the water is contaminated, the solids may adsorb and release the pollutant species. Because the specific surface area of the soils is large, the total effects on water quality can also be large. Strongly adsorbed pollutants will be retarded, and may be held while the water passes by. Less strongly adsorbed species may desorb, move, then readsorb over and over. These may move ten or a hundred times slower than the water mass, reducing the scope of the pollution problem. However, if an aquifer is to be decontaminated by flushing it with clean water, removal of the adsorbed pollutants may be difficult.

An important characteristic of solids is their relative immobility. Solid pollutants rarely move far from the site of disposal. Straining, adsorption, or coagulation and precipitation within the porous media remove them from the water (Chapter 5). This ability to remove even very small particles from water has been put to use. Filters containing sand or diatomaceous earth are widely used in water treatment and other purification processes. Septic tanks release highly contaminated water into the soil, relying heavily on the ability of the soil to remove the particles in order to prevent the spread of pollution.

For these reasons, the disposal of solid material (if it does not later dissolve) threatens groundwater quality less than the disposal of liquids.

Liquids

The movement of liquids is the most common kind of subsurface migration of hazardous wastes. Movement occurs in response to several forces. Gravity will cause liquids to move downwards where they exist as a film on soil particles in the vadose zone, or with the slope of the water table where the soil is saturated. In some cases, upward flow may result where water is released from a confined aquifer under pressure.

Water also moves in response to the "matric potential." Water has an affinity for dry soil, and will move from wet soil to dry just as it soaks into a dry sponge set against a wet one. Just above the water table this will create a "capillary fringe," a region where the soil is saturated even though the hydrostatic pressure is less than atmospheric. Higher in the vadose zone, water will be drained from the large soil pores, but will move as a microscopic film on particle surfaces.

The most common kind of subsurface pollutant is a contaminant dissolved in water. It moves as the water moves. Where it is not removed by interaction with the soil, the contaminant can pollute large volumes of soil and aquifer.

Contaminants will also move by transverse dispersion in a direction perpendicular to the average direction of water flow. Flow through porous media requires the water to move along torturous paths. As the flow divides and merges, the contaminant is mixed outwards. Because of this, a stream of contaminants moving away from a point source of pollution gradually broadens, just as a plume of smoke spreads after it leaves the chimney.

Other groundwater problems arise from the presence of a pure liquid. Gasoline and other petroleum hydrocarbons are common examples. If they are immiscible they will form a layer of pollutant on top of the water table and move according to their own pressure gradient. A layer of pure hydrocarbon contains a large mass of pollutants, and can contaminate correspondingly large amounts of water as it moves and dissolves.

As they flow, liquids may change. Solutes precipitate in response to changes in pH. The passage of crude petroleum through the soil may be limited by coagulation of the wax it contains.

Gases

The flow in soils is divided among millions of tiny pores. Moving fluids contact large amounts of solid surface, and tremendous amounts of shearing movement are generated. Resistance to shear in a fluid is called viscosity, and the viscosity of fluids has a large effect on flow rates in porous media. Gases move faster than water under comparable pressure gradients because gases are less viscous. They respond readily to a pressure gradient, and will be pushed outwards from a point of generation.

A volume of gas is much lighter than the same volume of water, and so is less affected by gravity. But it does respond. It moves up if it is less dense than air, or down if it is more dense. Transverse dispersion occurs as it does for contaminants in water. Molecular diffusion, from areas of high concentration to areas of low concentration, may also occur. Diffusion phenomena are more often significant for gases, because contaminant gases diffuse much more rapidly than solutes in water.

Methane generated by biological activity is commonly encountered in landfills which receive municipal waste or other decomposible organic matter. As the pressure builds up within the waste mass, the methane is driven outwards. It is a hazardous material because of its flammability. Houses near landfills have often been threatened by methane in basements or other enclosed areas. Accidents have even occurred at the exposed surfaces of landfills. In one case, children playing with matches ignited the gas and were burned.

Methane migration also has the undesirable secondary effect of carrying other gases with it. Just as soluble contaminants flow with water, gases mixed with methane will move with it. Vinyl chloride is a microbial breakdown product of the common solvent trichloroethylene, and so is often present in landfills. Transport of this gas with methane is particularly alarming because it is a known human carcinogen. Methane has carried it into the basements of houses near landfills. Benzene, xylene, and other toxic volatile compounds are also often detected.

Transport of metals in gases is rare because metal compounds with significant vapor presures are few. Only mercury has a significant vapor pressure in elemental form. A few others may form gaseous compounds, and the possibility of aerosol transport has been reported (Young and Parker, 1984).

A commonly encountered problem is the entrainment of odorous gases. Hydrogen sulfide and sulfur-containing organic compounds such as mercaptans are generated by the microbiological degradation of municipal waste. Young and Parker (1984) found methylmercaptan and butylmercaptan common in landfill gas. There were also a number of nonsulfurous but odorous compounds like 2-butanol and ethyl butanoate. Odor problems are considerably complicated by the importance of the atmospheric dilution which occurs as the gases leave the surface of the landfill. Releases which are a problem when a gentle breeze is blowing may not be noticed when the wind is stronger.

Often, the control of methane migration from landfills is instituted to prevent the release of these foul-smelling materials. Energy recovery is another benefit. Methane is a valuable fuel. Where large landfills are generating significant amounts, it can be collected. It can be sold or used for the generation of electricity.

Water vapor also moves in the pores of soil and waste, often driven by temperature differences. A mass of organic matter undergoing microbial decomposition will warm significantly. Water evaporating from the mass will spread to cooler regions and condense. Leachate may be generated at unexpected places.

CHEMICAL COMPOSITION

The most basic description of the composition of a waste, contaminated leachate, or polluted soil lists the amount of each element present. There are only 93 naturally-occurring elements, and most of these are rarely encountered. Thus an elemental composition requires relatively few tests and produces a well-defined, short data list. Convenient analytical technology exists. The trace metals, in particular, are readily quantified using atomic absorption spectrophotometry, neutron activation analysis, or other methods. Sample analyses such as "total lead" or "total copper," in parts of metal per million parts of sample, are widely determined. They have been given regulatory status as standards in pollution control.

Speciation

Elemental composition data ignore many important characteristics of wastes and water. Each of the elements present can exist in many forms, and be part of many compounds. The compounds have different solubilities, different toxicities, different adsorption coefficients, and a multitude of other different characteristics. The carbon in a sample may be coal, part of a chlorinated hydrocarbon, or the body of a living organism. Mercury can be a silvery metallic liquid, a dissolved ion, or the highly toxic organometallic methyl mercury.

Ionic State.   Elements are called metals because of the characteristics they display as the pure phase. In the "zero oxidation state," each atom is electrically neutral. It readily shares electrons with surrounding atoms. The resulting solid conducts electricity, has a shiny appearance, and often has the great strength which makes metals such useful materials. Metals are "oxidized" when electrons are removed from the atoms (often this is done by molecular oxygen, hence the name for the process). The resulting ion may become part of a crystalline oxide, or may dissolve in water. Charged species are often soluble in water because water molecules are electric dipoles: one end has a slight negative charge, and one is slightly positive. Thus a chunk of zinc metal is immobile in the soil, but will give rise to dissolved zinc ions which migrate easily to aquifers.

Other compounds may also generate ions. Acids and bases release positively charged protons or negatively charged hydroxide ions into solution. As each ion is released, a "counter-ion" of the opposite charge is also created. Acetic acid, present as a microbial decomposition product in landfill leachate, dissolves to produce a positively charged proton and a negatively charged acetate ion.

Compounds.   Assessment of the fate and effects of subsurface elements requires knowledge of the compounds of which they are part. The chemically stable combinations and permutations of the elements, however, are infinite. A sample of rich topsoil contains more compounds than could be listed in any readable report, including many for which no analytical techniques exist, and some which have never been described.

Analysis is thus limited to a small fraction of the number of compounds present. Investigators select those which are abundant, or which have particular environmental significance, as objects of study. No single analytical method is effective for all of the compounds likely to be present, or even for the majority of them. Characterization of subsurface soil and water requires a suite of methods, some tedious, and many expensive.

Complexes.   Ions and organic species are capable of forming associations called complexes. Complexes are formed by bonding mechanisms weaker than covalent bonds, often the nonspecific attraction of oppositely charged ions. While complexes may be less rugged than compounds, they still constitute entities that have characteristics different from those of their parts. They may also influence the fates and effects of subsurface contaminants.

The formation of metal ligand complexes is important for many of the toxic trace metals. Mott et al. (1987), in an investigation of leachate chemistry, have calculated the distribution of copper among various complexes (Figure 2–1). In water at pH 5, copper ions and copper sulfate are the dominant forms. In water at pH 9, however, 94.4 percent of the copper exists as carbonate and carbonate complexes.

Organometallic complexes are often formed with large organic molecules surrounding metal ions. The complex will present a less polar surface to the surrounding solution than the metal, and will prevent other reactions with the ion. Often, ions which are readily adsorbed and retained on soil particles form soluble organometallic complexes. The result may be much greater mobility.

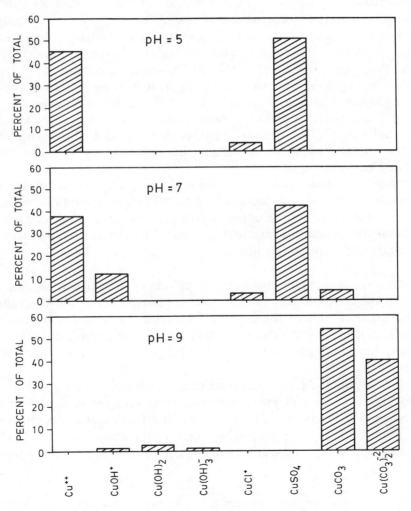

Figure 2–1. Speciation of copper at three values of pH.

Combinations of organic molecules also create complexes. Humic acid is a common constituent of soils. It is a relatively soluble, high molecular weight hydrocarbon of uncertain and probably variable composition, which results from microbiological decomposition of plant litter. It forms complexes with the highly toxic polychlorinated biphenyls (PCBs). PCBs are insoluble in water, and rarely move far when spilled on soil. The complexes are far more soluble, even to the point of resisting adsorption on activated carbon. It is possible that the complexes promote migration in soil which would not otherwise occur.

Trace Metals

The metallic elements from the middle of the periodic table are often hazardous wastes. These are sometimes called "heavy" metals to distinguish them from the lighter, less toxic species such as sodium. Examples include iron, nickel, copper, zinc, silver, cadmium, mercury, and lead. The Environmental Protection Agency lists 17 determinations that should be made to decide whether a waste contains toxic metals. Some, such as lead and mercury, are severely toxic in certain forms.

Mercury poisoning is sometimes referred to as Minimata disease. The city of Minimata, Japan suffered a poisoning epidemic which killed and maimed many people. The culprit in this case was methyl mercury, produced by microbiological transformation of ionic mercury discharged by industry. Metallic mercury, the silvery liquid, is far less toxic. It is a common accident that children break medical thermometers in their mouths and swallow the mercury. No ill effects result because the mercury passes directly through the digestive tract without entering the tissues of the body. Vapors from the same liquid, however, are very hazardous because they readily penetrate the tissues of the lung and enter the blood stream. Methyl mercury is worse yet because it accumulates in nerve cells, where the toxic effects are felt.

The toxicity of compounds is also highly variable for lead, cadmium, and others. But it is usual that wastes are simply analyzed for total metal concentration. This is partly because metals may be transformed into the toxic forms after discharge, as occurs with mercury. It is not safe to discharge the relatively nontoxic liquid, because it can evaporate to form deadly vapors, dissolve to form toxic ions, or be biotransformed into the extremely toxic organic form.

Another reason for measuring only totals is that analysis for the individual compounds is difficult, tedious, and expensive. A ready and inexpensive technology exists for the analysis of total metallic content. The total content for each metal is the criterion for determining whether a liquid waste is hazardous. For solids, either the total amount in the waste or the total amount which can be extracted in mild acid is measured.

Analysis of Trace Metals.   By far, the dominant method for analyzing metals contents in hazardous waste technology is atomic absorption spectrophotometry. In this method, the sample is vaporized and ionized. A light beam is passed through the vapor, and the amount of light absorbed at certain wavelengths is measured. Because each ion absorbs light only at specific wavelengths, it is possible to measure the amount of light absorbed by that metal. The concentration is proportional to absorption.

While the necessary instrument is expensive, it works well for rapid analysis of large numbers of samples, so the cost per sample is low.

Samples often require extensive preparation. Solids must be dissolved before the sample is analyzed. For most samples, this requires digestion in strong acid. Preparatory steps may also be done to determine the amounts of metals in various forms. Filtration, for example, will separate dissolved metals from those in or on the particles. Extraction of the sample in organic solvent, followed by analysis of the solvent and water phases, will isolate those metals which are parts of organic compounds or complexes.

A primary difficulty in atomic absorption spectrophotometry is matrix interference. Detection of the metal depends on its absorbance of light at a specific wavelength. If other substances in the sample also absorb at that wavelength, they will interfere with the determination of the metal. Again, elaborate chemical manipulation of the samples may be necessary to remove the interferences or to account for their magnitude.

Neutron activation analysis measures elemental concentrations by bombarding the sample with neutrons. The inner electrons of the atoms are raised to higher energy levels. When they return to their usual configuration, they release X-rays at wavelengths which are characteristic of the elements. Recording the spectrum of X-rays produced shows the amounts of the elements present.

Neutron activation analysis is valuable for its simplicity. It has the disadvantage, however, that it measures concentrations only relatively near the surface of the sample. For solid samples, this may produce results that are not characteristic of the whole.

Organic Chemicals

When scientists first began investigating the chemical world, they discovered that some chemicals could be converted to other forms. Compounds could be "synthesized." Many of the compounds produced by plants and animals, however, could not be manufactured. For many years, these chemicals were thought to have some special essence of life in them which required their manufacture by living organisms. They were called "organic."

We can now manufacture many of these, and know that they consist primarily of carbon, hydrogen, and a few other elements in the form of complex molecules. We have further learned to manufacture other chemicals with similar characteristics, and call them "organic," even though they never appear as products of living things. Some are even remarkable for their toxicity and for their resistance to attack by metabolic activity.

The list of organic chemicals is very long. It includes petroleum and its derivative products. Solvents and paints are usually organic, as are most pesticides. Biological products like DNA, proteins, carbohydrates, and oils are organic. Many of these contribute to groundwater contamination problems.

Organic chemicals may also exist as emulsions: small droplets of an immiscible fluid suspended in water (or occasionally, small droplets of water suspended in the immiscible fluid). Where this occurs, the droplets may behave almost like particles, being strained out of the water as it passes through the soil. Emulsions often display unusual physical characteristics, such as high viscosity.

A great many organic chemicals, however, are soluble in water. Some, like acetic acid, are completely miscible, and will form solutions of any composition with water. Others, like sugar, have a high but limited water solubility. Unfortunately, even relatively insoluble chemicals, like those in gasoline, dissolve in amounts sufficient to cause unacceptable degradation of water they contact.

Analysis of Organic Contaminants.   The great diversity of the world of organic chemicals has prompted development of many techniques for determining their identity and concentrations. No single technique is well suited to all species or all background solutions. All of the techniques require some degree of sample preparation before the instrumental determination, and often the preparation is elaborate and difficult. Interference from the sample matrix is common, and may preclude some methods.

Some organic chemicals, particularly large molecular weight species with very low solubility, are still beyond any analytical technique. Because these are commonly present in soils, this difficulty produces a significant gap in our knowledge of what occurs during the subsurface migration of hazardous wastes.

Despite these problems, a skillful analytical chemist can usually develop the data necessary to characterize soil and water contaminants. Often this requires initial extractions and separations to produce several solutions, each of which each contains only a portion of the chemicals present in the original sample. This "fractionation" reduces interference between determinations for compounds, and reduces matrix complications.

Vapor Phase Analysis.   Many of the compounds of interest, such as low molecular weight petroleum hydrocarbons and the chlorinated solvents, evaporate readily. If the soil or water sample is allowed time to reach equilibrium with the overlying atmosphere, a measurable amount of organic vapor will accumulate in the air.

Some special problems are associated with these volatile organics. Their release into the atmosphere at a spill site may endanger local residents or cleanup team personnel, or violate air pollution regulations. The analytical chemist must take care that the contaminants in a sample are not allowed to escape during the trip to the laboratory, making the determination erroneously low.

In some ways, however, it is easier to work with the volatile organics. Separation from the background matrix and the possible interference of less volatile organic chemicals occurs almost automatically. In "head space analysis," a sample of contaminated soil or water is placed in a vial and warmed to a standard temperature. After a suitable period of time, the gas is drawn off and analyzed. Collection of the vapors is readily performed in the field with fresh samples.

It is also possible to collect vapors directly from the soil, using holes drilled into the upper layers. This has been used to particular advantage for tracing gasoline contamination. In some cases, rather than drilling and taking fluid samples, it is possible to drive shallow probes by hand (or with light equipment), and collect and analyze the vapor samples on site. This provides data at a much lower cost.

Where the concentrations of the vapors are low, they may be collected on adsorbents. Commonly, this method employs a tube filled with granular activated carbon. A small pump is used to draw the vapors through the tube, and the organic materials are collected on the carbon. The tube is sealed and taken to the laboratory, where heat is used to drive the volatile hydrocarbons off the adsorbent and into a carrier gas. This gas is analyzed instrumentally. The method has the advantage that the sample is stable and sealed in a container for the trip to the laboratory. More important, large amounts of air can be drawn through the tube. Measurable amounts of sample are collected even when the vapor concentration is low.

Whether the vapors are first concentrated, or measured directly, the instrumental method of choice for detection and determination is usually gas chromatography. In this method, the sample is drawn through a column containing one of several granular or porous materials which have some tendency to adsorb organic vapors. An inert carrier gas flows continuously through the column. In each region of the column, the vapors are first adsorbed, because the concentration in the gas phase is high, then released as the following carrier gas without vapors comes by. This process occurs many times as the organic chemical is passed down the column.

Each of the chemicals has a different affinity for the column material. Each is delayed for a different amount of time as it passes through the column, and chemical species which enter the head of the column together will leave at different times. A detector (it may be a thermal conductivity detector, a flame ionization detector, an electron capture detector, a nitrogen-phosphorous detector, or a Hall effect conductivity detector) provides a signal as each species exits the column, generating a peak on a chart recorder.

Because the detectors respond to all organic chemicals in essentially the same way, the operator can distinguish the peaks only by their arrival time. Identification requires some detective work. A standard solution which contains a known amount of the species of interest is prepared, and comparisons with the unknown sample can determine which peak is associated with the standard chemical. The sizes of the peaks indicate the amount of chemical passing through the column, and thus the concentration of the species in the sample.

Primary among the numerous technical problems which can arise is overlapping peaks. If two species in the sample have similar detention times on the column, the detector will see both at the same time, and it will not be possible to determine the relative concentrations of each. In complex mixtures, interferences between the peaks become common, and many may overlap. The peaks and the problem may be resolved by using a different column material. Chemicals retained for similar times on one material may separate on others. Changing the column temperature may also help. Often, columns of different types are put in series, so that some species are separated by the first and others are separated by the second.

Obviously, the problems become more severe if there are more chemicals in the sample. Resolution of the compounds in complex mixtures may not be possible without an initial fractionation step.

Liquid Chromatography. Like gas chromatography, liquid chromatography passes specimens through an adsorbent column which detains different chemicals to different degrees. Again, a nonspecific detector records peaks as they appear. The time of arrival is used to identify the chemical, and the size of the peak is used to estimate its concentration.

As the name implies, the separation occurs in the liquid phase. The chemicals being determined are dissolved in a solvent. Liquid chromatography can identify many organic species which do not have vapor pressures sufficient for gas chromatography. The specifics of the adsorption process also differ, so that peaks which can not be resolved in gas chromatography may be amenable to separation by liquid chromatography. Samples are often prepared by simply extracting the soil in the appropriate solvent.

Modern instruments for liquid chromatography are "High Performance Liquid Chromatographs" (HPLC). This refers to the fact that the separations are carried out on columns with a very high pressure gradient. The great force moves the solution through columns with small pores (and high resistances to flow) within reasonable times. The columns of finely divided material, with their high active surface areas, provide maximum possible peak separation.

Gas Chromatography/Mass Spectrometry. Both liquid and gas chromatography suffer the limitation that the detectors used to define the peaks are nonspecific: they do not identify which chemicals are leaving the column. Identification depends on measuring the time of passage. Because a variety of subtle factors can affect column detention time, it is not possible to simply have a list of times for each chemical. Each column is calibrated for each experiment with standard solutions. This requires substantial additional effort. More important, it means that a peak often cannot be identified if the investigator has not anticipated its presence and prepared the appropriate standard solution. For complex unknown mixtures, this is often impossible.

Mass Spectrometry is a method that has great power in identifying unknown compounds. The sample in the vapor phase is passed through an electron beam, which breaks the molecule into fragments. The resulting ion stream is passed through a complex magnetic field which causes the trajectories of the ions to curve. Each ion follows a trajectory dependent on its particular charge and mass, and is thus separated from the others. A detector collects the ions arriving in different locations to produce a spectrum.

Each organic compound breaks up in a characteristic way, produces a characteristic group of ions, and creates a characteristic mass spectrum. The spectra have been compiled for hundreds of thousands of compounds, so it is possible to compare a spectrum from a sample with known standard spectra. (It is also possible to determine the compound directly from the characteristics of the spectrum, but this is often a challenging puzzle.)

Mass spectrometry is not commonly workable when used with complex mixtures of compounds. Ion fragments from several compounds are scattered in overlapping spectra, creating undecipherable patterns.

The method is powerful, however, when combined with gas chromatography. A complex sample passes through the chromatograph to separate the compounds. Each peak then goes to the mass spectrometer for identification. The ability of the chromatograph to separate is combined with the spectrometer's capability for identification. A computer automatically searches the library of recorded spectra, and determines the compounds.

The GCMS is an expensive instrument, and so is generally found only in laboratories which determine the composition of large numbers of samples. For soil contamination problems, it is most commonly used with the acid-base/neutral separation (U.S.E.P.A., 1982). In this procedure, the soil is first extracted with acid, then with base. The two solutions generated are then analyzed with GCMS. In most cases, the sequential extractions separate some compounds, the GC separates the remaining peaks, and the MS identifies them. The procedure is widely used for the characterization of unknown samples.

Gross Parameters

No instrument can completely characterize wastes and contaminated soils. Even partial characterization on a chemical-by-chemical basis is expensive. In some cases, tests which measure the bulk response of the sample have provided knowledge which is difficult to discern from a partial listing of chemical contents.

pH, Alkalinity, and Acidity.   The gross test most widely used is pH determination. pH is the negative log of the hydrogen ion concentration, and so is a measure of the concentration of a single chemical. But hydrogen ions participate in numerous reactions with a huge variety of soil constituents. Their pervasive influence allows many generalizations about the waste or soil when the pH is known. Water with a low pH (high hydrogen ion concentration), for example, will dissolve metals more rapidly than water with a high pH. Microorganisms will survive well at a moderate pH (6 to 9), but many will disappear if the pH becomes too high or too low. A moderately low pH (3 to 6) favors fungi over bacteria.

The pH meter measures the instantaneous concentration of hydrogen ions in solution. But many solutions contain acids or bases which have the capacity to exert a continuing influence on the solution. For example, if hydrogen ions are removed from a solution of acetic acid, more acid will dissociate, replenishing them. Removal of hydroxide ions may prompt the dissolution of solid hydroxides. Thus, the system will be capable of neutralizing far more than an instantaneous measure of hydrogen ion concentration would suggest. This acid- or base-neutralizing capacity of the waste or soil is an important characteristic, because the number of ions in solution at any one time is small. In situations where they are not continuously replenished, they will be depleted before very much reaction can occur.

An example can be seen in acid precipitation. Acid fog with a pH as low as 2 has been detected. People who were exposed to this fog, however, were not burned by the acid. The excess hydrogen ions present were almost instantly depleted when the droplets contacted skin, and there was no mechanism producing more ions. The hydrogen ion concentration was high, but the base neutralizing capacity was low.

The extent of the reactions which occurred was minute. In contrast, other solutions at pH 2 are hazardous. As reactions with the skin consume hydrogen ions, more are produced, so continuous reactions with the skin occur. An acid burn results.

This capacity of a solution to regenerate ions is measured as alkalinity or acidity. Alkalinity is the "acid neutralizing capacity" of the material, and can be directly measured by titrating a sample with acid. Acidity is defined similarly as the capacity of the sample to neutralize base.

Because wastes mixed with soils will inevitably be undergoing many acid-base reactions, the alkalinity or acidity of the materials is an important characteristic. Where they are low, the initial pH of the waste will change rapidly to a value determined primarily by the characteristics of the soil. The acid or base will have been readily neutralized. Where alkalinity and acidity are high, extensive reactions will occur and the waste will change the pH of the soil.

The relative alkalinity of the soils is also a determining factor. Where acid rain falls on soils of low alkalinity, streams and lakes are acidified. Where the same rain falls on soils high in natural alkalinity, such as that generated by carbonate minerals, it is quickly neutralized and has far less effect.

Redox Potential and Oxygen Demand.   As pH measures the concentration of protons in solution, the redox potential measures the availability of electrons. Where electron-donating chemicals are abundant, and electrons are readily available, the redox potential is low. Where electron acceptors are numerous and powerful, the redox potential is high. (This same characteristic is also measured as the "electrode potential." The electrode potential is equal to the redox potential multiplied by 0.059 volts.)

Because atmospheric oxygen is a strong electron acceptor, its presence raises redox potentials. In contrast, organic chemicals are usually electron donors, particularly during decomposition by microorganisms, and so may create environments with low redox potentials where they are abundant.

This close association with oxygen concentrations has lead to indirect characterization of the redox potential in terms of oxygen availability. "Aerobic" and "anaerobic" conditions are associated with high and low redox potential.

There are problems which arise, however, in assigning redox potentials to a waste or soil. The electron transfer, or "redox" reactions which must occur to establish chemical equilibrium are often slow. Often, equilibrium conditions are not established and the redox potential measured with respect to one set of electron donors and acceptors may be quite different from that measured with respect to another. This contrasts sharply to the situation for acids and bases. Protons are readily carried from acids to bases by water. One measurement of pH will determine the relative availability of protons for almost all of the species involved, because those species equilibrate rapidly with the protons in solution. For redox reactions, it is possible for strong electron donors like petroleum hydrocarbons to coexist indefinitely with a strong electron acceptor like oxygen, with no reactions occurring (as long as the temperature is not high!). A single redox potential will therefore not characterize all of the species present.

The kinetic hinderance which slows redox reactions is often overcome in the natural environment by biological catalysis. Microorganisms can consume petroleum hydrocarbons, for example, adding oxygen to create carbon dioxide and water, the equilibrium products.

Just as understanding acid-base reactions requires knowledge of the alkalinity and acidity of a waste, there are also capacity factors for redox reactions. The "chemical oxygen demand" (COD) of a waste is measured by titrating with a chemical oxidant to determine the total capacity of the material to donate electrons. The term "chemical oxygen demand" reflects the concept that the measurement indicates how much molecular oxygen could be consumed by the waste if the reactions went to completion. A waste with a high COD may also be said to have a "high reducing capacity."

Because oxygen-consuming reactions are often biologically catalyzed, it may be appropriate to measure "biological oxygen demand" (BOD) (microorganisms consume oxygen as they degrade organic chemicals). The test is done by putting a sample of the waste in a closed bottle with a seed culture of microorganisms and monitoring the disappearance of oxygen.

The biological oxygen demand is related to the chemical oxygen demand; in both cases, organic material is converted to carbon dioxide and water, so the amount of oxygen consumed may be much the same. But living organisms catalyze oxidation more gently. Often there are chemicals in the waste which they cannot utilize as food. Usually the BOD of a waste is lower than the COD. Organisms are also much slower. A typical BOD determination takes five days, and further calculation is necessary to predict the "ultimate BOD" from the five-day results. A chemical test takes only a few minutes. Finally, the microorganisms are delicate, and will confuse test results if growth is not vigorous. Toxic chemicals in the waste may inhibit the microorganisms. Further, the toxins may inhibit strongly at the high concentrations of waste used in the test, but not at the lower concentrations seen after dilution in the environment.

Given these problems, why is the biological test used? Most often, the oxygen demand of a waste will be exerted biologically. Microorganisms are common in soils, while powerful chemical oxidants are rare. Treatment systems commonly destroy waste by biological decomposition. If we are to predict the effects of the spill or the effectiveness of the treatment system, biological oxygen demand is the phenomenon of interest. A laboratory test which mimics the actual environmental processes is often the most useful.

A "total organic carbon" (TOC) test may also be used to characterize the organic matter in a water sample. It is different from the other tests, however, in that it does not directly measure the oxygen-consuming capacity of the material. Instead, it measures the amount of carbon present in the organic form. This difference is significant, because organic carbon may exist in several oxidation states, and thus consume different amounts of oxygen. The carbon and hydrogen in methane (CH_4), for example, consume four oxygen atoms in complete conversion to CO_2 and two water molecules. The carbon and oxygen in methanol (CH_3OH) consume only three because the molecule already contains an atom of oxygen. Both, however, contain one atom of organic carbon.

Total organic carbon can be determined by several techniques. Most begin with oxidation of the carbon to carbon dioxide, so that the distinction between the various forms of carbon is lost. However, for specific wastes, it is often possible to experimentally determine the relationships among total organic carbon, BOD, and COD. When this is done, the proportionality can be used to calculate approximate values for BOD from TOC or COD determinations. Because TOC and COD can be determined quickly and efficiently, this is an economically advantageous procedure.

Extractability

The degree to which chemicals in solid waste or contaminated soils may be extracted by liquids flowing through them is an important characteristic of the material. Easily extracted contaminants will migrate readily and pollute groundwater. Contaminants firmly held in the soil will not migrate.

Extractability is also an important characteristic when an aquifer is decontaminated by flushing. In such cases, the objective is to wash the soil by pumping water through it, then treat the water. The method is clearly impractical for tightly bound contaminants, and works best for chemicals which do not interact appreciably with the soil. For species of intermediate extractibility, the soil will require flushing by many "pore volumes" of water, and the method may not be economically practical.

Extractability is a key characteristic which determines whether a waste material is legally classified as hazardous. Federal regulations designed to prevent contamination of groundwater from land disposal, for example, consider only those metals which can be extracted by mild acid. The total concentration of metal in the waste or soil is not used to judge hazardousness.

For this reason, solidification is sometimes sufficient to convert a hazardous material to a nonhazardous material. Several processes are available which sequester trace metals in the soil or waste matrix to prevent leaching. In some cases, hazardous wastes have been used to make nonhazardous concrete.

Geochemists have characterized metals in soils according to their extractability. Easily releasable species include free or complexed ions, or adsorbed species which are rapidly released. Moderately releasable compounds include chlorides, sulfates, carbonates, and amorphous hydroxides. Less releasable are the oxides, sulfides, and organometallic solids. Crystalline rocks and clays are classified as inert, and will persist over very long periods of time without releasing their constituent chemicals.

In the case of soils or inorganic metalliferous wastes, the geochemical experience may be valuable guidance. For the majority of wastes and contaminated soils, however, the extractability cannot be reliably predicted from composition data. The material may contain many chemicals, the physical nature of the waste may be complex, and the chemical and physical constants describing the behavior of some of the constituents may never have been measured. Further, a detailed description of the composition of the waste may not be available.

Extractability is determined by the physical characteristics of the waste as well as its chemical nature. A finely divided material presents a large available surface area, and the extraction will proceed rapidly. Some particles have a different composition at the surface, so the material which reacts does not have the bulk composition of the solid. Soil particles, for example, may be covered by a layer of chemical precipitate which prevents extraction of underlying compounds.

Extractibility is also influenced by the nature of the extractant. Most metals are more easily extracted by acids than by bases or neutral solutions. Organic solvents will transport many kinds of hydrocarbons which cannot be extracted in significant amounts by water solution.

This complexity suggests development of a standard test to determine the extractability of constituents in wastes and contaminated soils. The United States Environmental Protection Agency has formulated such a test for the purpose of making the legal determination of hazardousness. The extraction procedure (EP) begins with grinding the waste to standard particle size. The material is mixed with weak acetic acid and the concentrations of metals in the extractant are measured and compared to standards established for each.

The rationale of the test is that it should mimic, to the best extent possible, processes which might occur when the waste is disposed of. An acid extractant was chosen because acids are generated by microbiological action in landfills. Acetic acid, in fact, is commonly found in masses of disposed waste.

Determining proper conditions for the test is not easy. Disposed wastes may encounter a wide variety of environments and thus various extractants and extracting conditions. It is not possible to determine which conditions are appropriate for each waste: if a waste is not declared hazardous, there is little or no control over how it is handled. On the other hand, if worst case conditions are assumed, many wastes will be classified as hazardous even though they represent no hazard under the actual conditions they will encounter.

The State of California has developed its own extraction test, and has chosen more stringent conditions than the federal government. The test is more likely to extract significant concentrations of metals. In particular, the extractant chosen was citric acid, which is generally more powerful. The citrate ion is a chelating agent, and so will hold greater amounts of metal in solution. While citric acid is not commonly encountered in landfills, other chelating agents are. The State of California therefore decided that the presence of a chelating agent was appropriate in a test of extractability. Many wastes not considered hazardous under federal law were declared so under California law.

Sample Fractionation

The organic matter in a soil or waste sample is often an extremely complex mixture of compounds. Petroleum products, like gasoline or motor oil, are distillate fractions rather than single chemicals. Each contains hundreds or thousands of different species which have only their boiling temperatures in common. Natural materials, or their derivative products, contain the myriad chemicals of the living organism. A soil sample from a contaminated site might, therefore, contain hundreds of petroleum organics, all of the compounds in twigs and leaves, all of the cellular components of a hundred species of bacteria and fungi, and numerous biodegradation and photochemical breakdown products for each of these. Characterization necessarily begins with a separation of the mixture into less complex fractions.

A commonly used scheme of analysis is to extract the sample with acid, then basic solutions. Chemicals which act as bases tend to dissolve in the acid, and acids tend to dissolve in the bases. The result is division of the chemicals into two groups, each more amenable to analysis. Similarly, it is common to expose a sample to water, then to an organic solvent. The result is one solution containing the water-soluble chemicals, a second with species which dissolve in organics, and a residue of insoluble chemicals. Each fraction is then analyzed separately. The fractionations can become tests in themselves: "oil and grease," for example, is defined as all the material which dissolves from a soil sample exposed to a solvent like trichloroethylene.

Especially challenging samples may be analyzed with elaborate fractionation schemes and many instruments. Yen (1988) has suggested a plan for the analysis of complex mixtures of low- and high-molecular weight hydrocarbons, such as those found at old town gas sites (Figure 2-2). It exemplifies the complex analytical techniques which can be used (and shows the degree to which the preceeding discussion has been simplified). Part of the sample is used for preliminary testing of bulk characteristics. The choice of analysis depends on nontechnical knowledge, such as the history of the contaminated site. The tests considered are cation exchange capacity, transmission electron microscopy, thermal gravimetric analysis, thermal chromatography, measurement of bulk density, soil textural analysis, energy dispersive spectrometry, and a determination of organic matter content.

A second portion of the sample is heated gently to drive off volatile materials. These are collected for gas chromatography, which may be done with a thermal conductivity detector, a flame ionization detector, a nitrogen-phosphorous detector, an electron capture detector, or a Hall electrolytic conductivity detector. The GC can also be coupled with a second instrument: the isolated peaks are passed to a mass spectrometer (GCMS) or to an infrared spectrophotometer (GCIR) for identification.

When the volatiles have been driven off, the sample is extracted with a solvent. With the volatiles gone, interferences are reduced in the determinations of the remaining species. The solution can be analyzed for the intermediate molecular weight species using high-performance liquid chromatography (HPLC), or HPLC followed by mass spectrometry of the individual peaks. High-resolution mass spectrometry may provide more power for identification. Low molecular weight species may be analyzed by heating the solution, and again applying gas chromatography along with an electron capture detector or a flame photometric detector. Again, the GC can be followed by infrared or mass spectrometry to identify the separated peaks.

The degassing and solvent extraction steps leave behind relatively resistant materials. Demineralization is accomplished by treating the sample with strong acid. The metals in the sample can then be analyzed by atomic absorption spectrometry, inductively coupled plasma, X-ray fluorescence, or ion chromatography. The organics may be further understood after Fourier transform infrared spectrometry, Fourier transform nuclear magnetic resonance, X-ray diffraction analysis, or electronic spectroscopy for chemical analysis.

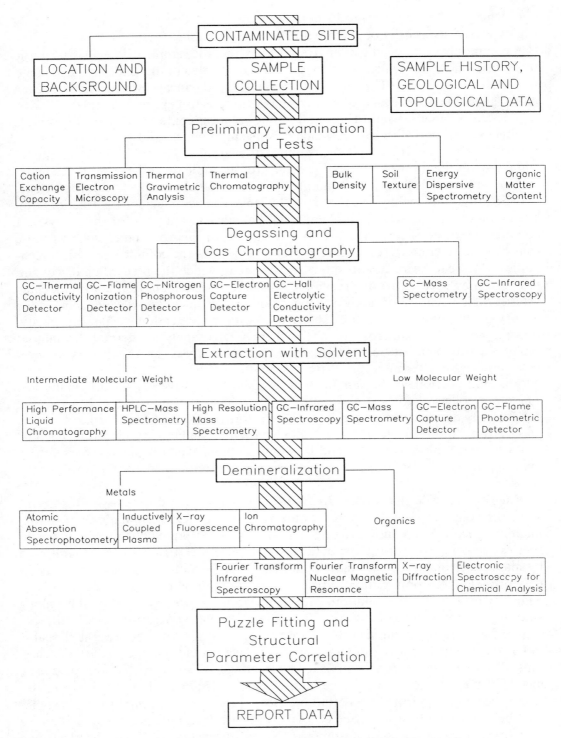

Figure 2–2. Analytical scheme for complex hydrocarbon mixtures in soil (Yen, 1988).

All of these results are considered together in an effort to get the most complete possible picture of the sample.

MICROBIOLOGICAL CONTENTS

Some wastes are hazardous because of the microorganisms they contain. Sewage can be a serious threat to public health. Landfills also receive materials containing pathogenic microorganisms.

Sewage sludge, the concentrated solids left after sewage treatment, contains many of the microorganisms which were present in the wastewater. It is often dumped in the same landfills which receive municipal wastes.

Disposable diapers carry disease organisms, and are disposed of in substantial numbers. As much as five percent of municipal waste may consist of diapers.

Hospital wastes also contain disease organisms. Culture dishes, used bandages, and other waste must be discarded. Regulations generally require special handling.

Pet feces are commonly disposed of in municipal garbage. While most diseases of animals are not passed to humans, there are a few of concern. Some canine intestinal worms infect people, so dog feces are a threat to public health.

Municipal garbage can carry disease, and so must not be allowed to contact humans. Of concern for the subject of this book, however, is whether pathogens might also be transmitted to groundwater with landfill leachate. Bacteria have been found in leachate, so it is a reasonable precaution to assume that the leachate is capable of transmitting disease. Certainly the leachate from septic tanks has caused disease outbreaks.

Viruses have been less studied. There is evidence that they are held by adsorption at the surface of soil particles.

Biological examination of wastewaters and leachates is beyond the scope of this book. The appropriate methods are like those widely used in sanitary engineering for the examination of domestic water supplies.

IS IT HAZARDOUS?

An obvious first question about the composition of a waste is "what dangers does it pose to public health?" The question is important in the legal and regulatory sense. Wastes which are not legally classified as hazardous are subject to very little regulation. Those which are deemed hazardous are heavily regulated, become the object of sensational news coverage and public outcry, and require expensive handling. It is one of the failures of present public policy that there is little recognition of degrees of hazard. A ton of soil contaminated with six parts per million of soluble lead causes the same outcry as a ton of pure pesticide. Materials which are considered hazardous because improper disposal might someday contaminate groundwater are lumped with those which present an immediate threat of death to anyone nearby.

Some efforts are being made to "fine tune" society's treatment of hazardous wastes. The State of California now recognizes "extremely hazardous" wastes, for which human exposure "may likely result in death, disabling personal injury or illness." A waste containing the pesticide toxaphene, for example, is extremely hazardous if the concentration is above 500 parts per million. If the concentration is between 5 and 500 parts per million, the waste is only hazardous.

Classification

Federal law considers wastes hazardous if they are ignitable, corrosive, reactive, toxic, or radioactive.

Classification of wastes as hazardous can be made on the basis of chemical content. A material which contains more than 2500 ppm of chromium is a hazardous waste under California law.

Classification may be made according to waste type. Battery acid, etching acid, plating waste, tanning sludges, and weed killer are hazardous.

Hazardousness can also be defined on the basis of specific tests applied to the waste. Several chemical characteristics are investigated.

Ignitability.   A waste is considered a fire hazard if it has a flash point of less than 60°C; or if it can ignite through friction, absorption of moisture, or spontaneous chemical changes; and if it burns when ignited.

Corrosivity.   Wastes with a pH above 12.5, or with the ability to corrode steel at the rate of a quarter inch per year, are classified as corrosive.

Reactivity.   A waste is hazardous if it spontaneously undergoes violent change, if it reacts violently with water, if it forms potentially explosive mixtures with water, if it forms toxic gases when mixed with water, or if it can explode.

Extractable Toxicity

Solid wastes may produce toxic leachates, a matter of prime importance for disposal. Federal law prescribes test procedures in which the waste is ground and exposed to mild acid. The concentrations of the organic and inorganic chemicals which appear in solution are compared to standards. A solid which releases 100 parts per million barium, or 0.4 parts per million lindane, for example, is a hazardous waste.

A summary of this information for some common chemicals has been prepared by the USEPA (Table 2–1).

Threat to Groundwater.   The U. S. Environmental Protection Agency has developed procedures for characterizing pesticides in terms of their potential for groundwater contamination (Creeger, 1986). Registration of a pesticide requires the maker to conduct many tests (Tables 2–2 and 2–3). Eight of the studies are considered in assessing the possible effects on groundwater (Table 2–4), and criteria have been chosen for classifying a pesticide as a groundwater threat (Table 2–5). Like the other tests for hazardousness, the tests are empirical, and each measures characteristics which are the complex result of many factors acting simultaneously.

Table 2-1.   Waste Characteristics Values for Some Common Chemicals.

REFERENCE CHEMICAL/COMPOUND	1 TOXICITY	2 PERSISTENCE	3 IGNITABILITY	3 REACTIVITY	1 VOLATILITY
Acetaldehyde	3	0	3	2	3*
Acetic acid	3	0	2	1	1
Acetone	2	0	3	0	3
Aldrin	3	3	1	0	0*
Ammonia, anhydrous	3	0	1	0	3
Aniline	3	1	2	0	1
Benzene	3	1	3	0	3
Carbon tetrachloride	3	3	0	0	3
Chlordane	3	3	0*	0*	0*
Chlorobenzene	2	2	3	0	1
Chloroform	3	3	0	0	3
Cresol-o	3	1	2	0	1
Cresol, m- & p-	3	1	1	0	1
Cyclohexane	2	2	3	0	3
Endrin	3	3	1	0	0*
Ethyl benzene	2	1	3	0	1
Formaldehyde	3	0	2	0	3*
Formic acid	3	0	2	0	2
Hydrochloric acid	3	0	0	0	3
Isopropyl ether	3	1	3	1	3
Lindane	3	3	1	0	0
Methane	1	1	3	0	3*
Methyl ethyl ketone	2	0	3	0	2
Methyl parathion in xylene	3	3	3	2	2*
Napthalene	2	1	2	0	1
Nitric acid	3	0	0	0	3*
Parathion	3	3	1	2	0*
Polychlorinated byphenyls	3	3	0**	0**	0**
Petroleum: kerosene	3	1	2	0	1*
Phenol	3	1	2	0	1
Sulfuric acid	3	0	0	2	1
Toluene	2	1	3	0	2
Trichlorobenzene	2	3	1	0	1
Trichloroethane	2	2	1	0	3
Xylene	2	1	3	0	1

1. Sax, N. I. 1975. *Dangerous Properties of Industrial Materials,* 4th Ed., New York: Van Nostrand Reinhold. The highest rating listed under each chemical is used.
2. JRB Associates, Inc., *Methodology for Rating the Hazard Potential of Waste Disposal Sites.* May 5, 1980.
3. National Fire Protection Association. 1977. National Fire Codes, Vol. 13, No. 49.
*Professional judgment based on information contained in the U.S. Coast Guard CHRIS Hazardous Chemical Data, 1978.
**Professional judgment based on existing literature.
Source: USEPA (1985).

LEACHATE COMPOSITION

Ultimately, the concern for subsurface migration of hazardous wastes is not for the composition of the waste but for the composition of the leachate or contaminated soils which result. If no escape occurs, the disposal system has been successful in protecting public health and natural resources. Because of the great number of wastes and their varying composition, it is not easy to predict what kinds of contamination will actually occur. Further, even where the composition of the waste is known, microbiological, chemical, and physical processes can greatly alter the na-

Table 2–2. Summary of Environmental Fate Data Requirements for Terrestrial Use Patterns for EPA Pesticide Registration.

DATA REQUIREMENTS*	DOMESTIC OUTDOOR	GREENHOUSE	NONCROP	ORCHARD CROP	FIELD AND VEGE. CROP	FORESTRY
Degradation						
Hydrolysis	R	R	R	R	R	R
Photolysis						
Water			R	R	R	R
Soil				CR	CR	CR
Metabolism						
Aerobic soil	R	R	R	R	R	R
Anaerobic soil					R	
Mobility						
Leaching	R	R	R	R	R	R
Aged leaching	R	R	R	R	R	R
Field Dissipation						
Soil	R		R	R	R	
Forest						R
Accumulation						
Rotational crop					CR	
Fish			CR	CR	CR	CR
Aquatic nontarget						CR

*Studies required only under certain conditions (such as photolysis in air, volatility studies, tank mix studies, and irrigated crop studies) are not included in this table.
R = Required; CR = Conditionally required; Underline (R, CR) indicates data requirements when an experimental use permit is sought.
Source: Adapted from Creeger (1986).

Table 2–3. Summary of Environmental Fate Data Requirements for Aquatic and Aquatic Impact Use Patterns for EPA Pesticide Registration.

DATA REQUIREMENTS*	AQUATIC USES	
	FOOD CROP	NONCROP
Degradation		
Hydrolysis	R	R
Photolysis (Water)	R	R
Metabolism		
Aerobic Aquatic	R	R
Anaerobic Aquatic	R	R
Mobility		
Leaching**	R	R
Field Dissipation		
Soil (Sediment)	R	R
Water	R	R
Accumulation		
Rotational crop	CR	
Irrigated crop	CR	CR
Fish	CR	CR
Aquatic nontarget		CR

*Studies required only under certain conditions (such as photolysis in air, volatility studies, and tank mix studies) are not included in this table.
**A batch equilibrium (adsorption/desorption) study.
R = Required; CR = Conditionally required; Underline (R, CR) indicates data requirements when an experimental use permit is sought.
Source: Adapted from Creeger (1986).

Table 2–4.   Environmental Fate Studies Used in Determining a Pesticide's Potential to Reach Ground Water for EPA Pesticide Registration.

Hydrolysis
Photolysis in water
Photolysis on soil
Aerobic soil metabolism
Anaerobic soil metabolism
Anaerobic Aquatic Metabolism
Leaching
Field dissipation (Terrestrial, aquatic, or forestry)

Source: Adapted from Creeger (1986).

Table 2–5.   Criteria for Potential to Reach Groundwater.

Water solubility: Greater than about 30 ppm
Distribution coefficient, K_d (ratio of amount of contaminant on soil
  to amount dissolved): Less than 5
Hydrolysis half-life: Greater than about 25 weeks
Soil half-life (field): Greater than about 2–3 weeks

Source: Adapted from Creeger (1986).

ture of the contamination which results (Chapters 5 and 6). Some surveys have been made at various sites to determine what materials are actually being encountered. Ghassemi et al. (1983) surveyed the composition of leachates produced by landfills. They found only a very general correlation of leachate composition with waste composition, confirming the assumption that the prediction of leachate quality is difficult. The most frequently encountered toxic organics were mono- and dichlorobenzenes. The organic species found in highest concentrations, however, were acetic acid, methylene chloride, butyric acid, 1,1-dichloroethane, and trichlorofluoromethane. Common trace metals were iron, copper, nickel, cadmium, chromium, zinc, and manganese. Those in highest concentration were iron, calcium, magnesium, and arsenic. Surprisingly, leachate quality from ordinary municipal landfills was not readily distinguishable from leachate generated at hazardous waste sites. Both represented substantial threats to groundwater quality.

More recent work has attempted to rank the threats presented by toxic discharges according to more complex criteria (Hallstedt et al., 1986). Each chemical was assigned a rating for toxicity and another for persistence. The combined rating factor ranged from 3 for slightly toxic, nonpersistent species to 18 for toxic, nondegradable chemicals. Aldrin, carbon tetrachloride, endrin, and PCBs, for example, received scores of 18. Acetone, which is less toxic and readily degraded, was scored at 6.

The threat associated with a given chemical also depends on the frequency with which it is found. Data on frequency of occurrence was gathered from reports of sampling programs at 32 sites. Benzene and toluene, which are components of gasoline as well as commonly used solvents, topped the list. Another common solvent, methylene chloride, was a close third (Table 2–6). Multiplying the frequency by the combined rating factor produced an overall ranking of threat to public health. When this was done, chloroform, trichloroethanes, methylene chloride, benzene, and tetrachloroethylene ranked 1 through 5 (Table 2–7). While it always remains

Table 2–6.   Toxic Compounds Listed in Order of Frequency of Occurrence.

COMPOUND	FREQUENCY
1. Benzene	.688
2. Toluene	.656
3. Methylene chloride	.594
4. Chlorobenzene	.531
4. Dichloroethanes (total)	.531
4. Tetrachloroethylene	.531
7. Chloroform	.500
7. Dichloroethylenes (total)	.500
7. Ethylbenzene	.500
7. Phenol	.500
7. Trichloroethanes (total)	.500
7. Trichloroethylene	.500
7. Xylenes (total)	.500
14. Dichlorobenzene	.438
14. Naphthalene	.438
16. Vinyl chloride	.375
17. PCBs (total)	.344
18. BHCs (total)	.313
18. Acetone	.313
20. Dialkylphthalates (total)	.281

Source: Adapted from Hallstedt et al. (1986)

Table 2–7.   Toxic Compounds Listed in Order of Frequency and Hazard Rating.

COMPOUND	FREQUENCY × COMBINED HAZARD RATING
1. Chloroform	9.00
1. Trichloroethanes (total)	9.00
3. Methylene chloride	8.91
4. Benzene	8.25
5. Tetrachloroethylene	7.97
6. Dichlorobenzenes (total)	7.88
6. Toluene	7.88
8. Trichloroethene	7.50
9. Chlorobenzene	6.38
10. PCB's (total)	6.19
11. Phenol	6.00
11. Xylenes (total)	6.00
13. BHCs (total)	5.63
14. Naphthalene	5.25
15. Dialkylphthalates (total)	5.06
15. Pyrene	5.06
17. Dichloroethanes (total)	4.78
18. Carbon tetrachloride	4.50
18. Ethylbenzene	4.50
18. Vinyl chloride	4.50

Source: Adapted from Hallstedt et al. (1986).

possible that any given toxic chemical can be dangerous in some situations, these lists may give an indication of where to look first for the problems of subsurface migration of hazardous wastes.

REFERENCES

Bitton, G. and Gerba, C. P., *Groundwater Pollution Microbiology.* New York: John Wiley and Sons.

Creeger, S. M., 1986. Considering pesticide potential for reaching ground water in the registration of pesticides. In W. Y. Garner, R. C. Honeycutt, and H. N. Nigg (Eds.), *Evaluation of Pesticides in Ground Water,* pp. 548–557. Washington, D.C.: American Chemical Society, 573 pp.

Ghassemi, M., Quinliven, S., Haro, M., Metzger, J., Scinto, L., and White, H., 1983. *Compilation of Hazardous Waste Leachate Data,* RW Energy and Environment Division, Torrance, CA 90505. U.S. Environmental Protection Agency, Office of Solid Waste Management, Washington, D.C. 20460.

Hallstedt, P. A., Puskar, M. A., and Levine, S. P., 1986. Application of the hazard ranking system to the prioritization of organic compounds identified at hazardous waste remedial action sites. *Hazardous Waste and Hazardous Materials.* 3(2):221–232.

Mott, H. V., Hartz, K. E., and Yonge, D. R., 1987. Metal precipitation in two landfill leachates. *J. Env. Eng.* 113(3):476–485.

United States Environmental Protection Agency, 1982. Test methods for evaluating solid waste: physical/chemical methods. 2nd Edition, U.S. EPA. SW-846. July.

United States Environmental Protection Agency, 1985. Uncontrolled hazardous waste site ranking system: A users' manual. Appendix A in 40 CFR Part 300, National Oil and Hazardous Substances Pollution Contingency Plan; Proposed Rule. *Federal Register* 50(29).

Yen, T. F., 1988. Personal communication.

Young, P., and Parker, A., 1984. Vapors, odors, and toxic gases from landfills. *Hazardous and Industrial Waste Management and Testing: Third Symposium, ASTM STP 851,* pp. 24–41. L. P. Jackson, A. R. Rohlik, and R. A. Conway, Eds. Philadelphia, PA: American Society for Testing and Materials.

3

Principles of Groundwater Flow

Robert L. Stollar
Principal Hydrogeologist
R. L. Stollar & Associates, Inc.

INTRODUCTION

This chapter presents a brief discussion of groundwater flow and contaminant migration. It has been written to introduce the subjects in a very general manner and to acquaint the reader with concepts and terminology. The discussion includes: the occurrence of groundwater, recharge and discharge, rock types comprising aquifers, mechanisms of contamination, distribution of contaminants, and site investigations to determine groundwater flow and contaminant migration patterns.

Groundwater presently supplies a large portion of our country's water requirements. At almost any location, groundwater may be tapped to provide a water supply. However, the distribution of groundwater reservoirs (aquifers) capable of producing large supplies is more limited.

Water continuously circulates in our environment through the hydrologic cycle. When water falls as rain or snow, a portion infiltrates the ground, moving through the unsaturated zone to the saturated zone beneath the water table. This is the area where the void spaces between the rock particles are completely filled with water. The water in this zone is termed groundwater.

The ability of the rock material, or aquifer, to store and transmit water is a function of its porosity and hydraulic conductivity. Porosity is the relative amount of void space (pores) in a rock or unconsolidated deposit. The amount of groundwater stored is related to the porosity of the aquifer. The hydraulic conductivity of an aquifer is an expression of the ability of a fluid to move through the aquifer, and is dependent on the properties of the aquifer material and the fluid within the aquifer.

Groundwater is constantly flowing from areas of recharge to areas of discharge. This flow is normally laminar (nonturbulent). Darcy's Law states that the quantity of water which will flow across a section of an aquifer, during any given period, is directly proportional to the hydraulic conductivity of the aquifer material, the hydraulic gradient within the aquifer, and the cross-sectional area. Simply stated, the hydraulic gradient is the head (energy) loss between two points divided by the dis-

40

tance between the points. These terms are discussed in more detail later in this chapter.

The rate of groundwater flow varies enormously because the hydraulic conductivities of aquifer materials range over nine or ten orders of magnitude. Thus, under hydraulic gradients commonly encountered in nature, velocities may be tens of feet per day or one foot in tens of thousands of years. Flow rates in most aquifers, however, are a few feet per day to a few feet per year.

In water-table or unconfined aquifers under natural conditions, the flow pattern is generally a subdued reflection of the topography. Changes in the natural flow patterns occur if the aquifer is stressed by means of pumping or artificial recharge. The degree of change depends on the rate and duration of pumping or induced recharge and aquifer characteristics.

Contaminants that have entered the groundwater can move horizontally or vertically, depending on contaminant density and the natural flow pattern of the water in the aquifer. They tend to travel as a well-defined slug or plume, but can be reduced in concentration with time and distance by such mechanisms as adsorption, ion exchange, dispersion, and decay. The rate of attenuation or reduction in concentration is a function of the type of contaminant and of the local hydrogeologic framework, but decades and even centuries are required for the process to be completely effective.

Under the right conditions, and given enough time, contaminating fluids invading a body of natural groundwater can move great distances, hidden from view but still toxic. The eventual point of discharge of the contaminated groundwater may be a drinking water well or surface-water body.

GROUNDWATER RESOURCES

Groundwater is widely available and more accessible than surface water. A low yielding domestic or farm well can be constructed almost anywhere; over one-third of the nation is underlain by groundwater reservoirs capable of yielding at least 75,000 gpd (gallons per day) to an individual well; and there are large areas where hundreds or even thousands of gallons per minute can be obtained from wells or springs. Groundwater is usually not subject to the rapid and sometimes great fluctuations in availability characteristic of a surface-water supply and is therefore more reliable. Still, the use of groundwater remains small compared to available supply.

At least one half of the population of the United States depends on groundwater as a source of drinking water (Miller, 1977). Of the total population, 29 percent use groundwater delivered in community systems and 19 percent have their own domestic wells. In addition, millions of Americans drink groundwater from wells which have been completed for industrial plants, office buildings, restaurants, gas stations, recreational areas, and schools. Few of the domestic wells in the nation are subject to routine or even initial evaluation of water quality. The several hundred thousand water systems supplying industrial and commercial establishments are rarely monitored.

Groundwater contamination inevitably results. The U.S. Environmental Protection Agency (1977) has reported that almost every known instance of aquifer contamination has been discovered only after a water-supply well has been affected. Often by the time the contamination is identified, it is too late to apply simple remedial measures. Either treatment at the well or very expensive aquifer restoration must be accomplished.

The problem facing scientists involved in the protection of groundwater resources is to identify the areas and mechanisms by which pollutants enter the groundwater system and to develop a reliable understanding of their movement. This is necessary to minimize the impact of industrial, agricultural, or municipal activities on the quality of the groundwater resource.

The maze of federal, state, and local regulations addresses many of the complex disposal issues and potential sources of contaminants, but does not provide comprehensive protection of groundwater. The national strategy of groundwater protection will require a better understanding of the environmental, legal, technical, and economic complexities of the resource.

OCCURRENCE OF GROUNDWATER

Water beneath the surface of the ground occurs in the void space of geologic material and can be in the liquid, solid, or vapor forms. In the unsaturated zone (or vadose zone), the void spaces range from being filled mostly with air to being nearly saturated with liquid. The unsaturated or vadose zone is usually divided into the soil moisture zone, the intermediate zone, and the capillary fringe (Figure 3–1). The vadose zone is characterized by pore spaces in which liquid water occurs at pressures less than atmospheric. In almost all areas, the vadose zone is underlain by a zone where all of the void spaces are filled with water which is at or above atmospheric

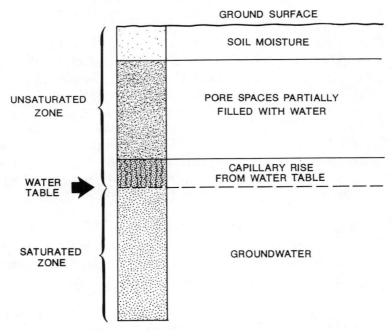

Figure 3–1.   Soil column.

pressure. This is the saturated zone. Until recently, studies of groundwater flow emphasized the saturated zone. In comparison, the movement of groundwater and contaminants in the vadose zone is much more complicated (Freeze and Cherry, 1979; Bear, 1972), and much research being carried out today is on the movement of water or contaminants through the vadose zone.

The soil zone supports plant growth and usually extends to less than ten feet below the land surface. It contains roots, voids left by the decayed roots of earlier vegetation, and animal and worm burrows. Beneath this is the intermediate zone. The thickness of the intermediate zone varies greatly and depends on the thickness of the soil zone and the depth to the capillary zone.

The occurrence of the capillary fringe is due to the molecular attraction between the liquid phase and the aquifer material, or solid phase, and from the surface tension of the air-water interface. As a result, water will rise in small diameter pores against the pull of gravity. Water in the capillary fringe and in the zones above are under a negative hydraulic pressure, that is, at a pressure less than atmospheric. In general, the smaller the diameter of the pore space, the higher the capillary rise. Near the bottom of the capillary fringe, the pore spaces are at or approach saturation.

The saturated zone starts where the pore spaces are filled with liquid and the water is at atmospheric pressure. The line dividing the capillary zone and the saturated zone is the water table (Figure 3–1). The water table is equivalent to the water level measured in a well which penetrates the saturated zone of an unconfined aquifer. Below the water table, the pore spaces remain saturated and the hydraulic pressure increases with increasing depth.

The saturated zone includes an extensive heterogeneous but interconnected geologic framework. This framework contains permeable water-bearing zones or aquifers, and less permeable units that act as confining layers, restricting the movement of groundwater.

Water-bearing rocks consist of saturated sedimentary, igneous, and metamorphic rocks. The sedimentary rocks are subdivided into clastics, evaporites, and carbonates. The clastic material includes consolidated or unconsolidated rocks. Igneous rocks may be plutonic (like granite) or volcanic (like basalt).

The consolidated sedimentary rocks, which are cemented or indurated, consist of fragments of other rocks transported from their source and redeposited. The unconsolidated materials, ranging in texture from clay (fractions of a millimeter) through silt and sand, to gravel and larger (on the order of centimeters), are not cemented or indurated. These unconsolidated deposits consist of material derived from consolidated sediment, igneous or metamorphic rocks, or other unconsolidated formations. The deposits consisting mostly of gravel and sand, with some silt and clay, form the most prolific aquifers.

Consolidated rocks, including sedimentary, igneous, or metamorphic rocks, consist of mineral particles of different sizes and shapes that have been welded by heat, pressure, or chemical reactions into a solid mass. Typical consolidated sedimentary rocks include limestone, dolomite, shale, siltstone, sandstone, and conglomerate. These may form aquifers.

Typical igneous rocks that are potential aquifers include basalt and fractured granite. The metamorphic rocks that may be used as aquifers include fractured schists and gneiss. Detailed discussions of the occurrence of water in geologic formations are given in Freeze and Cherry (1979), Davis and De Weist (1966), and Fetter (1980).

Water in any aquifer is contained in the voids within the rock. These are also called pore spaces or interstices.

Voids formed at the same time as the rock are called the primary porosity. For example, the pores in sand and gravel, and in other unconsolidated deposits, are primary openings (Figure 3–2). The voids occurring in lava tubes in basalt are also primary openings.

Voids created after the rock was formed are termed secondary porosity. Examples are the fractures in granite or consolidated sedimentary rocks (Figure 3–2).

Aquifers

Aquifers and confining beds are combinations of different geologic formations. Near the surface, few if any geologic formations are absolutely impermeable due to weathering, fracturing, and solution.

An aquifer is a saturated geologic formation, or group of geologic formations, which yields water to wells or springs in sufficient quantity to be of consequence as a water supply. Unconsolidated sands and gravels, sandstones, limestones and dolomites, basalt flows, and fractured rocks are examples of aquifers. A confining bed does not transmit water easily, and therefore restricts the movement of ground water either into or out of adjacent aquifers. A formation may be classified as an aquifer in one area but as a confining bed in another area depending on the availability of groundwater and the formation's ability to transmit water.

Aquifers act as storage reservoirs: during droughts, withdrawals can exceed recharge as the water table is drawn down. They are also conduits: water is transported from areas of recharge to areas of discharge.

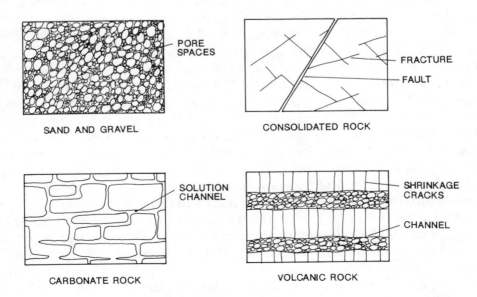

Figure 3–2. Types of void spaces.

Groundwater occurs in aquifers under two conditions. Where only the lower portion of the aquifer is filled with water, the upper surface of the saturated zone or water table is free to rise and fall. The water table is at atmospheric pressure and the water is unconfined. The aquifer is referred to as an unconfined or water-table aquifer (Figure 3–3).

Where the aquifer is completely filled with water under pressure and is overlain by a confining bed, the water in the aquifer is under confined conditions. Such aquifers are defined as confined or artesian aquifers (Figure 3–3).

In an unconfined aquifer under nonpumping conditions, the water level in a well is at the same elevation as the water table. The water table is responsive to changes in the amount of stored water, and fluctuates seasonally in response to variations in the rate of natural recharge.

Typically, the water table is highest in the late winter and early spring after natural recharge has occurred through precipitation, and before the irrigation season begins. Water levels are lowest in the fall, just after the high-use season.

The principal source of natural recharge to an unconfined aquifer is precipitation. Other sources are rivers, lakes, streams, and leakage from underlying aquifers.

Perched aquifers are local zones of saturation which occur within the unsaturated zone, above the regional water table. They are formed when water migrating downward reaches a localized confining bed. A zone of saturation forms above this layer (Figure 3–4). Although perched aquifers are sometimes tapped by wells, they are usually not sufficiently extensive to provide a significant supply of water. These units do, however, restrict and control recharge to the underlying aquifer and may be significant in contamination migration problems.

An alternative definition of perched systems relies on use criteria. For example, in California, if the aquifer used for water supply is deep and there is an overlying aquifer which is not used, the upper aquifer may be called a perched aquifer.

Confined aquifers are bounded above and below by geologic formations of rela-

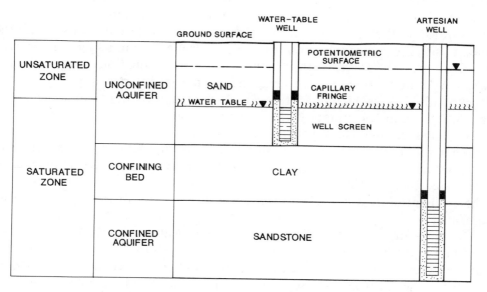

Figure 3–3. Confined and unconfined aquifers.

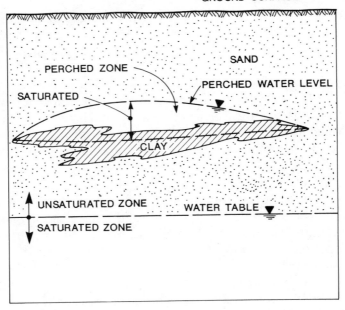

Figure 3-4. Example of a perched zone.

tively low hydraulic conductivity. In addition, the layer above the upper confining layer is often a water-table aquifer. The aquifer is completely saturated, and the upper surface is defined and fixed by the lower limit of the overlying confining unit. Under nonpumping conditions, when a well penetrates a confined aquifer, the water level stands above the top of the aquifer (artesian pressure). Where sufficient pressure is encountered, the water level may stand above the top of the well casing, causing the well to flow (artesian or flowing well). The hypothetical projection of the water levels is known as the potentiometric surface (Figure 3-3).

A confined aquifer does not receive recharge everywhere uniformly. Recharge is limited to specific recharge areas or areas where leakage from underlying or overlying formations occurs. Rather than being sensitive to volumetric changes, the water levels in wells in confined aquifers respond to changes in artesian pressure.

Rocks with identical characteristics may form an aquifer in one area, yet may act as a confining unit for a more permeable zone in another area. No confining unit is completely impermeable. Where a confining unit is sufficiently permeable to allow significant volumes of water to leak into or out of an aquifer, the formation is called a semiconfined or a leaky artesian aquifer. In addition, two artesian aquifers can be separated by a confining unit.

The mechanism of water release for a confined aquifer differs from that for an unconfined aquifer. In an unconfined aquifer, the water is removed from the pore spaces and the aquifer is dewatered (or drained). Water removal does not dewater a confined aquifer. Instead, water is released from storage by compaction of the aquifer and expansion of the water, while the void spaces remain filled. However, as water is removed from storage the pressure decreases and the potentiometric surface is lowered. An artesian aquifer becomes an unconfined aquifer when the potentiometric surface falls below the top of the aquifer.

The potentiometric surface may be at levels above or below the water table of an overlying unconfined aquifer. This will indicate whether water is leaking into or out of the aquifer (Figures 3-3 and 3-5).

The Hydrologic Cycle

Water moves continuously through aquifers, on the surface of the earth, and through the atmosphere. This endless process is the hydrologic cycle. The concept is central to an understanding of the occurrence of water and the development, management, and protection of water supplies.

Water evaporates from the oceans and other bodies of water, from exposed moist land, and from vegetation. This moisture rises and forms clouds, which will eventually return the water to the land or oceans in the form of precipitation. Precipitation includes rain, snow, and hail.

During a storm, infiltration rates vary widely, depending on land use, the character and moisture content of the soil, and the intensity and duration of the precipitation. When the rate of rainfall exceeds the rate of infiltration, overland flow or runoff occurs. Surface water flows from the land to streams, lakes, and ponds.

The first infiltrating water is absorbed by the soil in the unsaturated zone. After the soil moisture holding capacity is exceeded, the excess water migrates slowly through the intermediate zone and finally to the water table. The water flows through the aquifer to points of discharge.

The hydrologic cycle is illustrated in Figure 3-5. It also shows the mechanisms of groundwater movement, leakage between aquifers, and the concept of the hydraulic gradient. Groundwater is constantly moving from a point of recharge toward a point of discharge. If a particular region is a recharge area, the recharging water exerts a stress on the aquifer in the form of increased hydrostatic head. The concept of head is discussed in more detail later in the chapter. In areas which are designated as discharge areas, the hydrostatic head is usually lower. Thus, movement of groundwater is from regions of high hydrostatic head toward regions of low hydrostatic head. In practice, recharge and discharge areas of an aquifer are indicated by relative water levels. Within an aquifer, areas of high water-level elevations indicate higher hydrostatic head and areas of lower water-level elevations indicate lower hydrostatic head, so groundwater moves from areas of high water-level elevations toward areas of low water-level elevations. The hydraulic head difference, divided by the distance along the flow path, is known as the hydraulic gradient.

These concepts are illustrated in Figure 3-5. The water table is at the higher elevations at the recharge area in the center of the figure, and the water flows laterally downgradient to the pond and the stream. In addition, the figure shows the level to which the water rises in the two wells tapping a section of the artesian aquifer. In the well on the left, the water rises to a point below the water table. This elevation is the elevation of the potentiometric surface. The well on the right taps the same artesian aquifer and water rises to a point above the land surface. This well is then called a flowing or artesian well. The elevation of the water in this well is also

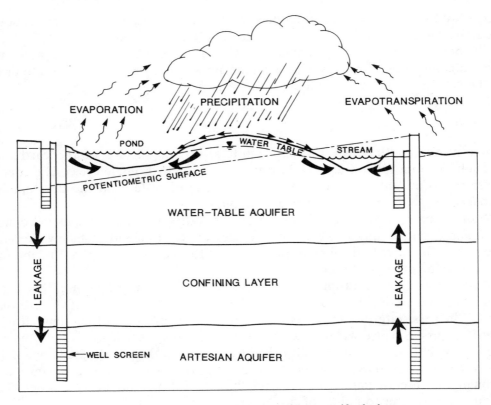

Figure 3-5. The hydrologic cycle and interaquifer leakage.

equivalent to the potentiometric surface. If the two points that represent the potentiometric surface are connected, the hydraulic gradient within the artesian aquifer can be determined. From this, groundwater in this artesian aquifer is determined to flow from right to left in Figure 3-5, down the hydraulic gradient.

In addition to precipitation, a water-table aquifer can be recharged where it is hydraulically connected to a surface-water source, such as a stream or a pond. A water-table aquifer can receive leakage through semipermeable confining beds of an underlying artesian aquifer. Artesian aquifers can receive recharge from or through confining beds or from precipitation and surface-water bodies in the outcrop area of the aquifer. Outcrops of geologic formations are where these subsurface formations are exposed at land surface or slightly below land surface.

Where the hydraulic head is greater in the water table than in the artesian aquifer (that is, the elevation of the water table is greater than the potentiometric surface), there is a potential for leakage to occur from the water-table aquifer to the artesian aquifer. This is shown in the pair of wells in the left side of Figure 3-5. If the gradient is reversed (that is, the elevation of the potentiometric surface is greater than the water table), leakage will occur from the artesian to the water-table aquifer. This is shown by the pair of wells on the right side of Figure 3-5.

Recharge locations can be points, lines, or areas. Artificial point recharge locations are very common, and are of major concern for groundwater contamination. These include waste disposal or recharge wells and individual septic tanks and cesspools. The beds of streams are commonly natural lines of recharge when the water levels are greater in the stream than in the aquifer. Reservoirs and large wastewater disposal ponds are examples of areas of artificial recharge.

Discharge locations for aquifers can also be points, lines, or areas. A spring is a natural point discharge, and a supply well is an artificial point discharge. Gaining streams (streams which receive water from the groundwater system) are line discharge areas. The water level in the stream is at an elevation below the water table. In this case, precipitation falling on adjacent upland areas infiltrates to the aquifer and flows to a nearby stream where it is discharged. Area discharge locations are swamps, ponds, lakes, and the sea.

It is important to understand that flow patterns are both regional and local (Figure 3–6). For example, water moving to the water table in a regional recharge area may infiltrate into the deeper section of the aquifer, move laterally, and discharge to a regional discharge area. However, water may also migrate to a local discharge area.

Water may also enter the system in a local recharge area. The difference must be understood when choosing the location of wells, to reliably determine aquifer flow conditions and contamination migration patterns.

Aquifer Geology

Groundwater occurs in all types of geologic formations. Knowledge of how these materials formed, and the changes that have occurred, is vital to understanding the hydrogeological system. The geologic formations are heterogeneous and anisotropic, which means that the hydrogeological properties of the formations are not uniform spatially or directionally. The hydrogeology of an area must be understood before the groundwater flow and contaminant migration patterns can be quantified. A hydrogeologic description of an area includes the review of previous geologic

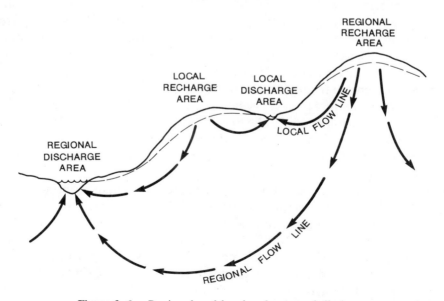

Figure 3–6. Regional and local recharge and discharge.

studies, site reconnaissance, field work, and analyses of data. Interpretation requires understanding of water storage and transmission (Fetter, 1980; Freeze and Cherry, 1979; and Davis and De Weist, 1966).

Aquifers may be composed of igneous, metamorphic, or sedimentary rocks. Igneous rocks are classified as plutonic (such as granite) or volcanic (basalt). Sedimentary rocks include clastic, evaporates, and carbonates. All of these may form either consolidated or unconsolidated deposits.

Clastic sedimentary rocks are composed of fragments of other rocks transported from their sources and deposited by water, glaciers, or air. Clastic rocks can be either unconsolidated or consolidated.

Consolidated Rocks. Hydrogeologically, consolidated rocks consist of mineral particles of different sizes and shapes that have been welded by heat, pressure, or chemical reactions. They may include sedimentary, igneous, or metamorphic rocks.

Determination of the hydrogeological characteristics of igneous and metamorphic rocks is not a trivial task. Extreme variations of lithology and structure, coupled with highly localized water-producing zones, make interpretation using the typical investigative techniques difficult. The storage and movement of groundwater occurs in small fractures. These fractures yield most of the water to wells.

It is difficult to determine the location of these fractures, and the groundwater flow patterns which result. In turn, prediction of the movement of contaminants through these fracture systems is very difficult.

Volcanic rocks display a wide range of hydrologic properties. Some, such as tuffs, have high porosity. Because the pores are not interconnected, the overall hydraulic conductivities are low and the tuffs produce very little water. The dense unfractured materials have very low porosities and hydraulic conductivities, and therefore, yield very little water to wells. Recent basalt flows, however, yield great amounts of water to wells.

If these formations contain water that can be pumped from wells, they are aquifers. Typical igneous rocks that may form aquifers include basalt and fractured granite. The metamorphic rocks serving as aquifers include fractured schists and gneiss. Detailed discussions of the occurrence of water in geologic formations are given in Freeze & Cherry (1979), Davis and De Weist (1966), and Fetter (1980).

Evaporites, which are formed by the precipitation of dissolved minerals, normally yield little water to wells. However, when such deposits come in contact with fresh groundwater, dissolution occurs. The dissolution can create highly permeable aquifers which produce water very high in dissolved solids. Eventually, underground caves, rivers, and sinkholes will form, generating a complex groundwater flow regime.

Carbonates, principally limestone and dolomites, originate from many sedimentary deposits. Organically precipitated limey muds, shell fragments, talus accumulations, calcite sand, reef masses, and deposits of the remains of small planktonic organisms are a few examples. Natural groundwater, which in some areas is reported to be slightly acidic, can slowly dissolve carbonate rocks along joints and fractures. The resultant porosity and hydraulic conductivity may range from low values, where the rock is slightly fractured, to extremely high values, where extensive fracturing and solution have taken place.

Sedimentary rocks can be partially consolidated by cementation. For example, a precipitated substance like silica or calcium carbonate may partially fill the pore space. The intergranular space or pore space available for groundwater is thus decreased. This decreases the available interconnected pore space and water transmission. In completely consolidated sedimentary rocks, the available pore space is further decreased by cementation and compaction.

In consolidated coarse-grained sedimentary rocks, like sandstone and conglomerate, pore space is usually small, and groundwater is stored and transmitted chiefly in fractures, joints between layers, and faults. There is virtually no pore space in fine-grained consolidated sedimentary rocks.

Unconsolidated Formations. The unconsolidated deposits are relatively uncemented and loosely compacted. These materials range from clay through fine sand to gravel and larger. These unconsolidated deposits consist of material eroded from distant sources, which may have been consolidated or other unconsolidated formations. The deposits consisting of mostly gravel and sand are important because they form the most prolific aquifers.

Where sand and gravel are homogenous, that is, the formation has many individual grains of the same size, the pore space will not be occupied by the grains. There is more available pore space that can contain water. Because of capillary and molecular attraction, some of this water is not available to wells.

In formations of silt and clay, even if the deposit is well-sorted (uniform in grain size), a significant percentage of the pore space is occupied by water held to the grains by capillary forces and molecular attraction, and is not available to wells. The quantity of water in storage is great, but the amount available to wells is small. Silts and clays are often confining layers in relation to aquifers of well-sorted sands and gravels.

The principal unconsolidated water-bearing formations in the United States are valley-fill, coastal plain, intermontane valleys, alluvium, and glacial drift aquifers. In areas where these aquifers are not in existence, the consolidated water-bearing systems may be aquifers.

Hydrogeology has always placed a major emphasis on the study of aquifers. Today, with the many contaminant-migration problems, more effort is being devoted to studying fluids moving through low-permeability beds. For example, although clays do not transmit water readily through pore spaces, contaminants do move through fractures in the clay. In fact, some chemicals increase the fracture permeability, allowing more of the chemicals to reach the aquifer below. It does not take large volumes of these contaminants to affect the groundwater in concentrations in the parts per billion range.

GROUNDWATER STORAGE

An aquifer is an underground storage reservoir which also serves as a conduit through which water moves to streams, wells, or other points of discharge. Fluctuations in the water table or potentiometric surface indicate that changes have occurred in the volume of water stored, just as changes in stage (or height) indicate changes in storage in a surface-water reservoir. The volume of water taken into or released from storage is proportional to the water-level change and to the specific yield for water-table aquifers and the specific storage for confined aquifers.

Porosity

The change in water storage is related to the total volume of void space available to hold water. The ratio of voids to the total volume of soil or rock is porosity. Porosity is expressed as either a decimal fraction or as a percentage. For example, if the total volume of soil is V_t, the volume of solids is V_s, and the volume of voids is V_v, then the porosity n is:

$$n = (V_t - V_s)/V_t = V_v/V_t.$$

If the volume of a saturated soil sample is one cubic meter and the volume of all the water within the sample (volume of voids) is 0.3 cubic meters, the porosity is:

$$n = 0.3/1.0 = 0.30 \text{ or } 30 \text{ percent.}$$

The porosity of unconsolidated deposits is related to the range in grain size (sorting), particle shapes, cementation, compaction, and fracturing. Fine grained materials tend to be better sorted and to have larger porosity. In well-sorted and well-rounded deposits, the size of the grains has no influence on porosity. For example, a deposit of boulders can have the same porosity as a deposit of clay. As small grains can fill the voids between larger grains, a formation that is not well sorted (has many different sizes of grains) will have a smaller porosity. Typical ranges of porosity for selected rocks are presented in Table 3–1.

Specific Yield

Only part of the water in an aquifer is available to wells, drains, springs, or seeps. Part is retained in the void spaces by the forces of tension, adhesion, and cohesion. This unavailable volume is retained water. The specific retention of the aquifer material is expressed as the percentage of the total volume of rock occupied by water that is retained in the interstices against the force of gravity. The portion available for withdrawal is the specific yield, or effective porosity, of the formation. It is the ratio of the volume of water that can be drained from a saturated rock by gravity to the total volume of the rock. Typical ranges of specific yield for selected rocks are presented in Table 3–1.

Thus, porosity is equal to the specific yield (S_y) plus the specific retention (S_r),

$$n = S_y + S_r \text{ where}$$

$$S_y = V_d/V_t$$

where V_d is the volume of water drained by gravity and

$$S_r = V_r/V_t$$

where V_r is the volume of water retained in the void space of the rock.

**Table 3–1. Representative Ranges of Porosity and Specific
Yield for Selected Rocks.**

ROCKS	RANGES FOR POROSITY (%)	SPECIFIC YIELD (%)
Gravel, coarse	20–35	10–25
Gravel, medium	20–35	15–25
Gravel, fine	20–40	15–35
Sand and gravel	20–35	15–30
Sand, gravelly	20–35	20–35
Sand, coarse	24–45	20–35
Sand, medium	25–45	15–30
Sand, fine	25–55	10–30
Dune sand	35–45	30–40
Loess	60–80	30–50
Peat	60–80	30–50
Till	25–45	5–20
Silt	35–60	1–30
Clay, sandy	30–60	1–30
Clay	35–55	1–20
Siltstone	20–40	1–35
Volcanic tuff	10–40	2–35
Sandstone (semiconsolidated)	1–50	1–48
Limestone	5–55	1–24
Igneous and metamorphic rocks, weathered	40–50	20–30
Fractured basalt	5–50	1–30
Dense crystalline rock	0–5	0–2

Source: Based on material from Morris and Johnson, U.S.G.S. (1967) and Davis and De Wiest, *Hydrogeology.* John Wiley & Sons, Inc., (© 1966).

An example illustrates the principle. A known volume of dry rock material is submerged in a known volume of water until saturated. The volume of water used to reach the saturation point can be calculated. If the rock material is allowed to drain, the volume of water drained can be measured:

Volume of rock: one cubic meter
Volume of water used to reach saturation: 0.3 cubic meters

Volume of water drained: 0.25 cubic meters then:

Total porosity = $n = V_v/V_t = 0.3 / 1.0 \times 100 = 30\%$
and the specific yield is

$$S_y = V_d/V_t = 0.25/1.0 \times 100 = 25\%$$

and the specific retention (S_r) is

$$S_r = (0.30 - 0.25)\ 100 = 5\%.$$

Typical values of specific yield and specific retention are shown in Table 3–2 and Figure 3–7.

With this information, the volume of water removed from an unconfined aquifer can be estimated by measuring water-level decline. For example, if the decline is one foot over an area of one square mile, the amount of water removed from aquifer storage can be calculated. For this example, assume that the aquifer comprises dune sand which has a specific yield of 30 percent.

Table 3–2.　Selected Values of Porosity, Specific Yield, and Specific Retention for Selected Rocks.

ROCK MATERIAL	POROSITY (%)	SPECIFIC YIELD (%)	SPECIFIC RETENTION (%)
Soil	55	40	15
Clay	50	2	48
Sand	25	22	3
Gravel	20	19	1
Limestone	20	18	2
Sandstone (semi-consolidated)	11	6	5
Granite	.1	.09	.01
Basalt	11	8	3

Source: Based on material from Morris and Johnson, U.S.G.S. (1967) and Davis and De Wiest, *Hydrogeology.* John Wiley & Sons, Inc., (© 1966).

Then:

$$1 \text{ foot of decline} \times 5280 \text{ ft} \times 5280 \text{ ft} = 27,878,400 \text{ ft}^3.$$

Multiplied by the specific yield (.30), indicates that 8,363,520 cubic feet of water (or 62,559,130 gallons) will be released.

The same principles allow calculation of water-table decline. Suppose 62.6 million gallons are uniformly removed from a dune sand aquifer over an area of one square mile, and the specific yield is 0.30 percent. The water-level decline (Δh) is:

$$(62.6 \text{ million gal} \times (1\text{ft}^3/7.48 \text{ gal}) \div (5280 \text{ ft} \times 5280 \text{ ft})S_y$$

$$= 0.30 \text{ ft} / S_y$$

$$= 0.30/0.30$$

$$= 1.0 \text{ ft of decline in the water table across the one square mile of aquifer area.}$$

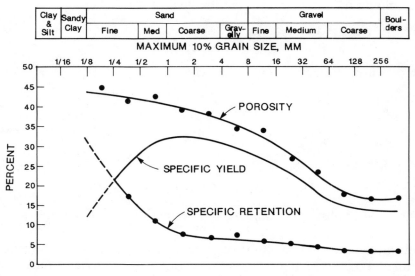

Figure 3–7.　Porosity, specific yield, and specific retention (adapted from D. K. Todd, *Ground-Water Hydrogeology.* John Wiley & Sons, Inc., © 1959).

It should be remembered that, in the part of the above solution resulting in 0.30 feet divided by S_y, the 0.30 feet indicates the decline in water level for a square mile in an equivalent open body of water. That is, if the decline occurred in a square lake, one mile on a side, and 62.6 million gallons were removed, the water level in that lake would fall 0.30 feet. However, in the aquifer, the water-level decline would have to be greater because the only space available for the water is within the pore or void space. Again, the percentage of available void space for adding water to or subtracting water from the aquifer is equivalent to the specific yield (S_y). Conceptually, it is realized that, if the area of and the decline or buildup within the aquifer are known, the volume of water can be computed. If the volume was just water being held in a tank or lake, the above volume (area × rise or decline of water level) would be the correct number. However, with the same areal extent as the above, and as the water is held only within the pore space, the thickness or height of the aquifer affected (filled or emptied) with the same volume of water would be greater than the height in the tank. An example is shown in Figure 3–8. Therefore, the volume would be divided by the specific yield (S_y). Similarly, if the volume of water taken from the aquifer is computed by measuring the decline or rise in water levels and the area affected is known, the volume of aquifer affected by this rise or decline can be computed. However, remembering that the water can only be derived from the void or pore spaces, and that the volume computed above is for the entire section of aquifer, the volume should be multiplied by the specific yield (S_y) to determine the actual volume of water released from or added to storage.

Specific Storage

For confined aquifers, calculations of water yield reflect different phenomena. The *storativity* of an aquifer is defined as the volume of water released or taken into storage per unit surface area of the aquifer per unit decline or rise in head. It is a dimensionless coefficient synonymous with *storage coefficient:*

  S = Volume of water / Unit area × Unit head change

   = Cubic feet / Square feet × feet

   = Cubic feet / Cubic feet.

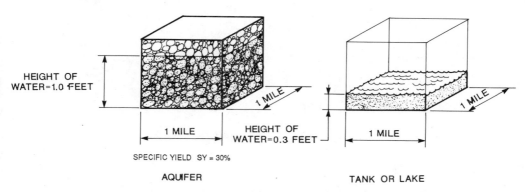

HEIGHT OF WATER=1.0 FEET

1 MILE

1 MILE

HEIGHT OF WATER=0.3 FEET

1 MILE

1 MILE

SPECIFIC YIELD SY = 30%

AQUIFER

TANK OR LAKE

Figure 3–8. Storage and volume of water held within an aquifer.

For unconfined aquifers, the storativity is equivalent to the specific yield (assuming gravity drainage is complete). In a confined aquifer, the storage coefficient reflects the water released from storage within the aquifer due to the expansion of water and the compression of the aquifer.

When water is discharged from a confined aquifer, the potentiometric surface remains above the top of the aquifer. (If it drops below the top of the aquifer, the aquifer becomes unconfined.) Therefore, the mechanism of removing water is not through the actual dewatering of the aquifer, as in the unconfined situation. Detailed discussions of these mechanisms are found in Bear (1979) and Freeze and Cherry (1979).

In simple terms, withdrawal of water produces a change in pressure. The change in pressure produces only small changes in storage volume. The hydrostatic pressure within an aquifer partially supports the overburden while the solid structure provides the remaining support. Withdrawal of water reduces the pressure, producing a small decline in storage volume. The reduction of hydrostatic pressure increases the load on the aquifer.

The compression of the aquifer causes a release of water. In addition, the lower pressure allows the water to expand. The water released in a confined aquifer is related to the water released due to the expansion of water and compaction of the aquifer. This water released is also equivalent to the storage coefficient. The range of the storage coefficient for a confined aquifer is from 0.00005 to 0.005 (the coefficient is in cubic feet of water released per square foot of aquifer per foot of decline in potentiometric head, and so is dimensionless (Bear 1979; Freeze and Cherry, 1979)).

Water expansion and aquifer compression occur in unconfined aquifers as well. However, the contribution to the volume of water released is negligible compared to the volume of void drainage. For example, in a unconfined aquifer, the volume of water released from storage due to the expansion of water and compression of the aquifer is about 0.00005, while that due to drainage is 0.2; therefore the total storage coefficient is 0.20005, very close to 0.2.

The difference between withdrawals from unconfined and confined aquifers may be illustrated. Assuming that the respective storage coefficients are 0.15 and 0.0015, it is possible to compute the volume of water released from an aquifer of one square mile if the respective water level surfaces decline five feet:

$$V = S \times A \times \Delta h$$

where

S is the storage coefficient;
A is the area of the aquifer; and
Δh is the change in head.

In the unconfined unit this would be:

$$(5 \text{ ft}) \times (5280 \text{ ft}) \times (5280 \text{ ft}) \times (0.15) = 21 \times 10^6 \text{ ft}^3.$$

In the confined aquifer this would be:

$$(5 \text{ ft}) \times (5280 \text{ ft}) \times (5280 \text{ ft}) \times (0.0015) = 0.21 \times 10^6 \text{ ft}^3.$$

One hundred times less water is released from storage for the same decline in head in the confined aquifer.

For another example, if the 21×10^6 cubic feet of water were released from one square mile of an unconfined aquifer, it is possible to compute the area of a confined aquifer needed to release an equivalent amount of water:

$$21 \times 10^6 \text{ ft}^3 = 0.0015 \times (5 \text{ ft}) \times (\text{area}),$$
where 5 ft is the water level decline;

$$2.8 \times 10^9 \text{ ft}^2 = \text{area} = 100 \text{ mi}^2.$$

Note that the ratio of 21×10^6 to 0.21×10^6 is 100. There are great differences between the storage coefficients of unconfined and confined aquifers. But where potentiometric levels in a confined aquifer are reduced below the top of the aquifer, the aquifer is changed from confined to unconfined conditions. Therefore, the storage coefficient increases from values representing a confined aquifer to values representing an unconfined aquifer. The differences in the storage coefficient of confined and unconfined aquifers are very important when determining the response of aquifers to pumping.

GROUNDWATER FLOW

Groundwater is constantly flowing from areas of recharge to areas of discharge. This flow is laminar, and the quantity of water flowing is directly proportional to the hydraulic conductivity of the aquifer material, the hydraulic gradient, and the cross-sectional area.

Heads and Gradients

Movement of groundwater is from points of higher energy or fluid potential to points of lower energy. In hydrology, the energy level is usually called the energy head or head, and is expressed in units of length (e.g., feet). Total head is the sum of elevation, pressure, and velocity head. Because velocities in porous media are usually low, velocity heads may be neglected without appreciable error. Therefore, the potential energy equation can be reduced to:

$$h_t = z + h_p$$

where h_t is the total head, z is elevation head, and h_p is pressure head.

If the elevation of the measuring point is known and the depth to water is subtracted from the measuring point, the result is the total head as measured at the well (Figures 3–9 and 3–10). The elevation head (z) is measured relative to some standard elevation (usually sea level) to the center of the well screen. The pressure head (h_p) is usually measured from the well screen to the water table. If the water flow is horizontal, the total head is the same in any vertical section of the aquifer. If the screen were near the bottom of the aquifer, h_p would increase and z would decrease; however, the total head (h_t) would remain the same.

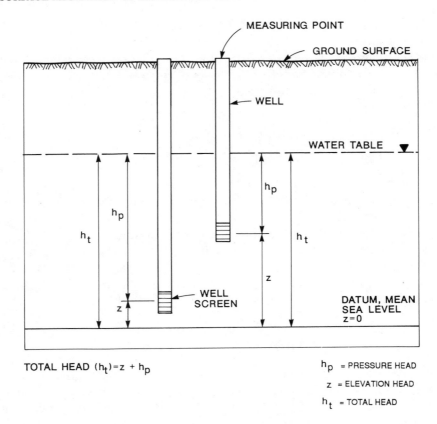

Figure 3–9.   Definition of total, elevation, and pressure head.

Examples of the relationship of pressure, elevation, and total head are shown in Figure 3–10. Figure 3–10A and B indicate what is generally thought to be a typical case, that is, water flows "downhill" or downgradient. The bottom of this aquifer is either tilted from left to right, as in Figure 3–10A, or is horizontal, as in 3–10B. The total head (h_t) is greater on the left side of the diagram, and water flows from left to right. Figure 3–10C is different. Here, the bottom of the aquifer is tilted upward away from left to right and it appears as though water is flowing "uphill." In actuality, the total head is greater on the left and water is flowing from higher head to lower head, or downgradient. Groundwater moves in the direction of decreasing total head, which may or may not be in the direction of decreasing pressure head.

The direction and rate of movement of groundwater is controlled by the hydraulic gradient. The hydraulic gradient is the change in total head per unit distance in a given direction. In groundwater terminology, the term total head is usually shortened to head. The change in head (Δh) is divided by the distance between measuring points. In Figure 3–11, $h_{t1} - h_{t2} = 110 - 100 = 10$ feet. The distance between the wells is 2500 feet; therefore, the hydraulic gradient is 10 feet / 2500 feet or 0.004 foot/foot.

Graphic depiction of groundwater flow routinely employs equipotential and flow lines. Equipotential lines connect points of equal head, and thus represent the water-table surface of an unconfined aquifer or the potentiometric surface of a confined

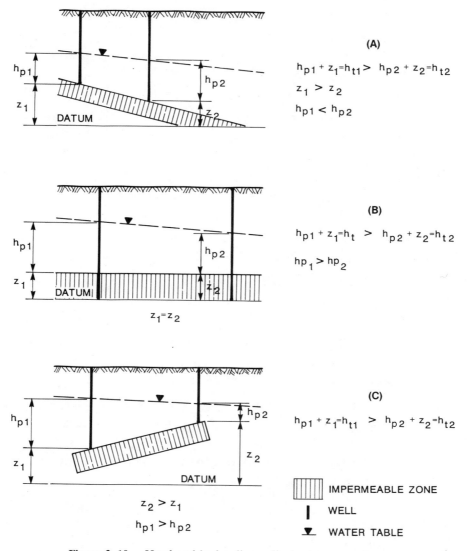

Figure 3–10. Head and hydraulic gradient measurements.

aquifer above the same data plane. The representation of the equipotential surface (or water-table surface or potentiometric surface) is similar to a topographic map of the land surface. They illustrate the three-dimensional nature of the water table or potentiometric surface. For example, where the lines are close together, the surface slopes steeply.

Flow lines show idealized paths followed by particles of water as they move through the aquifer. Because groundwater moves in the direction of the hydraulic gradient, flow lines are perpendicular to equipotential lines in homogeneous and isotropic aquifers.

In actuality, there are an infinite number of equipotential lines and flow lines describing flow in an aquifer. Only a few need be drawn to illustrate groundwater flow. Equipotential lines are drawn so that the change in head is the same between adjacent pairs of lines. Flow lines are drawn so the flow is equally divided between

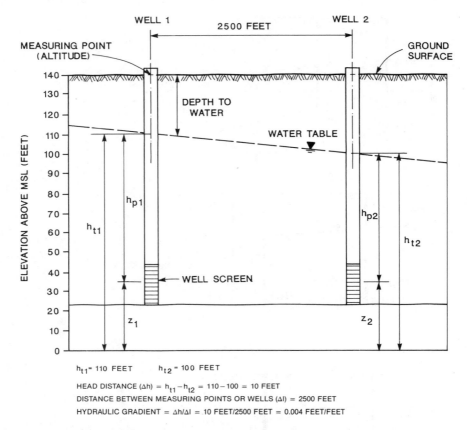

$h_{t1} = 110$ FEET $h_{t2} = 100$ FEET

HEAD DISTANCE (Δh) = $h_{t1} - h_{t2}$ = 110 − 100 = 10 FEET

DISTANCE BETWEEN MEASURING POINTS OR WELLS (Δl) = 2500 FEET

HYDRAULIC GRADIENT = $\Delta h / \Delta l$ = 10 FEET/2500 FEET = 0.004 FEET/FEET

Figure 3–11.   Cross section illustrating hydraulic gradient.

pairs of flow lines. If this is done exactly, a flow net is created (Bear, 1979; Freeze and Cherry, 1979).

Flow nets conveniently illustrate the direction of groundwater flow and can be used to approximate the quantity and velocity of water moving through an aquifer. Examples of flow diagrams shown both in map view (areal view) and cross-sectional view are illustrated in Figure 3–12.

Preparation of groundwater flow diagrams requires water-level and elevation data. Well construction and geologic data must be analyzed. All of the water-level data used must represent the same section of one aquifer, or the data will be misleading. Wells tapping different aquifers should not be used to construct flow diagrams; different wells tapping the upper and lower sections of an aquifer and those that fully penetrate the same aquifer, should not be used to construct a flow map. Figure 3–13 is an illustration showing these relationships. Failure to distinguish between water levels of different aquifers and to identify wells which connect more than one aquifer are common mistakes in preparing potentiometric maps.

There are studies utilizing water levels from different aquifers and different sections within the same aquifer. Properly understood, this data can be used to determine vertical movement of water, and interaquifer exchange of water.

In general, if the elevation of the measuring point (for example, a notch cut in the top of the casing) is determined and the depth to water is measured from this point, the water-table elevation, or potentiometric surface, of the aquifer is ob-

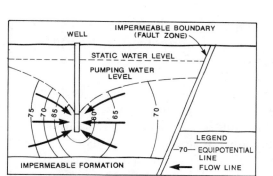

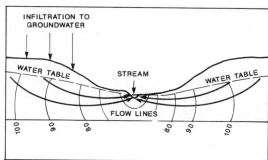

CROSS SECTION NEAR PUMPING WELL SHOWING THE FLOW PATH FOLLOWED BY WATER MOVING TOWARDS THE WELL

CROSS SECTION THROUGH A STREAM VALLEY SHOWING FLOW LINES IN THE GROUNDWATER SYSTEM

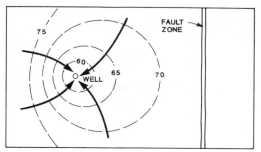

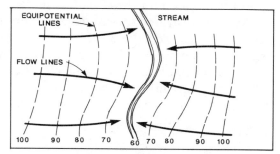

MAP OF POTENTIOMETRIC SURFACE DURING LONG-TERM PUMPING SHOWS THAT THE IMPERMEABLE BOUNDARY CAUSES GREATER DRAWDOWN IN THE DIRECTION OF THE FAULT

AERIAL VIEW OF WATER-TABLE CONFIGURATION FOR STREAM VALLEY REPRESENTED IN THE UPPER FIGURE.

Figure 3–12. Examples of flow maps (adapted from Heath, U.S.G.S. 1984).

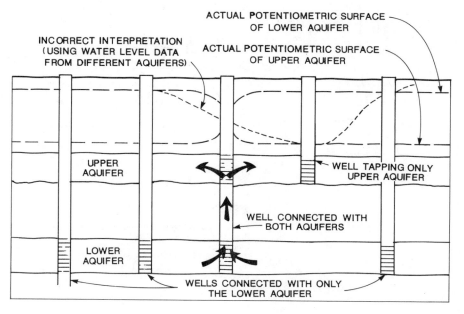

Figure 3–13. Water levels in relationship to well completion (adapted from Davis and De Wiest, *Hydrogeology*. John Wiley & Sons, Inc., © 1966).

tained at the location of the well. With enough wells, an equipotential map can be constructed, which is the elevation of the water table, or potentiometric surface.

An example of a water-table map is given in Figure 3–14. In area A of Figure 3–14, the depth to the water table is greater than the bottom of the stream, and the geologic material is permeable. Water from the stream is recharging the aquifer. The stream is a losing stream. The downstream bend of the contours indicates this. Flow lines are drawn from the stream to the aquifer. In area B, the equipotential lines are perpendicular to the stream. This indicates that the stream is at the same elevation as the water table, neither gaining nor losing, and groundwater flow is parallel to the stream at this point. Further downstream, at C, the equipotential lines bend sharply upstream. The flow lines indicate that groundwater is discharging into the stream and the stream is gaining.

Area D illustrates an area where pumping has affected the contours. The contours completely encircle the well. The lowest elevations are near the well. The flow lines show that the water is flowing toward the well from all directions. In addition, there are two contours with the same elevation (175) near the well. This indicates, as we move downstream of the well, water-level elevations increase (showing flow toward the well) until a certain point and then the elevations begin to decrease, showing

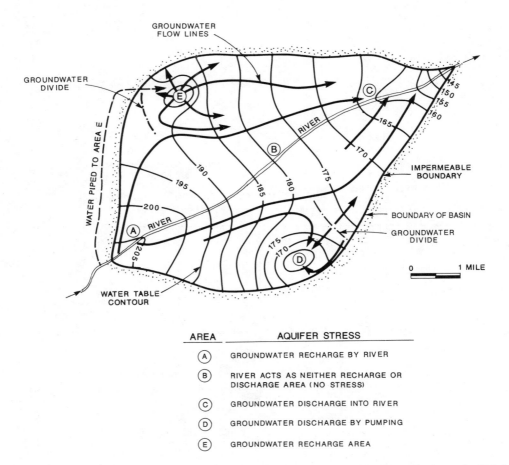

AREA	AQUIFER STRESS
A	GROUNDWATER RECHARGE BY RIVER
B	RIVER ACTS AS NEITHER RECHARGE OR DISCHARGE AREA (NO STRESS)
C	GROUNDWATER DISCHARGE INTO RIVER
D	GROUNDWATER DISCHARGE BY PUMPING
E	GROUNDWATER RECHARGE AREA

Figure 3–14. Hypothetical water table map (adapted from Davis and De Wiest, *Hydrogeology.* John Wiley & Sons, Inc., © 1966).

flow is away from the well. This area is called the groundwater divide: a point of stagnation where the groundwater velocity is zero. Groundwater on opposite sides of the divide flow in opposite directions.

The limiting flow line is also an important concept (Figure 3-14). It is the approximate location of the last flow line that would be intercepted by the cone of depression caused by pumping the well. Any flow lines on the opposite side of the limiting flow line from the well may be influenced by pumping, but will continue to flow downstream of the cone of depression following the regional or natural flow paths. Any flow line between the limiting flow line and the well will illustrate that water will flow toward the pumping well. The concepts of limiting flow lines and groundwater divides become very important when determining the movement of contaminants and attempting to control the movement of contaminants.

An area of recharge is shown at area E. Here, water is being added to the groundwater system, and the groundwater is flowing in all directions away from the point of recharge. Note that a groundwater divide and limiting flow lines also exist for point sources of groundwater recharge.

From this short exercise, it can be seen how important flow maps can be to interpret the groundwater flow system. In addition to the obvious changes that occur to the shape and elevations of contours near points of groundwater recharge and discharge, the shape and spacing of contours also allows interpretation of aquifer thickness and hydraulic conductivity. An example of this is given in Figure 3-15. The upper diagram is a groundwater contour map. The two lower diagrams are interpretations that can be inferred from the contour map.

The contours in Figure 3-15A represent flow through a confined aquifer. The abrupt change in the spacing of the contours is explained by Figure 3-15B. The steepening of the gradient is caused by a significant decrease in hydraulic conductivity. That is, if the quantity of flow through the system remains constant, more head is needed to push the water through a material with a lower hydraulic conductivity. Therefore, the gradient steepens, as shown by the closer spaced contour lines.

Gradient changes can also be caused by changes in the area through which the water must flow (Figure 3-15C). If the width of the aquifer decreases or is constricted, more energy is needed to push the water through, and the gradient increases. Again, this is shown by closely spaced contour lines.

To determine the hydraulic gradient from an equipotential map, the head difference (Δh) is measured along a flow line between the two equipotential lines (Figure 3-16). The length of the flow line (Δl) is also measured. From the figure, the change in head (Δh) between points A and B is 99 feet − 96 feet, or Δh is 3 feet. The length along the flow line is 1,000 feet. Therefore, the gradient ($\Delta h/\Delta l$) is 3/1000, or 0.003.

Preparation and use of groundwater contour maps is fundamental to understanding the groundwater flow systems and contamination migration patterns (Bear, 1979; Fetter, 1980; Freeze and Cherry, 1979; Ferris et al. 1962; McWhorter and Sunada, 1977; Todd, 1980; and Walton, 1970).

Hydraulic Conductivity

Aquifers act as porous conduits transmitting water from areas of recharge to areas of discharge. A measurement of how well an aquifer transmits water is the hydraulic conductivity. This term replaces the term field coefficient of permeability or perme-

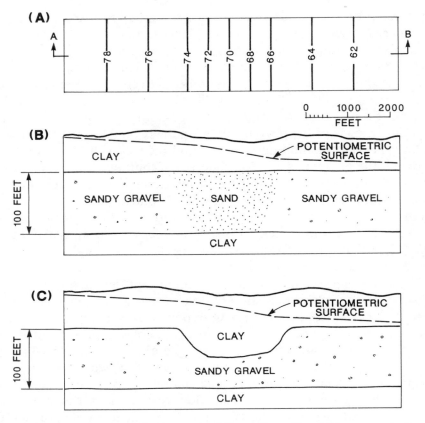

Figure 3–15. The effect of hydraulic conductivity and aquifer thickness on hydraulic gradient (adapted from Heath and Trainer, *Introduction to Ground Water Hydrology.* John Wiley & Sons, Inc., © 1968).

ability, and should be used to define the water-transmitting characteristic of the geologic material in quantitative terms. However, in qualitative terms, geologic materials are still described as being permeable and impermeable.

Quantitatively, hydraulic conductivity (K) is a measure of both the geologic material and the fluid moving through the material. It is directly proportional to the specific weight of the fluid and inversely proportional to the dynamic viscosity of the fluid. For geologic materials, hydraulic conductivity is influenced by the mean grain size, distribution of grain size, sphericity and roundness of the grains, and the nature of their packing.

Conceptually, the hydraulic conductivity is the volume of water per unit time that moves through a unit area of aquifer with a unit decline in hydraulic gradient. Although the units are the same as those for velocity, hydraulic conductivity is not a measurement of velocity. It is one of the parameters used to calculate the velocity of groundwater. In practice, hydraulic conductivity is stated in units of feet per day (ft/day), gallons per day per square foot (gpd/ft^2), centimeters per second (cm/sec) and meters per day (m/day).

Hydraulic conductivity is variable for different types of rock material, and varies within a rock unit from place to place (Table 3–3). If the conductivity is the same in different areas of the aquifer, the aquifer is homogeneous; if the hydraulic conductivity varies from place to place, the aquifer is heterogeneous.

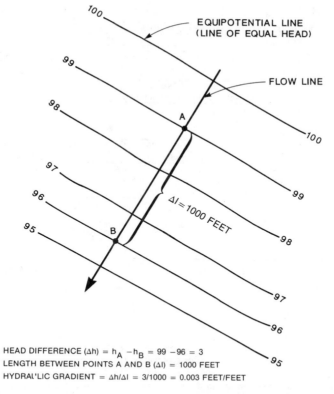

HEAD DIFFERENCE (Δh) = $h_A - h_B$ = 99 − 96 = 3
LENGTH BETWEEN POINTS A AND B (Δl) = 1000 FEET
HYDRAULIC GRADIENT = $\Delta h / \Delta l$ = 3/1000 = 0.003 FEET/FEET

Figure 3-16. Calculating hydraulic gradient from an equipotential map.

If the values of conductivity are the same in all directions, the aquifer is isotropic. If the hydraulic conductivity differs with direction, the aquifer is anisotropic. The hydraulic conductivity is generally greater for horizontal movement than for vertical.

Another measure of the ability of aquifers to transmit water is transmissivity. The transmissivity (T) of an aquifer is equal to its hydraulic conductivity (K) multiplied by the saturated thickness of the aquifer (b):

$$T = Kb.$$

The units of transmissivity are gallons per day per foot (gpd/ft), square feet per day (ft²/day), or square meters per day (m²/day).

The transmissivity is the parameter computed during field tests to determine aquifer coefficients, and in computations to solve groundwater problems. As transmissivity is proportional to saturated thickness, it is a variable for an unconfined aquifer and a constant for a confined aquifer that is isotropic and homogeneous. In an unconfined aquifer, the water table defines the upper boundary of saturation. As the water table rises or declines, the saturated thickness either increases or decreases, respectively. Therefore, the transmissivity also increases or decreases in proportion to the saturated thickness. In a confined aquifer, the saturated thickness is a constant. Therefore, the transmissivity remains constant. More details on transmissivity are given by Bear (1979), Freeze and Cherry (1979), and Todd (1980). It should be realized that even changes in saturated thickness due to natural water-

Table 3-3. Ranges of Hydraulic Conductivity for Selected Rocks.

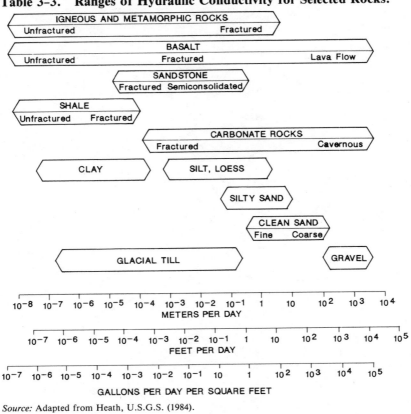

Source: Adapted from Heath, U.S.G.S. (1984).

level fluctuations will cause the transmissivity of an unconfined aquifer to vary, even if the aquifer is homogeneous and isotropic.

Darcy's Law and Groundwater Velocity

The velocity and movement of groundwater can be estimated using Darcy's Law (Figure 3-17). Darcy's Law indicates that the discharge of water is proportional to hydraulic conductivity, hydraulic gradient, and area of flow:

$$Q = KA\Delta h/\Delta l, \text{ where:}$$
K is the hydraulic conductivity
A is the area of flow
$\Delta h/\Delta l$ is the hydraulic gradient.

The units of discharge, $Q(L^3/T)$, are (for example) in gallons per day (gpd) or cubic feet per second (ft³/sec).

The specific discharge through the aquifer is derived by dividing both sides of Darcy's equation by the area of flow, A. Thus,

$$q = Q/A = K\Delta h/\Delta l, \text{ units are in } L/T$$

The units of specific discharge are feet per day, or meters per day (Figure 3-17). This is equivalent to the discharge measured as water flows through a pipe filled

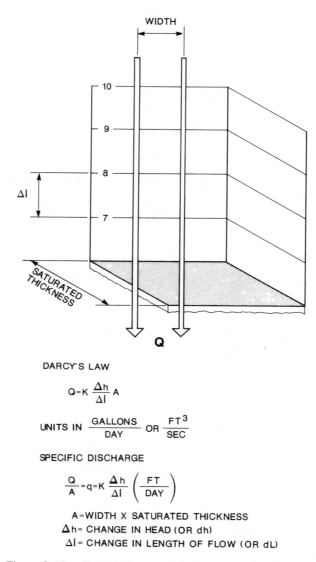

DARCY'S LAW

$$Q = K \frac{\Delta h}{\Delta l} A$$

UNITS IN $\frac{\text{GALLONS}}{\text{DAY}}$ OR $\frac{\text{FT}^3}{\text{SEC}}$

SPECIFIC DISCHARGE

$$\frac{Q}{A} = q = K \frac{\Delta h}{\Delta l} \left(\frac{\text{FT}}{\text{DAY}} \right)$$

A = WIDTH X SATURATED THICKNESS
Δh = CHANGE IN HEAD (OR dh)
Δl = CHANGE IN LENGTH OF FLOW (OR dL)

Figure 3–17.   Darcy's Law and flow through an aquifer.

with sand. It is really a discharge per unit area (Figure 3–18). The specific discharge is not equivalent to the velocity of groundwater.

The cross-sectional area of flow for a porous medium is actually much smaller than the dimensions of the aquifer. Groundwater can only flow through the pore spaces. In addition, part of the pore space is occupied by stagnant water (related to specific retention) or water which clings to the aquifer material. The only pore space that is available for saturated flow is that measured by the specific yield (S_y) or effective porosity (n_e) of the aquifer.

The effect is like a narrow constriction in a pipe (Figure 3–18). If the flow in the pipe remains the same, and the diameter remains the same, the velocity throughout the pipe is constant. At a point where the pipe is constricted, the velocity must increase in order to move the same amount of water. In porous media, the pores

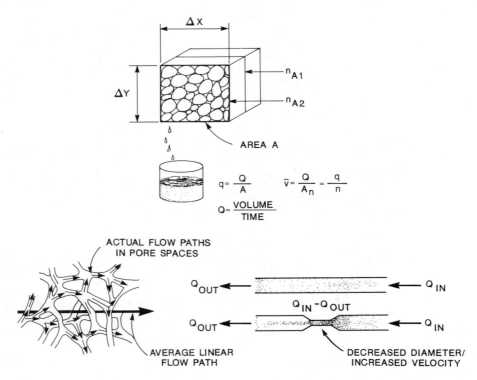

Figure 3-18. Relationship between specific discharge and average velocity.

represent a similar reduced cross-section for flow (Figures 3-17 and 3-18). The velocity of water through an aquifer is:

$$v = q/n_e = (K\Delta h/\Delta l)/n_e$$

The units are feet/day or meters/day. For example, if the hydraulic conductivity is 180 feet/day, the gradient is 0.001 foot/foot, and the effective porosity is 0.2, the specific discharge (q) and the velocity (v) are:

$$q = Q/A = (K\Delta h/\Delta l) = 180 \text{ ft/day} \times 0.001 \text{ ft/ft} = 0.18 \text{ ft/day}$$

$$v = q/n_e = 0.18/0.20 = 0.9 \text{ ft/day}$$

A more rigorous approach to velocity considers the sinuosity of torturosity of the flow paths. Water must flow around sand grains and only through pore spaces that are continuous and interconnected (Figure 3-18). However, these exact paths cannot be measured. The velocity computed through a porous media must be considered an average linear velocity. Some drops of water flowing through the aquifer will move faster and some will move slower than the average linear velocity.

The flow paths within porous media converge and diverge constantly. Thus, even with laminar flow, there is some intermingling of flow. This produces transverse dispersion: mixing at right angles to the direction of groundwater flow (Figure 3-19). There are also differences in velocity caused by friction between the water and the grains. The water velocity is smaller near the grain and greater in the center of

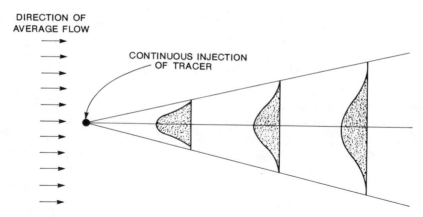

Figure 3-19. Effects of dispersion on a groundwater contaminant (adapted from Freeze and Cherry, *Groundwater,* © 1979, pp. 76, 396, 399. Reprinted by permission of Prentice-Hall, Inc., Englewood Cliffs, NJ).

the pore. This causes longitudinal dispersion: mixing in the direction of flow (Figure 3-20).

Dispersion becomes more important when dealing with the movement of solutes, the substances dissolved in the water. Solutes can be contaminants, tracers, or natural substances. In general, they are transported with flowing groundwater. But as the solute moves dispersion tends to spread it. It is diluted by the mixing.

Solutes are also diluted by molecular diffusion, which carries them from regions

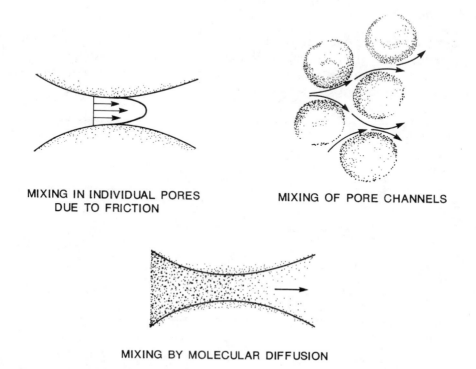

Figure 3-20. Some causes of dispersion (Adapted from Freeze and Cherry, *Groundwater,* © 1979, pp. 76, 396, 399. Reprinted by permission of Prentice-Hall, Inc., Englewood Cliffs, NJ).

of high concentration to regions of low concentration (Figure 3–20). Diffusion is more important where velocities are very small.

The effects of longitudinal dispersion may produce a breakthrough curve (Figure 3–21). This is a plot of the ratio of the measured concentration to the original concentration of a solute versus time. A hypothetical experiment illustrates the concept. Suppose the solute is injected at a constant rate and concentration. At some point downstream, the concentration of the water is measured. The concentration rises slowly at first, as only the fastest flowlines arrive at the point of measurement. The arrival of the bulk of the solute causes a rapid rise. Finally, the measured concentration gradually approached the input value as the medium is saturated. The spread and delay of arrival times will be much greater where adsorption is occurring (Chapter 5).

The phenomena of dispersion is important in the study of groundwater pollution. However, recent research indicates that its importance in defining the movement and shapes of plumes of contamination is not as great as was once thought.

Dispersion is difficult to measure in the field because the movement of contaminants is also affected by the aquifer heterogeneity, stratification, soil-water-solute relationships, ion exchange, filtration, and other conditions and processes. Stratification, heterogeneity, and anisotropic properties of aquifers and confining beds have a much greater effect on the movement of solutes than does dispersion. Indeed, the measured values of dispersion in the field are probably more a measure of the heterogeneity of the aquifer material than true dispersion.

Recharge and Discharge

In water-table aquifers under natural conditions, the flow patterns are generally a subdued reflection of the topography. Water flows from high to low areas. Changes in the natural flow patterns occur if the aquifer is stressed artificially by means of pumping or recharge.

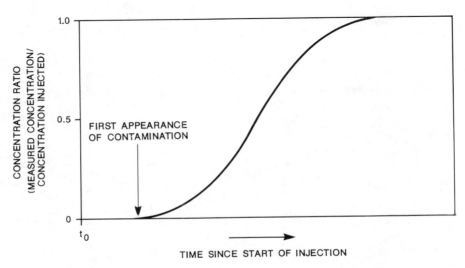

Figure 3–21. Arrival of a contaminant at a point with time.

In unconfined aquifers, recharge occurs in topographically high areas and discharge generally occurs in the topographically low areas. The water table in recharge areas is usually deep (the unsaturated zone is thick) and the water table in discharge areas is very close to the surface (Figure 3-6).

The potentiometric surface for confined aquifers is similar to the water table. This surface has a gradient from areas of recharge to areas of discharge. In relatively shallow confined aquifers these regional recharge and discharge areas can be shared with the unconfined aquifers. In deeper confined aquifers, the recharge area is usually in the outcrop area and the discharge areas are located where the top of the aquifer is near the bottom of major streams, lakes, estuaries or seas.

Aquifers also receive recharge and discharge through the process of leakage. Leakage rates are controlled by the vertical hydraulic conductivity and thickness of the deposits through which leakage occurs, the head difference between the source formation and the aquifer being recharged, and the area through which leakage is occurring. The leakage can be a major source of either recharge to or discharge from an aquifer over large areas (Bear, 1979; Freeze and Cherry, 1979; Freeze and Witherspoon, 1966; Toth, 1962; and Walton, 1970).

The key to understanding leakage, recharge, and discharge is proper interpretation of data from properly located monitoring wells. The example of Figure 3-22 is used in the following paragraphs to illustrate the important concepts.

Monitoring wells A and B are located in a recharge area, while wells C and D are located in a discharge area. Each is only opened to the formation at the bottom of the well, so the measured head only represents this point within the aquifer. The water-level elevations in wells A and B are equal, as both wells intercept the same equipotential lines. However, A is located at the water table and B is located at a depth below the water table. The water level in A is at the water table, as expected. However, the water-level elevation in B is below the elevation of the water table. In this case, if a shallow well were located next to B, the water level would be at the water table, which is higher than the water level in B. This indicated that ground

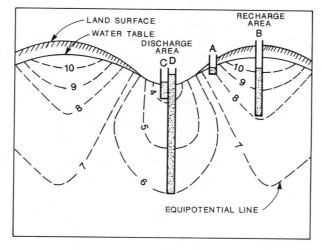

Figure 3-22. Water level comparisons in recharge and discharge areas.

water has a downward flow component and the well is located in a groundwater recharge area.

In a discharge area, the relationship is different. This is shown in adjacent monitoring wells C and D. Assume that the wells are located next to each other. However, well C is located at the water table and well D is located at depth. The water level in well D is higher than the water table, and therefore, higher than the water level in well C. The water level in well D may be misinterpreted as representing a confined aquifer.

Stagnation points (Figure 3–23) exist naturally in complex regional and local discharge and recharge systems. At the stagnation point, the groundwater has a velocity of zero.

Care must be taken when analyzing chemical samples taken from wells tapping different sections of the aquifer. Figure 3–23 shows that the samples may represent water originating from different areas within the aquifer. For example, contaminants entering the aquifer in the area around A would migrate vertically, pass under the local discharge area and move toward the regional discharge area. Freeze and Witherspoon (1966) and Toth (1963) give very interesting discussions on regional flow patterns.

In addition to regional flow and natural processes, artificial discharge and recharge also have large effects on the groundwater flow system. These include pumpage and point sources of recharge such as recharge wells, spills, and recharge ponds and pits.

Effects of Pumping

Pumping groundwater from wells is a means of artificial discharge. The response of aquifers to these withdrawals influences contaminant migration. As the water is pumped, the water table near the well is lowered and the groundwater from the aquifer begins to flow towards the well. As pumping continues, the local water table continues to fall, and the rate of flow toward the well increases, until the flow equals

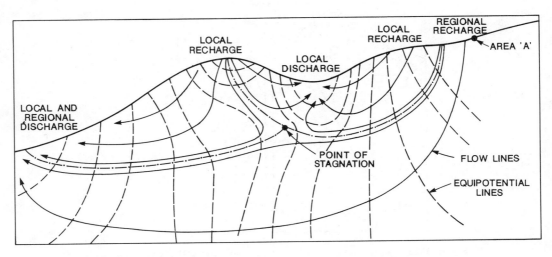

Figure 3–23. Profile of regional and local flow systems.

the withdrawal. If withdrawal continues, the water levels in the aquifer will continue to decline (assuming recharge is not occurring).

The rate of water level decline within the aquifer or the movement of water from the aquifer into the well results in the formation of a cone of depression. The shape of the cone of depression is shown in Figures 3–24A and 3–24B, for water table and confined aquifers respectively.

The cones of depression in water table and confined aquifers are very different. Pumpage from a water-table aquifer actually drains and dewaters the formation. Because the storage coefficient is larger than those for confined aquifers, the amount of water drained is relatively large and the cone of depression expands slowly.

Within a confined aquifer, withdrawals cause a decline in the pressure surface and do not dewater the aquifer. Because the storage coefficient is so small, very little water is derived from each unit area of the aquifer, and therefore the cone of depression expands very rapidly and can cover large areas. If the water level declines to a level below the top of the aquifer, the response of the aquifer to continued pumping is that of a water-table aquifer.

The effects of pumping are usually illustrated as a cone of depression that occurs in an aquifer without any regional gradient. However, it is important to conceptually understand that, in actual field situations, the cone of depression is superposed over a regional gradient, as shown in Figure 3–25. The lower section of the figure can be compared to the profile of the cone of depression shown in Figure 3–24A. The difference is that this profile is superposed over a flow system with a gradient. Therefore, a groundwater divide develops downstream of the well. Upstream of the divide, water flows towards the well. Downstream of the divide, water flows away from the well. The location of this divide up- or downstream depends upon a balance between the demand placed on the well and the ability of the regional ground-

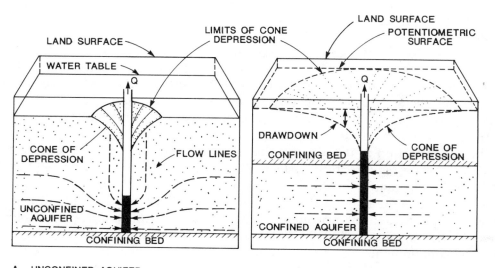

A. UNCONFINED AQUIFER   B. CONFINED AQUIFER

Figure 3–24. Profile of cones of depression in confined and unconfined aquifers (adapted from Heath, U.S.G.S. 1984).

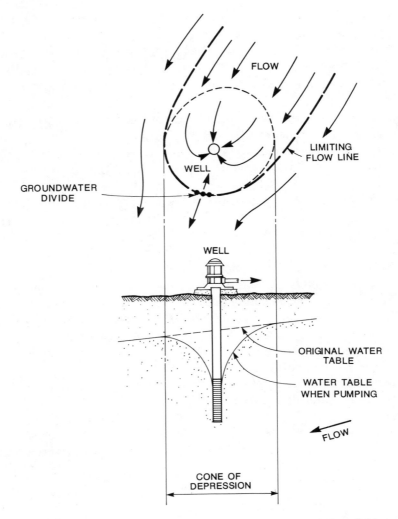

Figure 3–25. Effects of pumping in a uniform regional flow field.

water flow to satisfy the demand. Some of the conditions which can cause the location of the divide to change include: a change in the pumping rate, a change in the flow conditions in the aquifer, or the presence of other pumping or recharge features.

The upper portion of the diagram shows an areal view of the cone of depression superposed on a regional flow system. The important aspect is again the limiting flowlines and groundwater divide. As shown in Figure 3–25, all water within the area bounded by the limiting flowlines and divide will be captured by the well. All water outside the limiting flowlines and downstream of the divide will not be captured by the pumping well. This is obviously important when describing the movement of contaminants or when planning a pumping scheme to capture contaminants.

It is also important to conceptualize the effect of pumping large amounts of water from production wells located near a local plume. This is shown in Figure 3–26. When a well is pumping, the flow of water from the aquifer is from all directions

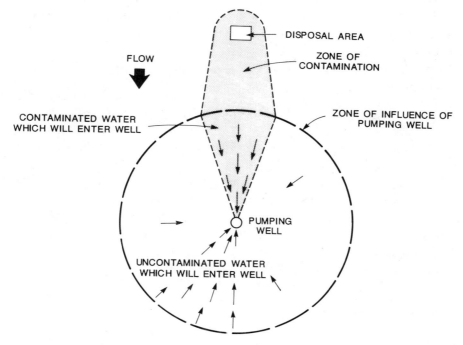

Figure 3–26. Dilution effects due to pumping from a production well on groundwater contaminants [Adapted and reproduced with permission from O. C. Braids, G. R. Wilson, and D. W. Miller, "Effects of Industrial Hazardous Waste Disposal on the Ground Water Resource," in R. B. Pojasek, ed., *Drinking Water Quality Enhancement Through Source Protection*. Stoneham, MA: Ann Arbor Science (an imprint of Butterworth Publishers, 1977.)]

(360 degrees) around the well to the well. If a contaminant plume is entering the affected area, it will be captured by the well. However, the volume of the plume entering the capture area will be very small in comparison to the entire volume of clean water. Therefore, concentrations of the contaminant measured in a sample taken from the production well will be very diluted. For this reason, constituent concentration data from large water supply wells may lead to underestimation of the true concentrations or location of contaminants.

In order to predict or compute the effects of pumping from an aquifer, including the location of groundwater divides and limiting flowlines, the yield of aquifers, and the movement of groundwater, the hydraulic gradient and the aquifer parameters such as hydraulic conductivity and storage coefficient must be known. In order to determine these coefficients, aquifer tests are carried out. An aquifer test comprises pumping a controlled volume of water from a properly designed well at a constant rate, usually for several hours to several days, and measuring the water-level decline in the pumping well and in especially designed monitoring wells located at different distances from the pumping well. Aquifer tests are described in detail in Freeze and Cherry (1979), Fetter (1980), Todd (1980), and Walton (1984, 1970, 1962).

The hydraulic gradient is usually determined by measuring water levels in a series of wells, converting the water levels to elevations, preparing contour maps with these elevations and calculating the gradient by measuring the change in water-level elevations per unit length. The importance of the hydraulic gradient to define the locations of groundwater divides and limiting flowlines (capture zones) is described by Todd (1980).

In addition to discharge, recharge also affects the water levels and flow system in an aquifer. Figure 3–27 illustrates the effect of a recharge pond superposed on a regional flow system. The lower portion of this figure is a profile of the effect of a recharge pond. The water leaking from the lagoon allows the groundwater levels to rise to the height of the bottom of the lagoon. Therefore, the water table rises to the bottom of the lagoon, and the profile of the new water table is shown in the figure. Water from the lagoon will flow upstream of the original gradient until it reaches the groundwater divide; at this point, the water will enter the regional system and will begin to flow downstream within the regional system. Therefore, there is a high degree of vertical movement in a system such as this. The upper portion of the diagram shows the areal view of this system. Again, the flow area affected is shown by the groundwater divide and limiting flow lines. The location of these are dependent on the amount of recharge and the regional flow conditions.

The importance of conceptualizing these types of systems occurs when the re-

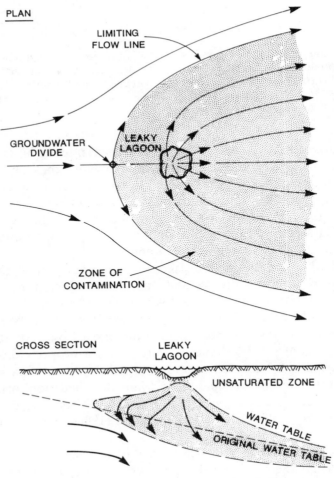

Figure 3–27. Effects of artificial recharge in a uniform regional flow field (adapted from Deutsch, U.S.G.S. 1963).

charge water is contaminated. Examples of these occurrences are given in the next section. More detailed discussions of natural and artificial recharge and discharge are given by Freeze and Cherry (1979), Fetter (1980), and Walton (1970 and 1984).

GROUNDWATER CONTAMINATION

The principal processes that influence the transport behavior of contaminants in groundwater are basically advection, dispersion, sorption and transformation (Roberts and Mackay, 1986). Advection and dispersion describe the rate of movement and dilution of a contaminant or solute. Sorption, or the partitioning of the contaminant between the liquid and solid phases, results in a decrease in concentrations in the water without changing the total mass of the compound, and also in the retardation of its movement relative to groundwater flow. Due to sorption, it usually appears that the velocity of movement of the contaminant is slower than that of the water. This relative difference in velocity is measured by the factor, retardation. Roberts and Mackay (1986) and Freeze and Cherry (1979) give excellent discussions on this topic. In fact, Mackay (1989, personal communication) is presently carrying out field experiments to determine the retardation of certain constituents and will also determine information on sorption.

The following discussions are limited to conceptualization of contaminant or plume movement in the groundwater flow system. These discussions should allow for a better understanding on how to determine the nature and extent of groundwater contamination problems, how to initiate remedial actions, and how to monitor the problem.

First, consider the shape of a plume of contamination within the groundwater. If it is a continuous source, superposed on a regional flow system, without any other major stress on the aquifer such as pumping or injection, the typical cigar-shaped plume may be expected, as shown in Figure 3–28A. However, if the source of contaminants is a one-time event or an intermittent source, its shape may appear as shown in Figure 3–28B. That is, segments or pieces of a plume separated over space. Therefore, trying to map the size of this plume would be difficult, as analyses from samples from wells placed between the slugs would not detect contaminants, or segments where the concentrations of the contaminants are dilute.

Commonly, the plume appearance is somewhere between the two descriptions

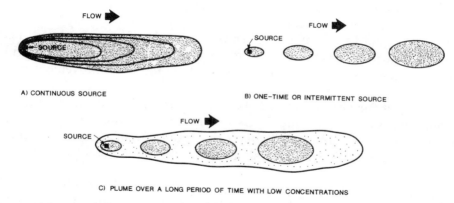

A) CONTINUOUS SOURCE

B) ONE-TIME OR INTERMITTENT SOURCE

C) PLUME OVER A LONG PERIOD OF TIME WITH LOW CONCENTRATIONS

Figure 3–28. Examples of shapes of plumes in an aquifer.

above, as in Figure 3–28C. This is especially true where the concentrations of contaminants within the plume are relatively low. The plume generally has the regional cigar shape. That is, if very low concentrations of contaminants are plotted, a wide, elongated plume can be mapped. Within this plume, there are slugs of relatively high concentrations of contaminants. Therefore, if the actual source location is not known, it is difficult to determine where the source is located by mapping the plume.

This relationship also exists during long-term remedial action and monitoring programs. The monitoring wells usually show high levels of contamination during early phases of remedial action and monitoring. After time, however, the concentrations become reduced and interpretation of the geometry of the plume becomes more difficult. The decision as to whether to drill additional wells to continue to map the plume geometry over the years of remediation must be made.

The actual shape and size of the plume depends on the geologic framework, the heterogeneity of the aquifer, local and regional flow patterns, the types and original concentrations of contaminants, variations in the rates of leachate production, and on local stresses occurring within the aquifer. Typical examples are shown in Figure 3–29A and B. Figure 3–29A represents a plume being monitored in Idaho. The plume occurs in a fractured basalt which appears to be very heterogeneous. The plume appears to have spread as far in a lateral direction as in a longitudinal direction. The more typical plume, represented by a more uniform geology, is the cigar-shaped plume shown in Figure 3–29B. This plume was mapped in a sand and gravel aquifer.

Another important impact on the shape of plume is the local stresses on the aquifer. An example is seen in Figure 3–30, in an aquifer where pumpage from several production wells is occurring in a lateral direction from the source of contamination. The pumpage will tend to pull the plume laterally, which has the effect of spreading the plume. Therefore, after many years, the plume may appear wider than if its movement were just affected by the regional flow system. Even after the pumpage ceases in the production wells, the plumes will continue to appear wider than originally expected.

The effect of the aquifer's heterogeneity is very important for the movement of contaminants. An example of this is shown in Figure 3–31. In Figure 3–31A, the formation is homogeneous and the contaminants move through the medium as a uniform slug. However, Figure 3–31B illustrates the movement of contaminants through a formation containing stringers of material with both high and low hydraulic conductivities. Obviously, the contaminants will move much more rapidly in material with high hydraulic conductivities than in material with low conductivities. This is equivalent to a formation with coarse sand and very fine sand, respectively. This is important, in that contaminants may move rapidly in sand stringers that occur in a fine-grained material where rapid contaminant movement was not expected.

This is described further in Figure 3–32. This figure illustrates the movement of contaminants in an aquifer having layers of different hydraulic conductivities. It also shows the relative breakthrough curves that would occur if each section of the aquifer was tapped by an individual well and by a well tapping all three sections of the aquifer.

The wells are tapping three sections of the aquifer containing silty sand, clean sand, and sand and gravel. The silty sand (K_1) has the lowest hydraulic conductivity,

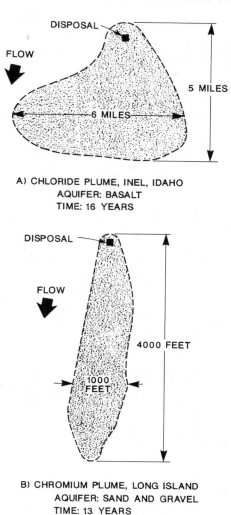

A) CHLORIDE PLUME, INEL, IDAHO
AQUIFER: BASALT
TIME: 16 YEARS

B) CHROMIUM PLUME, LONG ISLAND
AQUIFER: SAND AND GRAVEL
TIME: 13 YEARS

Figure 3-29.   Effects of heterogeneity on the shapes of contaminant plumes.

the clean sand (K_2) has an intermediate hydraulic conductivity, and the sand and gravel (K_3) has the highest hydraulic conductivity.

Well A is opened to all three sections of the aquifer. The graph showing the breakthrough curve for Well A is shown on Curve A in the figure. The breakthrough curve illustrates the relative concentrations (C is the concentration measured; C_0 is

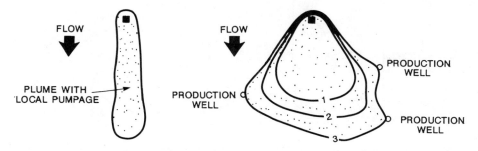

Figure 3-30.   Effects of pumping wells and the shapes of contaminant plumes.

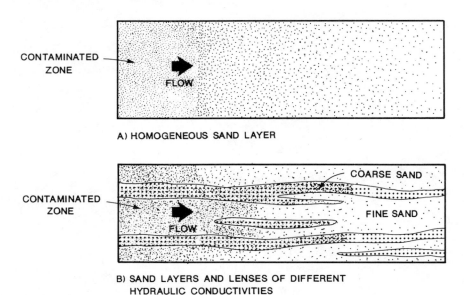

Figure 3–31. Effects of heterogeneity on the movement of contaminants (adapted from Freeze and Cherry, *Groundwater,* © 1979, pp. 76, 396, 399. Reprinted by permission of Prentice-Hall, Inc., Englewood Cliffs, NJ).

the original concentration injected) versus time. As shown by Curve A, the spread of the contaminants reaches the well quickly (T_1) and the relative concentration rises relatively slowly. Curve B represents the breakthrough curve for the section of aquifer with the lowest hydraulic conductivity. Here the contaminants take the longest period of time to reach the well, T_3. However, the rise in relative concentrations,

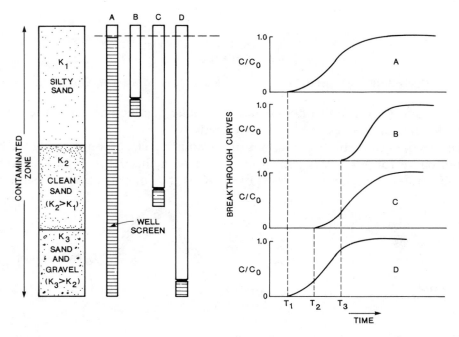

Figure 3–32. Effects of hydraulic conductivity on the movement of contaminants (adapted from Grisak et al., 1978. Reprinted by permission of the American Water Resources Association).

C/C_o, is faster than for the more permeable formations. Curve C shows the breakthrough for the formation with the intermediate conductivity. The plume reaches this well at T_2 and the curve is less steep than in Curve B. Finally, Curve D represents the well tapping only the material with the highest conductivity. Again, the breakthrough curve shows that the contaminants reaching the well the quickest, at time T_1. However, the relative concentration increases more slowly with time.

Another variable that can have a large effect on the shape and location of the plume within the aquifer is the density of the contaminants. If the density of the contaminant is less than that of water, the contaminant or plume of product will appear to float on top of the water table. If the density of the contaminant is greater than water, the plume will sink towards the bottom of the aquifer. This is shown in Figure 3–33. It should be realized that, in the concentration range of parts per billion, density differences will be insignificant compared to the intermolecular forces and the contaminant would neither float nor sink. For example, let's assume that trichloroethylene (TCE) were the contaminant migrating within the aquifer. If the concentration of TCE were 100 ppb, the plume would move with the groundwater it is dissolved in, and would not sink. However, if a tank of TCE would rupture, and the product was dumped directly into the aquifer as 80 to 100 percent TCE (800,000 parts per million); the contaminant would sink to the bottom of the aquifer, as shown in Figure 3–33B.

The movement of solutes of different densities through the subsurface is complicated, and care must be taken in determining the geometry of the system, in developing remedial action programs, and in monitoring the results of these programs. The movement is not only a function of the hydrogeological framework, but due to high adsorption and volitilization rates, the volume of the spill also becomes important for substances such as gasoline, halogenated and nonhalogenated organics, solvents, and other petroleum-related products.

If the spill is of a relatively small volume, all of the material may be sorbed by the soil before it reaches the water table. An example of a larger spill is shown in Figure 3–34. Here, much of the product remains in the soil, although some of the product reaches the water table and dissolves in the groundwater. The rate of dissolution is as great or greater than the volume reaching the water table. Therefore, a product layer of gasoline or oil does not form.

Figure 3–35 shows an example where the volume of product entering the subsur-

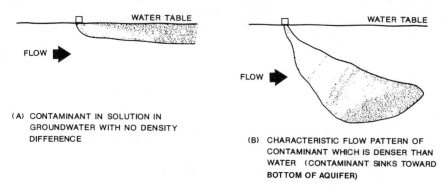

Figure 3–33. Effect of density on the movement of contaminants (adapted from Freeze and Cherry, *Groundwater,* © 1979, pp. 76, 396, 399. Reprinted by permission of Prentice-Hall, Englewood Cliffs, NJ).

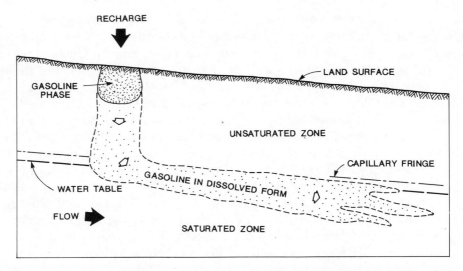

Figure 3–34. Movement of gasoline in the subsurface (adapted from Schwille, 1975).

face is large enough so that a product layer forms. In this case, there is the phase held in the soil, a phase dissolved in groundwater, and a product phase floating on top of the water table (assuming the product density is less than that of water). Where the product reaches the water table and enters the flow system as a separate floating phase, the product layer actually depresses the water table. In order to determine the true gradient of the water table, density corrections need to be made. In this case, there are definitely two plumes that can behave very differently from one another. Monitoring programs and clean-up programs for multiphase flow systems are very different than for just a dissolved phase system. An excellent discussion of the multiphase flow system is given in Schwille (1975).

An example of another complicated situation relative to the migration of different density fluids is shown in Figure 3–36. In this case, two hypothetical tanks contain-

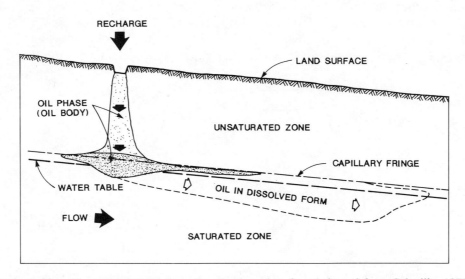

Figure 3–35. Example of multiphase flow in the subsurface (adapted from Schwille, 1975).

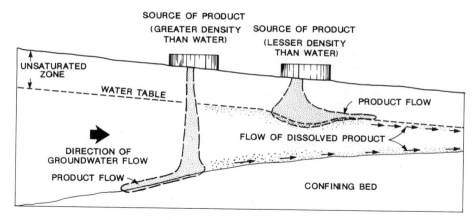

Figure 3-36. Example of the subsurface movement of fluids with densities greater than, and less than, that of water.

ing different density products rupture. One tank contains fluids of a density that is less than that of water, and the second tank contains fluid whose density is greater than water.

The product with the lower density migrates vertically through the unsaturated zone, reaches and slightly depresses the water table, moves downgradient as a product, dissolves in the groundwater, and also flows downgradient in solution.

The tank on the left contains a product where the density is greater than water. Enough of the product leaks so that the product itself reaches the water table. Because the density is greater, the product sinks vertically through the aquifer. At the bottom of the aquifer, some of the product does dissolve but a portion remains undiluted. The dissolved product flows with the groundwater, downgradient. However, the product at the bottom of the aquifer flows with gravity which, in this case, is in the opposite direction of regional groundwater flow. Obviously, designing a program to detect the geometry of this system would be complicated and probably be based on learning as the program is continued, and on hindsight. As is probably realized, the remedial program would also be very complicated.

The shape and movement of the plume is also affected by the solute being sorbed or transformed, as discussed in Chapter 5. The effect of these processes appears to retard the rate of movement of the contaminant. That is, it appears as though the contaminant is moving slower than the groundwater. An example of this is shown in Figure 3-37. In this illustration, the relative rate of movement of four contaminants that react (sorption, transformation, etc.) with the geologic formation, or are (in general) retarded at different rates, are compared. The rate of movement, as shown by the figure, obviously affects the shape of the plume. For example, contaminant A may be a pesticide that is highly sorbed on soil particles. Contaminant B may be benzene, which is also highly sorbed, but not as much as the pesticide. Contaminant C can represent trichloroethylene (TCE), which moves more rapidly through the formation, when compared to benzene and pesticides, but is still retarded when compared to the velocity of water. The apparent slower velocity, in this case, is probably due to sorption on the soil and the volatilization of the TCE. Contaminant D may be chloride, which is conservative, and migrates through the formation at velocities equivalent to water.

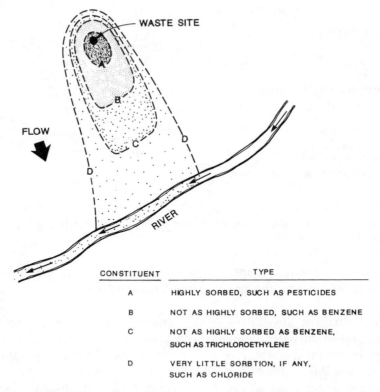

CONSTITUENT	TYPE
A	HIGHLY SORBED, SUCH AS PESTICIDES
B	NOT AS HIGHLY SORBED, SUCH AS BENZENE
C	NOT AS HIGHLY SORBED AS BENZENE, SUCH AS TRICHLOROETHYLENE
D	VERY LITTLE SORBTION, IF ANY, SUCH AS CHLORIDE

Figure 3–37. Effects of retardation on the movement of contaminants (adapted from LeGrande, 1965. Reprinted by permission of the American Geophysical Union).

Let's assume that all of these contaminants were stored in a waste tank and were released to the environment at exactly the same time. The relative position of the plumes at a specific time in the future might appear as in Figure 3–37.

Contaminant D already is discharging to the river. Contaminant C may be many hundreds of yards downstream of B, yet upstream of D; the front of Contaminant A is still very close to the source.

The effects of retardation and the relative movement of different contaminants are important when designing investigative groundwater programs or monitoring programs. For example, if wells were placed only between the edge of Contaminant B and the river, the nature and extent of Contaminant A may not be discovered for many years.

Understanding the behavior of flow beneath a landfill is also helpful when designing or monitoring cleanup programs. In many cases, the landfill area becomes a recharge area. This causes the water table to rise beneath the landfill, as shown in Figure 3–38. The leakage into the ground and the rise in the water table causes a downward vertical component of flow to occur. Therefore, near the center of the landfill, a groundwater mound occurs and the contaminants would move vertically downward, whereas at the edges of the landfill, the flow would be more lateral. If the water table rises significantly, springs (or leachate seeps), which are groundwater discharge areas, may form on the sides of the landfill.

Recharge ponds and lagoons can have a less severe impact on the regional water

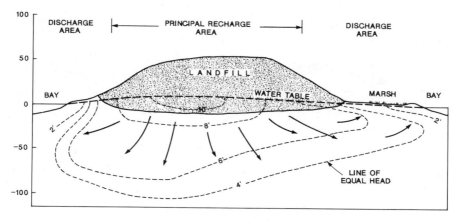

Figure 3-38. Movement of contaminants below a landfill.

table than large landfills. The effects of lagoons and ponds on the water table are shown in Figure 3-27. The effects on the movement of contaminants in the areal and cross-sectional views are shown in Figures 3-39 and 3-40, respectively. Figure 3-39 illustrates the migration of contaminants from disposal ponds into and through the regional flow system, to a creek, which is the discharge area. Figure 3-40 illustrates the vertical component of flow. The high heads within the recharge basins cause a large vertical component of flow near the source area. Therefore, contaminants move vertically into the aquifer and then horizontally along with the regional flow system. In this case, the groundwater in the regional flow system also moves upward in the aquifer and discharges into the creek. Therefore, the contaminants also moved vertically and discharged into the creek. This is very similar to the hypothetical diagram shown in Figure 3-6.

Again, this illustrates that the movement of contaminants is controlled by the natural flow system and local stresses on the system. Therefore, to design a program to detect contaminants in the subsurface, the regional and local systems need to be understood. An example is shown by a more hypothetical case in Figure 3-41. In

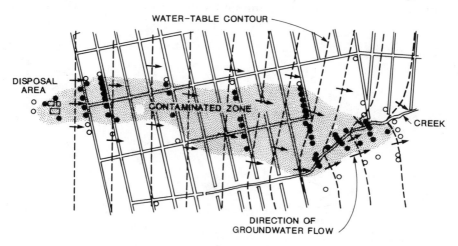

Figure 3-39. Map view of the movement of a plume (adapted from Perlmutter and Lieber, U.S.G.S. 1970).

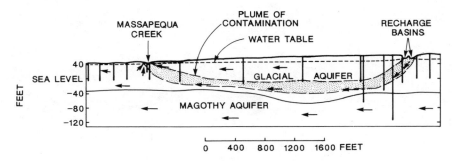

Figure 3–40. Cross-sectional view of the movement of a plume (adapted from Perlmutter and Lieber, U.S.G.S. 1970).

this case, the shallow and deep monitoring wells would have intercepted clean water, whereas only the middle well would have indicated the presence of contaminated groundwater, again illustrating that proper monitoring well placement and interpretation of water-quality data depends on a thorough understanding of the local and regional flow systems.

As mentioned in the previous sections, in addition to the effects of the regional flow system and recharge, pumping and/or natural discharge affects the vertical extent of contaminant migration within the aquifer. Figure 3–41 illustrates that, in recharge areas, the contaminant has the potential to move deeper into the formation. Also, Figure 3–41 shows that the river may be a discharge area, and that the regional flow system discharges to the river. Therefore, as the contaminants are migrating within the regional system, they will also move vertically upward through the aquifer, and discharge to the river.

Figure 3–42 illustrates the effect of pumping from a well that is screened near the bottom of the aquifer. If the well were not pumping, the contaminants would have moved into the aquifer (probably remaining near the top) and migrated downstream. Due to the pumping, however, the gradient of the flow system is toward the well. In addition, as pumping from the well has caused the water to migrate toward the bottom of the aquifer, the contaminant will move both vertically and laterally

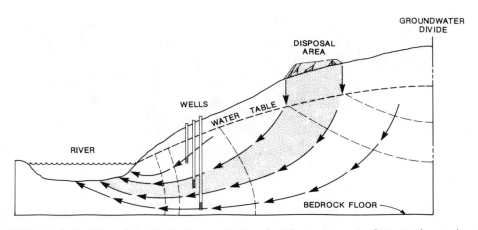

Figure 3–41. Relationship of screened zones to measuring the movement of contaminants in a regional flow system (adapted from Miller, USEPA. 1977).

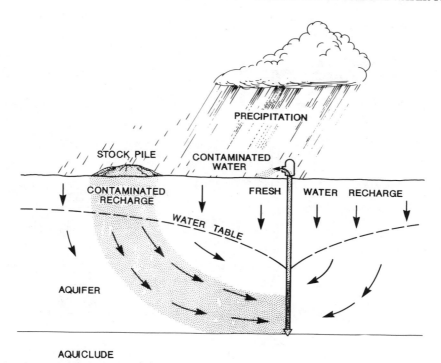

Figure 3-42. Influence of pumping on plume migration (adapted from Deutsch, U.S.G.S. 1963).

towards the well screen, thereby spreading the contaminants throughout the aquifer. If pumping in the well ceases, the contaminants would probably migrate away from the source nearer to the surface of the water table. If the well pumped intermittently, the contaminants may be spread throughout the thickness of the aquifer. If this control on contaminant migration is coupled with changes in stratigraphy, the complex nature of the flow of contaminants can be realized.

Manmade conduits have a large effect on the spread of contaminants within an aquifer. An example of this could be in uncased, multiscreened, or gravel-packed wells, in which the gravel pack taps more than one formation. Figure 3-43 represents a well which is uncased or opened to multiple layers of permeable formations, or can also represent gravel-packed wells where the gravel pack cross-connects multiple aquifers. In Figure 3-43, a tank containing hazardous materials ruptures and the contaminants flow through the unsaturated zone. In this case, the fluids reach a confining bed (very low permeability) within the unsaturated zone and a perched water table begins to form. As the mounding occurs, water and contaminants in this zone begin to flow laterally. If the contaminants reach the uncased well (which is just an open hole through the formation), the contaminants will enter the well, cascade down the open hole and reach the water table. The contaminants would then enter the aquifer and begin to contaminate the regional groundwater flow system. Thus, the uncased well becomes a source of contamination.

If the leaking tank was discovered and a drilling program was initiated, the contaminants in the perched zone may be found and a remedial action program can be carried out. However, if the existence of the uncased well was unknown, then the plume beneath the water table forming downstream may not be discovered. Also,

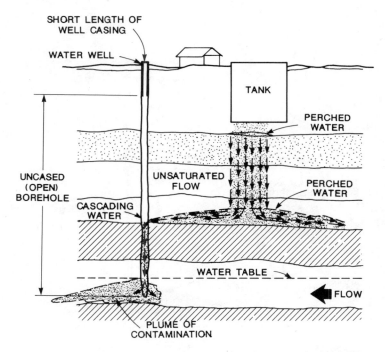

Figure 3–43.   Potential contamination of groundwater from perched water entering an uncased well.

if it is discovered due to other sampling programs within the aquifer, the original source of the contamination may never be determined.

This type of complication is common in areas such as Silicon Valley, California. This area consists of unconsolidated geologic materials of alternating relatively thick layers of permeable and nonpermeable materials. Therefore, there are multiple aquifers and multiple confining layers. In addition, this valley was largely an agricultural area with many irrigation wells before it was urbanized. For efficiency, and to produce the greatest amounts of water, these irrigation wells tapped multiple aquifers. In addition, the wells were gravel-packed across each aquifer. Therefore, these wells allowed each aquifer to be interconnected.

As the contaminants entered the upper aquifer and migrated away from the source, the potential for the contaminants to reach and enter the gravel pack of an irrigation well was high. Because the gravel pack is permeable, and the heads in each aquifer decrease with depth in this area, the contaminants migrated down the gravel pack and entered multiple aquifers. Therefore, the contaminants entered many aquifers in the valley to depths of tens to hundreds of feet. This is illustrated in Figure 3–44. This figure shows the relative equipotential lines and flow lines. The well tapping multiple aquifers is a source of recharge to the lower sand. Therefore, if contaminants reach this well in the upper sands, they would flow into the lower sand and migrate downstream.

The next series of figures represent a few examples of the relationship of the vertical location of the screen of a monitoring well tapping an aquifer and the migration of contaminants within the aquifer. Figure 3–45 is similar to some figures shown before. The contaminants migrate vertically downward to the water table. The area of the source is in a recharge area, allowing for a component of downward

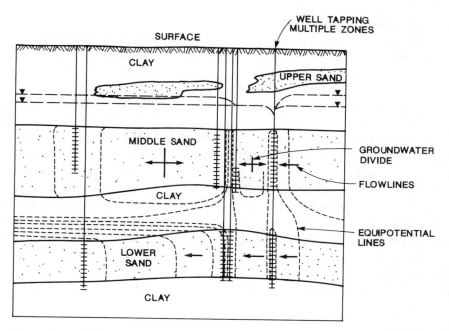

Figure 3–44. Flow section showing effect of wells screened in multiple aquifers.

vertical flow; the contaminants then move downstream horizontally within the regional flow system, and then move upward through the aquifer to a stream which is a groundwater discharge area.

The example shows four monitoring wells, one tapping the full saturated thickness of the aquifer. Two of these wells tap either the upper or lower sections of the aquifer and intercept uncontaminated water. Samples from the fourth well, which taps the middle of the aquifer, are representative of the contaminated water. The well tapping the full saturated thickness of the aquifer does intercept the contaminated water. However, it also captures clean water; therefore, the sample will be diluted and will indicate lower concentrations of the contaminant than a well tapping just the plume.

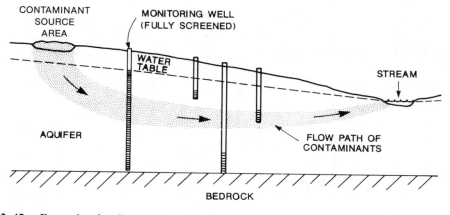

Figure 3–45. Example of wells screened in different sections of an aquifer, and their relationship to the measured concentrations of contaminants.

Figure 3–46 again illustrates the effect of pumping, superposed on the regional flow system, on the migration of contaminants. Figure 3–46A illustrates the flow without pumping and 3–46B illustrates the effect with pumping. In this diagram, when the well is not pumping, the flow in the regional system is towards the stream. Therefore, the plume would move from the source towards the stream. When the well begins to pump, a groundwater divide will form between the contaminant source and the stream. This indicates that groundwater flow on the production well side of the divide flows toward the production well and, on the downstream side of the divide, water flows toward the stream. Therefore, contaminants on the upstream side of the divide will flow toward the production well and, on the downstream side of the divide, the contaminants will flow towards the stream.

Under nonpumping conditions, the two monitoring wells located in Figure 3–46 would probably intercept the plume. However, when pumping, the plume changes direction and migrates toward the pumping well. This demonstrates the importance of determining the location of pumping wells within the regional system when trying to map contaminant plumes.

Figure 3–47 illustrates locating monitoring wells in situations where there are mul-

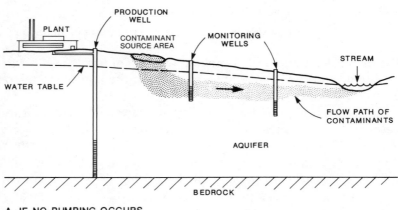

A. IF NO PUMPING OCCURS

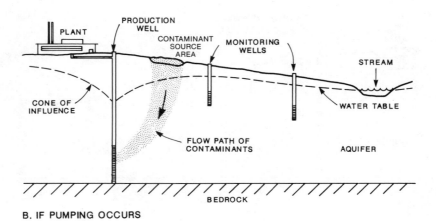

B. IF PUMPING OCCURS

Figure 3–46. Locating monitoring wells to map contaminants in an area affected by pumping wells.

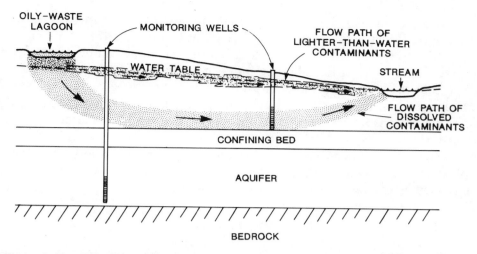

Figure 3-47. Effects of monitoring well location on measurements of multiphase plumes.

tiphase contaminants and multiple aquifers. The well tapping the lower aquifer obviously does not detect any of the contaminants; the well tapping the lower portion of the upper aquifer intercepts the dissolved plume; and the movement of the products that are lighter than water migrate downstream undetected.

Occurrence of groundwater flow in a fractured rock system can be very complicated. Figure 3-48 illustrates monitoring wells which tap a fractured system. Wells tapping the interconnected fractures which are connected to the source may contain contaminants. However, wells tapping fractures not connected to the source, or wells not tapping fractures, will probably remain uncontaminated. The fractures containing contaminated fluids and fractures containing clean water can be located very close to one another.

The natural phenomena that complicates the understanding of the hydrogeologic environment, the groundwater flow system, and the contamination migration pat-

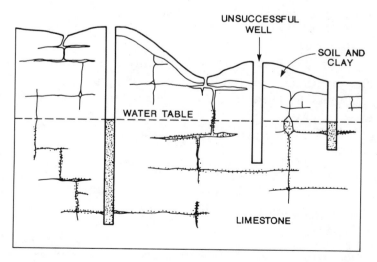

Figure 3-48. Location of wells in fractured rock in relationship to measuring contaminants.

terns, is further complicated by problems that are posed by the practices occurring at waste disposal facilities. At these waste disposal facilities, a variety of materials can be discharged from many different sources, thereby there can be numerous distinct plumes of contamination, moving independently. Extraction of groundwater by wells within the area modify the natural flow paths of the groundwater and the contaminant plumes. In some cases, monitoring programs that have years of history illustrate fluctuations in concentrations of some constituents, while others remain fairly constant.

This overview has shown that the movement of groundwater and contaminants is complex, and the investigation of groundwater contamination can require extensive and costly assessments over a considerable amount of time.

SITE INVESTIGATIONS

In a typical assessment of groundwater quality at a hazardous waste disposal facility suspected of causing contamination, the first step is a thorough survey of potential contaminant discharges to groundwater within, and in the vicinity of, the site. This includes reviewing the detailed history of the facility, including records of inventories, spills, leaks, and abandoned operations. Background information is extremely important, because contaminants that may have entered the groundwater system many years before can still be in the study area and be intercepted by monitoring wells installed as part of the investigation.

The key steps to carrying out a groundwater contamination assessment program include a study of the:

- Plant history and operation.
- Inventory of potential contaminants, historical and present.
- Regional inventory of other sites or sources of contamination.
- Regional geologic system, including identification of aquifers and confining beds.
- Regional hydrogeologic system, including the groundwater flow system, potential or existing contamination migration patterns, and recharging and discharging boundaries, such as streams and lakes.
- Regulatory criteria and standards that will affect and guide the program.
- Nondrilling investigation activities, such as sampling existing wells, streams, and waste sources, and geophysics and soil gas studies[1] to evaluate the potential groundwater quality at the site.
- Design of a groundwater assessment program to meet site specific and regulatory objectives.
- Carrying out of well drilling and sampling programs.
- Interpretation of groundwater flow and contaminant migration system.
- Design and carrying out of remedial actions, if necessary.

---

[1]Gas removed from the unsaturated zone and analyzed can be used, at times, to determine potential sources of contamination or to map high concentration plumes in the saturated zone. The constituent of interest needs to be a volatile constituent.

- Establishment of long-term monitoring program objectives and protocols necessary to solve site specific problems and regulatory requirements.
- Management of the collection, tabulation, interpretation, and reporting of data during monitoring program.

If the site is under litigation, or is a Superfund site, the above process is different, and the investigation is no longer an "informational" study; instead it is a "litigatory" study. In general, this includes the Remedial Investigation and Feasibility Study (RI/FS). Before any work can be carried out at the site, Quality Assurance Program Plans (QAPP), Sampling and Analyses Plans (SAPS), and Health and Safety Plans (HASP) must be approved by the regulatory agency (which can be a state agency or the U.S. Environmental Protection Agency, EPA). How the data are collected, analyzed, and reported may be different for an RI/FS than for an informational study. Therefore, it is suggested that the many publications and manuals that have been prepared by the EPA for this work be reviewed (USEPA, 1985, 1987, 1988).

The EPA can be contacted through their regional offices, which are listed below:

EPA Region 1
JFK Federal Building
Boston, MA 02203
617/565-3715

EPA Region 2
26 Federal Plaza
New York, NY 10278
212/264-2525

EPA Region 3
841 Chestnut Street
Philadelphia, PA 19107
215/597-9800

EPA Region 4
345 Courtland Street, NE
Atlanta, GA 30365
404/347-4727

EPA Region 5
230 S. Dearborn Street
Chicago, IL 60604
312/353-7000

EPA Region 6
1201 Elm Street
Dallas, TX 75270
214/655-6444

EPA Region 7
726 Minnesota Avenue
Kansas City, KS 66101
913/236-2800

EPA Region 8
1860 Lincoln Street
Denver, CO 80295
303/293-1603

EPA Region 9
215 Fremont Street
San Francisco, CA 94105
415/974-8071

EPA Region 10
1200 Sixth Avenue
Seattle, WA 98101
206/442-5810

The first step is to review the available data at the site, to determine if a potential contamination problem exists. This would include a historical study of plant operation. This usually starts with interviews related to present and historical waste disposal practices. Also, an inventory of raw materials that may be considered hazardous materials should be carried out. This inventory should not only include purchased material, but also estimated amounts of usage, and records of disposal of these materials. On-site investigations of present and past practices should include record searches for underground tanks, buried pipelines, sewers, cesspools, septic tanks, dry wells, other types of sumps, product and waste storage areas, loading areas, storm water collection areas, and, of course, records of product losses and spills.

This type of study should include the use of present and historical maps and aerial photographs and interviews. Foundation borings or other soils investigation logs should be reviewed. Sometimes these boring logs record observations such as odors, oily substances or products, or soil discoloration. These logs also can be used to interpret the geology.

Data from water or existing production or monitoring wells also aid in hydrogeologic interpretation. For example, driller's logs and the construction details of supply wells can provide valuable information on the interrelationships of aquifers and confining layers. In addition, results of previous monitoring programs can provide historical data on ground and surface water quality.

Similar regional information needs to be included in this type of assessment, because the groundwater quality beneath the facility can be affected by off-site sources. Degraded water quality may be migrating from off-site to on-site. Therefore, a regional inventory of potential contamination sources should be attempted through field inspection, both present and historical land-use maps, and aerial photographs. Old aerial photographs are especially useful because they may aid in locating abandoned hazardous material facilities such as landfills, lagoons, and other industrial facilities. Very important information can be gathered from the EPA and state and local regulatory agencies. There may be work already occurring at other sites under the authority of these agencies. These should be available to the public or through the use of the Freedom of Information Act.

Unsewered residential areas should be mapped. If these areas use septic fields, they may be potential sources of organic contamination. For example, historically, dry cleaning facilities using septic systems have been sources of organic solvent contamination in the groundwater.

Additional sources of contamination can be municipal sewer lines, or the trenches dug for these lines or for similar construction. Historical and present surface drainage patterns should be reviewed using topographic maps and aerial photographs. These drainage-ways may have carried contaminants away from industrial facilities. Highway and railroad rights-of-way are also subject to spills that may have been flushed or leached into the groundwater system. Sometimes records of these incidents are kept by the agencies that are involved.

During the regional assessment of potential sources, regional geology and hydrology data should also be collected and interpreted. This information on geology, hydrology, and soils is important in evaluating potential sources of contaminants in the subsurface and their potential for vertical and horizontal migration through the groundwater system. These data, which include information about geology, groundwater, climate, water quality, and water use, are available for many areas in the nation from state and local agencies, the EPA, and through the U.S. Geological Survey publications.

Published area-wide and site-specific topographic maps can be used to estimate the probable direction of groundwater flow for shallow systems, and areas of high water table, groundwater discharge/recharge areas, and ground and surface water interrelationships.

Regional groundwater pumpage is also very important. This pumpage can change regional flow directions. Old wells, both used and abandoned, can also affect the flow. As mentioned in the previous section, old wells that are uncased, have cracks in the casing, faulty construction, or have gravel packs that connect multiple aquifers, can lead to cross-contamination of aquifers. Therefore, efforts to locate pumping centers, irrigation wells, domestic wells, and even monitoring wells should be made. Maps and tables should be compiled with information about these wells, including their construction details, years of construction, years of use, water use, and aquifers and confining beds penetrated. This information can be collected from agency publications and records, and possibly from interviewing water-well contractors.

Groundwater quality assessments are usually being carried out because a problem has occurred at a facility. The activities are usually being reviewed by regulatory agencies. Therefore, the agencies' needs should also be recognized. For example, some laws and regulations specify minimum requirements for the number of wells, their locations, their construction, the constituents to be analyzed, and the frequency of sampling. The agencies that may be involved (sometimes there are multiple agencies) should be interviewed before any field programs are formulated to determine the requirements of the particular agency.

Again, it should be realized that there is no consensus on what constitutes an appropriate groundwater assessment program or an optimum monitoring program, or even optimum procedures for drilling and sampling. The regulatory standards for groundwater, drinking water, and soil should also be outlined and integrated into the program design for determining what analyses are to be done or constituents are to be analyzed.

To complete the initial assessment, on-site background water-quality data can be obtained by sampling springs, supply wells, standing surface-water bodies, streams, storm sewers, and waste products in disposal areas, such as holding tanks, sumps, and ponds. Analyses of samples from existing sources of contamination should be carried out, but it should be realized that these analyses may not be representative of the actual contaminants currently beneath the site.

Some of the initial tasks can use methods other than direct sampling or drilling. For example, geophysics, including seismic refraction, resistivity, very low frequency (VLF) and ground-penetrating radar, can be used to map the geometry of aquifers, pipelines, and other manmade underground structures. Geophysics can also be used to map plumes of contamination (Stollar and Roux, 1975). In addition, to indirectly map plumes of volatile organic contaminants, soil gas surveys are commonly used. Results of many of these studies have been published in articles in the National Water Well journal, *Ground Water*.

All of the site-specific and regional information collected in this discussion should be summarized and interpreted in a manner that responds directly to the needs of the site assessment. This information should include maps and tables to summarize the data. Finally, a work plan should be completed. The work plan should discuss the objectives of the assessment program, the data collection and sampling program, with techniques and protocols, the quality assurance plan, the health and safety plan, and schedules.

The work plan should contain the overall objectives of the program. With a groundwater contamination program, these objectives should include, at a minimum:

- Collection of the necessary data to verify and characterize the presence of contamination
- Assessment of the nature and areal and vertical extent of groundwater and soil contamination at the site
- Evaluation of the specific risks and hazards to public health and the environment that result from the contamination problem
- If necessary, identification of the appropriate cleanup criteria
- Conducting field investigations to collect the site-specific data necessary to meet all of the above objectives, and to provide all of the hydrogeological information for the design and evaluation of the final remedial action or remedial action alternatives for removing the contaminants

Again, the purpose of the assessment of hydrogeologic data is to evaluate regional and site-specific information related to the geologic and hydrogeologic characteristics of the site. The objectives should be to:

- Determine the geometry of the groundwater flow system
- Determine the groundwater flow directions, velocities, and travel times
- Determine contaminant migration patterns, both horizontally and vertically
- Determine the quantity or quality of water withdrawn from existing wells
- Determine data deficiencies

As part of the work plan, a Health and Safety program should be developed to ensure the protection of the project personnel, observers, and visitors. The plan should describe the personnel protection and safety considerations to be followed during any type of field activity that involves risks of potential exposure to contaminants.

The work plan should also include sampling and quality assurance plans. These should cover all proposed field activities, including well drilling, soil and groundwater sampling, aquifer testing, groundwater level monitoring, and surface water flow and quality measurements. These written descriptions will ensure that field work objectives or protocols are well planned and will provide field personnel with specific guidance for each field activity. At a minimum, the sampling plan should include:

- Sampling objectives
- Locations of sampling points (wells, borings or other facilities to be constructed, sampled, tested, or monitored)
- Rationale for sample locations, numbers, and analytical parameters

- Schedule of sampling, monitoring, and testing
- Detailed procedures for monitoring, testing, and sampling collection and preservation
- Procedures for sample labeling, sealing, storage, packing for shipment, and chain-of-custody
- Analytical methods to be employed for sample analyses
- Description of methods to be employed for data analysis

Choosing the best method for drilling and sampling is dependent on access to the property, depth to the water table or formation of interest, formation thickness, and the type of geologic materials. The various methods of drilling are described and compared in Chapter 6. Today, a common method for drilling monitoring wells is the hollow-stem auger. With this method, because fluids are generally not added during drilling, the depth to the water table can be determined, the subsurface formations can be sampled for geologic analyses, and also, if the hollow section of the auger is of large enough diameter, the monitoring well can be installed within the hollow stem. The borehole can then be gravel packed and sealed with bentonite and grout, and finally the auger can be removed.

Although there are field programs that have used the hollow stem auger to install wells to depths of 120 feet, and some actually report depths to 300 feet (Driscoll, 1986), it should be realized that the optimum depths to drill and install monitoring wells with the hollow stem auger is probably less than 75 feet. If the water table is 150 feet below the land surface, the use of the hollow stem auger would not be recommended. The common drilling method for these depths would be mud rotary. However, if the depth to water was not exactly known, it would be difficult to use a mud rotary and to detect the top of the water table. Therefore, in this case, an air rotary may be used. It becomes obvious that the method of drilling should be dependent on local hydrogeologic conditions, the type of data that is to be collected, and the objectives of the program. Whatever the method of drilling is chosen, it is mandatory to attempt to avoid cross-contamination of aquifers.

The materials to be selected for well construction, whether steel or plastic casings and screens, are also dependent on the method of drilling, the type of contamination expected, and the depth of the well. This is discussed in Chapter 6. The diameter of wells will also be dependent on the objective of the program. The nature and objectives of the program will also determine the number and depth of wells and the length of well screens, and whether well clusters tapping different sections of the aquifer or different aquifers are necessary.

The drilling contractor should be selected on the basis of competence, cost, experience, and reputation as to the level of cooperation to be expected when facing difficult field problems. Contracts for drilling wells into contaminated groundwater are usually let on a time and materials basis, and not on the cost per foot of drilling. This is usually to be fair to the driller and to be able to slow the drilling down at times when it is necessary to collect the data to describe detailed information about the hydrogeologic or contaminant migration conditions.

During the field program, data must be collected carefully. The geologic samples are collected to define the difference in lithology that can affect vertical and horizontal groundwater flow, to estimate hydraulic conductivity of the aquifer and confining bed material, to investigate the unsaturated zone, and to collect soil samples for chemical analyses.

Cores, split-spoon samples, or other sampling techniques are used to collect the geologic material. These are described in Chapter 6. There are many methods available for sample collection, depending on the type of geologic material, whether the material is in the saturated or unsaturated zone, the type of drilling technique being used, and whether the sample is being collected for physical or chemical analysis.

Depending on the drilling method, different types of geophysical logs also can be carried out in the borehole. These logs, including resistivity, self-potential, natural gamma ray, neutron and others, can aid in interpreting the type of geologic formation, boundaries of confining beds and aquifers, estimates of hydraulic conductivity and porosity, borehole characteristics, and water quality. Therefore, in combination with the geologic samples collected, it allows for better design of the monitoring or production wells.

The cores and geologic samples can be used for describing the geologic materials, estimating grain-size distribution for designing the well, estimating porosity and hydraulic conductivity, for chemical analyses, and for head space analyses. Head space analyses are carried out by placing the soil sample in a jar, closing the jar so it is airtight, and then, after a set time, measuring the relative amount of volatile organic constituents, if any, emanating from the soil. If volatiles are detected, the remainder of the core can be forwarded to the laboratory for complete analysis.

While collecting soil samples during the drilling process, there are also methods of collecting water samples with depth for chemical analyses. This will allow estimates of changes in water quality with depth to be made. If temporary screens are placed in the formation to collect the samples, head measurements can also be made. In this manner, estimates of the major vertical components of flow can be made during the drilling process.

After the well is completed, in order to determine the aquifer coefficients, various aquifer tests can be carried out. These include slug, tracer, and borehole dilution tests, as well as detailed controlled aquifer tests. These tests are described in Freeze and Cherry (1979), and Walton (1970 and 1984).

Once the field work and interpretation of data are complete, the groundwater flow and contaminant migration patterns are described satisfactorily, and the remedial action is initiated, a long-term monitoring program needs to be established. Initiating the monitoring programs are described in Chapter 6. Some of the elements of a monitoring program include the following:

- A systematic approach to data collection and quality control
- Tabulation of well construction details
- Tabulation of water-level and quality data, so that the data are readily available to interpret and to guide the field crews during the next round of sampling
- Tabulation of the elevation of measuring points and depths to the water table and the bottom of the wells. These data should be readily available for interpretation and for the field crews. (These data allow field crews to determine if there are anomalies such as wells filling up with fine-grained material)

- A permanent well numbering system and well-marked measuring points established for wells
- Location maps with usable scales
- Sampling and purging procedures must be established on a well-by-well basis, and written protocols must accompany field personnel
- A sampling order of wells should be established, from the least contaminated to the most contaminated well
- A uniform sampling protocol is essential
- Measurement of water levels and sampling should occur in the shortest period of time, one day if possible
- Frequency of sampling should be established and based on both technical rationale and regulatory constraints

The sampling protocol and pumping equipment chosen depend on such factors as the types of contaminants, depth of water table, well construction, screen submergence, and well yield. Most of these are discussed in Chapter 6.

Some of the analyses of groundwater should be carried out at the well head. These include, if necessary, pH, specific conductance, dissolved oxygen, temperature, Ph, and alkalinity. Decisions on whether to filter the sample in the field, especially when the constituents to be analyzed are metals, is always a point of discussion. The decision as to whether to filter or not should be based on technical considerations and the requirements of the regulatory agency. If filtered, the analyses for metals represents metals dissolved in water only. If the sample is not filtered, the results of the analyses for metals would represent concentrations of metals dissolved in the water and total metals. The total metals includes the metals dissolved from soil particles, which may be natural. To determine the natural metal concentrations of the soil is not a trivial problem. In fact, in some cases, the concentration of natural metals can approach, or be greater than, regulatory standards. This, then, can present problems in deciding whether or not the water or soil is contaminated. In addition to filtering, the protocol for preservation, packing, and shipping of samples, along with the chain-of-custody, must be established and rigidly controlled. This is discussed in more detail in Chapter 6, groundwater monitoring.

The chemical analyses are a key element in the monitoring program, and therefore choosing the laboratory should be based on the laboratory's experience and reputation; quality assurance and control programs; if necessary, the laboratory's certification with state or Federal agencies; turn-around time; cooperation; communication procedures; reporting procedures; and costs. Changing laboratories during the monitoring program should be avoided, if possible. It should be remembered that chemical variations can be a reflection of different laboratory procedures, rather than changing groundwater quality conditions.

Another key element to long-term monitoring is managing, interpreting, and reporting on the data being collected. Within a relatively short time after the monitoring program has been initiated, hundreds, if not thousands, of individual bits of information will be generated. Assuring the quality of the data and storing, retrieving, and manipulating the information being collected is a major task. It is recommended that the data be stored on a computerized data management system.

As mentioned previously, the data from the laboratory has to be reviewed through a quality assurance (QA) program, and validated. (If it is a Superfund site, the QA program will be established by the EPA). The validation process can be complicated and costly. The program should be worked out with the regulatory agency and laboratory before the data are collected. These data should be interpreted quickly in-house before it is submitted to the regulatory agencies.

The interpretation of the data should be in the form of preparation of tables, maps, and hydrographs of both chemical and water-level data. The data should also be compared to the historical data, to determine if there are changes occurring in the flow system or contaminant migration patterns. In addition, it should be determined whether the changes are areal or vertical in nature, or in individual wells. The changes should also be checked to determine if they are physically and statistically meaningful, or are related to errors in the data or errors in the reporting of the data.

The inventory of on-site and off-site sources should be updated periodically. Also, the entire monitoring program should be evaluated at least annually to determine if the monitoring frequency or analyses need to be changed, or if additional wells are needed or should be removed to meet the monitoring objectives.

Finally, a strategy should be developed to deal with problems that may be revealed by the monitoring program. The strategy would include a system of sample replication to verify results, assigning responsibility for decision making and liaison with the regulatory agency, and defining alternatives for cleanup and abatement.

It should be realized that plume shapes, lengths, rates of movement, and concentrations over time cannot always be predicted accurately, even with extensive study. The lack of adequate information or disposal history makes it difficult to determine the length of time the contaminant has been in the environment. In fact, usually so little data are available related to the mass of contaminants that entered the system that efforts to use predictive models is difficult. Also, when multiple sources contribute to degrading the water quality, the cause and effect relationships are difficult to develop.

As of yet, there is no consensus on drilling methods, sampling frequency or protocol, standard quality assurance procedures, or numbers of wells needed to define problems. Some drillers are still unfamiliar with the specialized techniques necessary to construct small-diameter wells, developing wells in low permeability formations, the prevention of cross-contamination, or protecting themselves from hazards. In addition, there appears to be a shortage of groundwater professionals who are truly familiar with carrying out site surveys and how to interpret the hydrogeologic data.

Failure to analyze a flow system properly can lead to the improper design of monitoring and remedial action programs. Therefore, the problems can lead to an improper design of the entire program due to the improper understanding of monitoring results.

GROUNDWATER MODELS

A groundwater model is a simplified representation of the actual hydrogeological system. In order to simulate groundwater flow and/or solute transport in an aquifer, analytical or numerical modeling techniques are used to approximate aquifer conditions. Any system that can duplicate the response of a groundwater reservoir is a model of the reservoir. The operation of the model and manipulation of its inputs and results are termed simulation. A variety of models may be used. Historically, these included electric analog models, which have been replaced by numerical and analytical models. The models are used to simulate groundwater conditions and to forecast head, flow, and water-quality changes that can occur under different stresses on the aquifer.

Models can be as simple as using Darcy's Law or so very complicated that a computer may be needed to solve thousands of complicated mathematical expressions. These mathematical models are, in reality, a set of equations which, subject to certain assumptions, describes the physical process within the aquifer. The models used most commonly include the analytical approach. An example of an analytical flow model is the Theis equation. These are usually exact solutions to the differential equations using many simplifying assumptions. The second approach is the numerical model. Examples of numerical flow models are the Prickett-Lonnquist Aquifer Simulation Model (1971) (PLASM) and the U.S. Geological Survey Modular Flow Model (McDonald et al., MODFLOW, 1984). In this approach, the system is complicated and the simplifying conditions can no longer be made. Therefore, the partial differential equations need to be approximated numerically. The finite difference and the finite element techniques are two methods used to solve these equations. When using these approaches, the continuous aquifer system is replaced by discrete variables that are defined by grids or nodes. That is, the aquifer system is subdivided into parcels. To define the equipotential surface everywhere within the aquifer, the hydraulic head is replaced by a number of algebraic equations that define hydraulic head at specific points. The system of equations is solved using matrix techniques. A computer program is used to solve the equations. There are many computer programs available to simulate groundwater flow and solute-transport problems. An excellent primer on the subject of modeling is given by Mercer and Faust (1981). Their book also has excellent case histories and a reference list. Javandel et al. (1984) is another excellent reference.

The model can be used as a tool to attempt to interpret the hydrogeological and solute-transport systems, as well as to predict changes in the flow or transport system due to the addition of stresses to the system. Initially, a conceptual model of how the system operates is formulated by the hydrogeologist. The first step is to understand the physical behavior of the system and the cause and effect relationships. These concepts are then translated into the numerical model. The second step is to input the aquifer system description data and to calibrate the model. This is done by completing the simulations and by comparing the results to measured data. In a flow model, the results are usually in the form of head changes, and in a solute-transport model, the results are usually in the form of concentrations of contaminants. A more detailed discussion of model calibration, sensitivity analyses, and verification of models is given by Mercer and Faust (1981).

Before selecting the modeling approach, the objectives of the modeling effort should be outlined in detail, along with the knowledge of the aquifer system, and whether additional data will be collected. Selection of a particular approach or model should be based on the nature of the specific aquifer problem. For example, if the problem involves water-level drawdown near a well, a radial flow model with small grid spacing is preferable to a regional model where local effects are lost due to large grid spacings.

A good approach would be to initiate the model at the time when the field program is in the early stages. In this manner, the data collected can be integrated with the model study. The steps in the modeling approach include data collection, data preparation for the model, history matching, and predictive simulation. These tasks are iterative processes. This procedure is shown in Figure 3–49.

The first step in constructing the model is to determine the boundaries of the region to be modeled. This can include impermeable or no flow boundaries, recharge or specified flux boundaries, or constant head boundaries. Once the boundaries are established, the modeled area is subdivided into a grid with either rectangu-

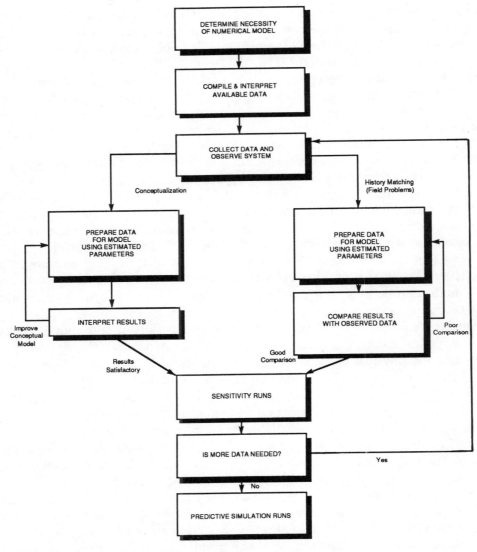

Figure 3–49. Building a groundwater model (adapted from Mercer and Faust, 1981).

lar grids for the finite difference technique or irregular polygon subdivisions for the finite element technique. Examples of these are shown in Figure 3–50.

After the grid is established, the aquifer parameters must be entered into each grid. The necessary data for the flow and solute-transport models are listed in Table 3–4.

Once the initial parameters are input, the model needs to be calibrated. For exam-

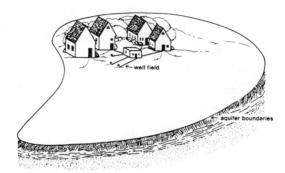

MAP VIEW OF AQUIFER SHOWING WELL FIELD
AND BOUNDARIES

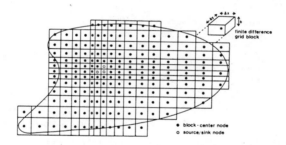

FINITE–DIFFERENCE GRID FOR AQUIFER STUDY, WHERE
ΔX IS THE SPACING IN THE X–DIRECTION, ΔY IS THE
SPACING IN THE Y–DIRECTION AND b IS THE AQUIFER
THICKNESS

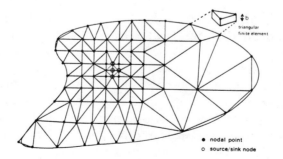

FINITE–ELEMENT CONFIGURATION
FOR AQUIFER STUDY WHERE b IS THE AQUIFER THICKNESS

Figure 3–50.   Examples of finite difference and finite element grid systems to represent a model of an aquifer (adapted from Mercer and Faust, 1981).

Table 3-4. Preparation of Hydrogeologic System Description for Building a Groundwater Flow and Solute-Transport Model.

Groundwater flow model:

- Map showing areal extent and lateral and vertical boundaries of aquifer.
- Topographic map showing surface-water bodies.
- Maps of water table, potentiometric surfaces, and saturated thicknesses.
- Maps displaying transmissivity variations of aquifers and confining beds.
- Map showing variations in storage coefficients.
- Data, graphs, and maps relating saturated thickness to transmissivity.
- Data and maps on interrelationships of surface water and groundwater.
- Data and maps on recharge from irrigated areas, recharge basins, recharge wells, etc.
- Data and maps on surface-water diversions.
- Data and maps on groundwater pumpage distributed in time and space.
- Data and maps on streamflow distributed in time and space.
- Data and maps on precipitation, evapotranspiration, and relationships to head, recharge, or discharge.

Solute-transport model:

- All data inputs needed for groundwater flow model.
- Estimates of parameters comprising hydrodynamic dispersion, retardation, sorption, and transformation.
- Distribution of effective porosity, velocity, and total porosity.
- Information on natural background chemical concentrations.
- Estimates of fluid density variations and relationships to concentrations.
- Data on boundary conditions for concentrations.
- Areal and temporal distribution of water quality.
- Streamflow quality distributed in time and space.
- Sources and strengths of pollution.

Source: Adapted from Mercer and Faust (1981).

ple, the simulated heads need to be matched with the measured heads. If a match does not occur, a trial and error procedure is initiated to refine estimates of aquifer characteristics, boundary conditions, and aquifer stresses.

According to Mercer and Faust (1981), there are no exact rules on when a satisfactory match is obtained. In general, it depends on the objectives of the model, the complexity of the system, and the length of history of observed data. Once the model is calibrated, it can be used to predict changes in the groundwater flow system and contaminant migration patterns. The confidence in the predictive results are dependent on the model limitations, the accuracy of the calibration, and the reliability of the data.

Models are, at times, misused. Prickett (1979) describes three common and related misuses of models. These misuses are overkill, inappropriate prediction, and misinterpretation. For example, if a complicated three-dimensional model is constructed from very little data, when a one- or two-dimensional model could have been used satisfactorily, the effect can be considered overkill. Another form of overkill that is commonly carried out is designing a very small grid system over a very large area.

When modeling a thin confined aquifer, the hydrogeologist should be aware of the consequences of pumpage. That is, large rates of pumpage can cause the head in the aquifer to fall below the top of the aquifer. This would cause the aquifer conditions to change from confined to unconfined conditions. The decline in water levels due to pumpage in an unconfined aquifer is much different than that in a confined aquifer. Therefore, if the aquifer was modeled as only a confined aquifer, very large errors in predicted head relationships would occur. This error is considered to be an inappropriate prediction.

The worst misuse of a model, misinterpretation, is using the results as fact without interpreting the results for reality. That is, the results contradict the conceptualized hydrogeologic conditions. Proper application of the model requires a conceptual understanding of the hydrogeologic conditions.

In general, the numerical models are important tools for simulating the behavior of aquifers. The models can simulate complicated boundaries, heterogeneous aquifers, groundwater flow, and contaminant migration. In order to construct a model, the hydrogeologic system is formalized into a conceptual understanding of the system. The model can, at times, be used to gain a better understanding of the system, especially if it is used in an iterative process with the data collection program. At times, using the model to understand the system can decrease the costs of field programs and make the data collection program more efficient. If the hydrogeologic system is understood, along with the modeling techniques, limitations, and sources of errors, the modeling effort can be very effective.

CASE STUDY

A long-term groundwater monitoring program aided in the design of a remedial action program and is presently illustrating the success of an industry's contaminated groundwater abatement and control program (Stollar, 1983). The monitoring program began in the mid-1970s, and therefore could not take advantage of many of the new monitoring and sampling techniques. However, with sound hydrogeologic judgment, and good investigative techniques, the monitoring program is successful and is ensuring the completion of the corrective action program.

The industry produces phosphoric acid, phosphates, and, most recently, phosphorus pentasulfide. Wastes from these processes (except the phosphorous pentasulfide) were disposed in in-ground-pits—a method acceptable to regulatory agencies in years past.

After it was determined that the shallow groundwater beneath the plant contained relatively high concentrations of contaminants, a program was funded to determine the extent of the problem, a description of groundwater flow and contaminant migration patterns, and to develop a groundwater management and monitoring program.

An important part of every investigation of groundwater pollution is to locate and define the extent of the contaminated body of groundwater. The usual method for accomplishing this is to install and sample numerous test wells, a costly and time-consuming procedure. A much faster and less costly method, which has proven to give accurate results in certain cases, is the earth-resistivity survey (Stollar, 1975). Earth resistivity is simply the measurement of an electrical current passing through the ground from one point to another. Because earth resistivity is inversely proportional to the conductivity of groundwater, the location of the groundwater that has been contaminated by a relatively high concentration of conductive industrial wastes, for example, may be quickly and accurately traced. Based on the resistivity values obtained near the control points, an apparent resistivity of 200 ohm-feet or less was considered to be representative of contaminated groundwater. Using the 200 ohm-foot value, the extent of the contaminated groundwater body was estimated. Subsequent drilling and water sampling of wells on either side of the approximate 200 ohm-foot boundary confirmed the validity of this interpretation.

To determine the vertical and lateral distribution of contamination and groundwater flow within the alluvial aquifer, as well as to verify the boundaries of the contaminated water body inferred from the resistivity survey, a drilling program was initiated at the plant site. During this program, 23 observation well sites were completed under the supervision of a hydrogeologist. Each site contained from one to three wells. Where well sites contained clusters of two or three wells, they were completed to different depths. Such a well cluster is illustrated in Figure 3–41.

After all wells were completed, they were sampled by pumping or bailing approximately three times the standing water held within the casing or, until the well was pumped dry. If the well was bailed or pumped dry, it was allowed to recover for a short time and then sampled.

Water samples collected from each well were measured on-site for specific conductance, pH, and temperature, and split into three sub-samples for subsequent analysis. All samples were collected, preserved, and analyzed in accordance with standard EPA protocol. From these data, the present pattern, direction, and rate of groundwater flow in both the horizontal and vertical direction were determined. Contour maps of water-level altitudes were also constructed. Each contour on this map represents a line of equal energy head. Groundwater flow lines cross these contour lines at right angles, the direction of flow being from higher to lower head. Regionally, the natural groundwater flow is towards a river located south of the plant.

In addition to the general horizontal flow components, the vertical components of groundwater flow can also be interpreted from data collected in the well clusters. The distribution of heads near and under the areas that were the sources of contamination is very complicated.

Many of the groundwater characteristics examined during the study, such as temperature, conductivity, pH and total dissolved solids, are indicators of the contaminated plume. However, the constituent that best represents the body of contaminated groundwater is phosphate.

To illustrate the extent of the plume in the subsurface, a north-south cross section of the phosphate constituent was prepared, and is shown in Figure 3–51. The contour lines join points of equal concentration. This is an example of the cross-sectional distribution of phosphate.

The movement of the contaminated groundwater plume is controlled by parameters that describe the aquifer, the transport of the contaminant, and by stresses on the aquifer. It appears that, over time, the artificial stresses such as pumping and the recharge ponds have controlled the direction of movement of the plume to a large extent.

Recharge of the contaminated fluids to the groundwater system at the site of the abandoned wastewater pits created a recharge mound. A recharge mound is a high point in the water table where flow is away from the source in all directions. Thus, after the creation of these pits, the flow direction is southerly towards the river. Also, at this time, Wells 1 through 5 (Figure 3–52) were pumping, which caused a

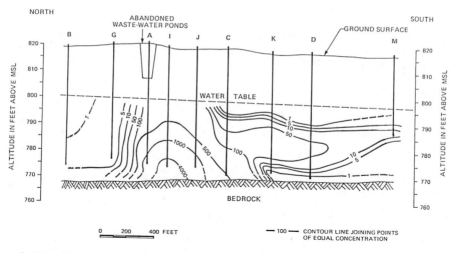

Figure 3–51. Cross section of phosphate concentrations (adapted from Stollar et al., 1983).

cone of depression to form in the area along Maple Street. The gradient created by a combination of the cones of depression and the recharge mound enabled the contaminant to move rapidly towards Maple Street. However, because the gradients created by the pumping had probably caused all flow paths north of Maple Street to be intercepted by a production well, the plume could not move past the wells to the south side of the street. The plume, therefore, was probably fanshaped, beginning at the pit and spreading towards the wells along Maple Street from Well 5 in the southwest to Well 1 in the southeast. With the abandoning of wells over a period of time, the groundwater flow pattern changed significantly.

From 1966 to the present, the pumping pattern at the southern portion of the plant has remained the same; the only well pumping is Well 5. Therefore, the portion of the plume that had reached Well 1 is now moving with the natural gradient in a south to southwesterly direction. The main direction of the plume is now in a southwesterly direction with an area extending from just west of Well 1 to Well 5.

Data did not indicate that the plume reached Well 5 until 1975. The change in the phosphate content from 1966 to 1988, as determined from samples collected from Well 5, is shown on Figure 3–53.

This indicates that, with continued pumping, the phosphate concentration will continue to rise in the well. Presently, the rate of increase of the maximum concentrations of phosphate has not been projected in time. Of course, if all sources of contamination are stopped, the maximum concentrations probably will never exceed the highest concentrations which are shown in Cluster I (Figure 3–54).

Part of the problem in attempting to determine and to predict changes in concentration with time, and maximum concentrations, are the changes that occur in the mobility of the constituents as they move through the aquifer. Phosphate is absorbed on or reacts with the soil particles. When the reactive sites of the solids within the aquifer become fully occupied with these constituents, the concentration of the plume will no longer be reduced by this mechanism. Therefore, it will move as a concentrated slug. The information needed to delineate the physics and chemistry of these phenomena is still experimental, and therefore some of the present answers

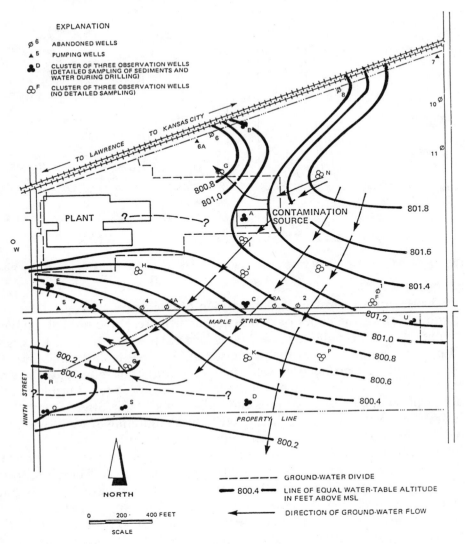

Figure 3–52. Water table configuration, April 1978 (adapted from Stollar et al., 1983).

depend upon estimates and empirical relationships. In order to carry out such predictions, it will be necessary to build a solute-transport model.

After the well drilling program was completed, a groundwater monitoring program was designed and initiated.

As the results of the drilling program only represented a synoptic picture of the plume of contamination, movement of the plume and changes in constituent concentration over periods of time could not be predicted with the data. However, if, over a long period of time, observations of changes in, or movement of, the plume are made frequently, observed data will answer some of the above questions, and predictions for future changes may be made. To accomplish this, a formal monitoring program was established.

When the program was initiated in 1976, the sampling for water levels and quality was carried out on a quarterly basis. As expected, rapid changes in movement of

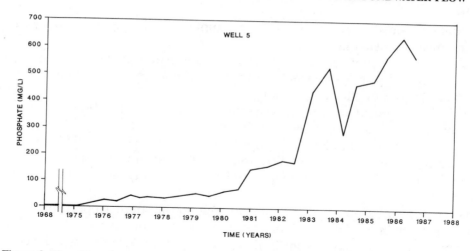

Figure 3–53. Phosphate concentrations in Well 5 (adapted from Stollar et al., 1983).

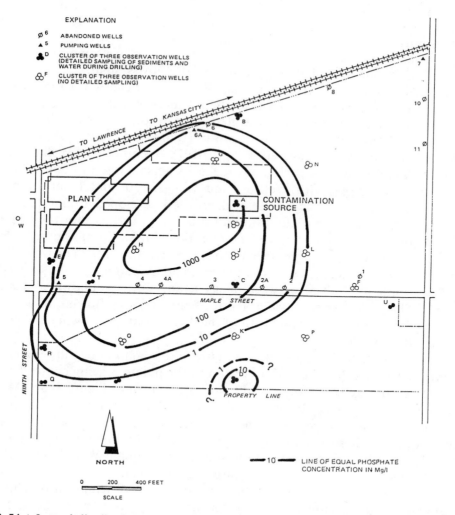

Figure 3–54. Lateral distribution of phosphate concentration, April 1978 (adapted from Stollar et al., 1983).

the contaminant did not occur and the sampling frequency was reduced to semi-annual measurements in 1978.

As in the solving of any groundwater contamination problem, the groundwater flow system is of primary concern, and the actual concentration changes are secondary. That is, the quality of groundwater, or changes in quality, or movement of plumes, is dependent on the flow system or changes in the flow system, and on sorption and attenuation of the constituents.

Results of the present monitoring program illustrate this point. This can be shown with changes of phosphate concentration in Well 5, located in the southwest portion of the plant site, water-table maps, and maps showing the lateral extent of the plumes.

As mentioned before, Well 5 is the last well on the southern portion of the property that is still in use and is the furthest from any of the sources of pollution along this southern line of wells. From 1967 through early 1975, the phosphate concentrations determined from water samples in this well were representative of background quality. In 1975, it appears that the plume reached the well and concentrations of phosphate within the groundwater began to increase. From 1975 to the present, concentrations have increased steadily. This is shown on Figure 3–53. As Well 5 is a major supply well and is pumped relatively heavily, it has a major influence on the movement of the phosphate plume and on the flow system.

Whenever pumping rates from this well change significantly for long periods of time, the impact on the flow system at the southern portion of the plant also is significant. An example of this is illustrated on maps of the water-table configuration for April, 1978 and May, 1979, shown in Figures 3–52 and 3–55, respectively.

When water levels were measured in April, 1978 (Figure 3–52), Well 5 was being pumped for a relatively long period of time at its normal rate of about 125 gallons per minute (gpm), along with Well 6A and Wells 7, 10, and 11 in the northern portion of the property. A large cone of depression developed around the well and the resulting water-table configuration indicates the effect, as shown on Figure 3–52. The figure also contains representative flow lines. From Figure 3–52, the groundwater is shown flowing in a south to southwest direction. The flow direction is being controlled by the regional flow pattern, and by pumping wells such as Wells 5 and 6A. South of Well 5, between Wells R and Q, a groundwater divide had developed. This indicates that water south of the divide is flowing in a southerly direction, away from Well 5; and north of the divide, water is flowing in a northerly direction, towards Well 5.

Comparing the flow patterns on this figure with a map indicating the extent of the plume, it appears that the major portion of the plume is being controlled by the pumping from the well. The map of the plume is shown in Figure 3–54. As shown by this figure, the regions of the plume containing high concentrations of phosphate are still very near the source areas, and the concentrations decrease away from the sources. In the southern portion of the property, the phosphate plume appears to be controlled by the effects on the flow system of pumping Well 5. The effects of attenuation and sorption of phosphate are not discussed.

However, changes in the flow system can affect the distribution of phosphate. An example of this is shown by the water-table configuration for May, 1979, as shown in Figure 3–55. During this period, problems with well efficiency were occurring in Well 5. As the well became inefficient, pumping rates decreased.

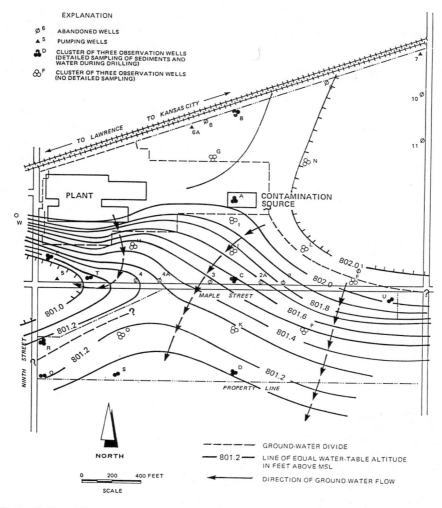

Figure 3-55. Water table configuration, May 1979 (adapted from Stollar et al., 1983).

Finally, the well was shut off, cleaned, and redeveloped. Due to the decreased pumping in Well 5, water levels in the entire southwestern portion of the property began to recover. The data also indicate that, during the actual time of measurements, pumping began in Well 5 and water levels fell slightly.

The intermittent pumping at Well 5 probably has caused the change in the groundwater flow pattern. The groundwater divide shown for May 1979 is closer to the well and does not extend as far in an easterly direction.

This indicates that, as shown in the figures, the groundwater flow is southerly. If this water-table configuration were to persist, the plume of phosphate also will continue to move in a southerly direction away from Well 5 and off the property. In other words, the plume would not be controlled. To prevent this, a second production well has been installed in the vicinity of Well 5.

Interpretation of the data collected during the monitoring program has indicated the importance of controlling the water-table configuration in the southwest part of the plant property. This, in turn, will have a major effect on controlling the movement of the plume. As Well 5 was the only pumping well in this area, a new well has been drilled just a few hundred feet to the northwest of Well 5. Pumpage from this new well will increase the amount of stress on the groundwater system, cause a larger amount of flow to move towards the two pumping wells and will increase the amount of plume area to be intercepted by the wells. In other words, with pumpage from the new well, very little of the plume will move past the property boundary.

The location and pumping rates of the new well was determined by using an analytical model. The map showing the computed water-table (analytical model) configuration, Figure 3–56, indicates that the combination of pumpage from the two wells will aid in controlling the plume. As shown on the map, the groundwater divide has shifted further south. This indicates a larger area where groundwater will flow to-

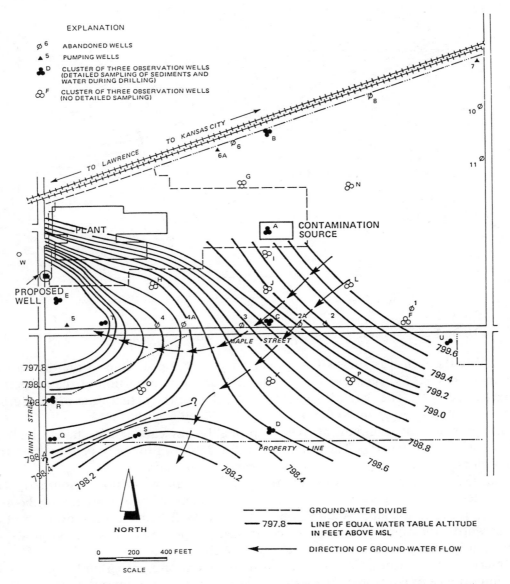

Figure 3–56. Computed water table configuration for proposed abatement program (adapted from Stollar et al., 1983).

wards the pumping wells. Almost the entire intact phosphate plume is within the area of influence of the pumping wells. Therefore, the plume will be controlled by the wells and the concentration of phosphate will increase steadily in Well 5 as the high concentrations in the central core of the plume move in the direction of the well. Groundwater elevation data collected recently have substantiated the above predictions, as shown in Figure 3–57. It is interpreted that increased phosphate concentrations will occur in the new well, but should be a very slow occurrence, as the most concentrated part of the plume is being intercepted by Well 5.

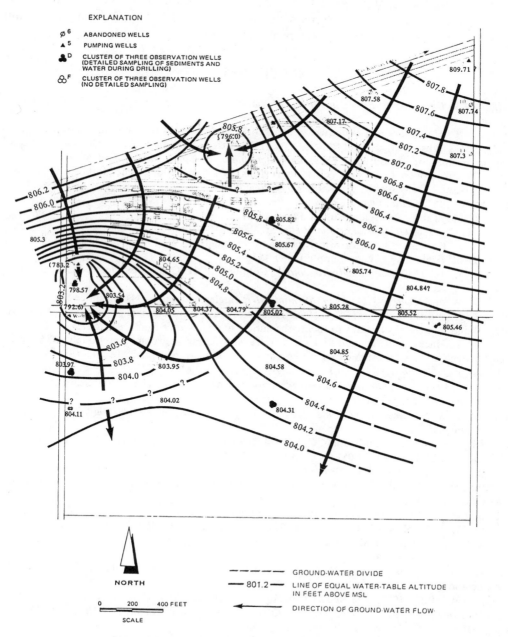

Figure 3–57.   Water table configuration, May 1987.

In conclusion, the use of resistivity studies and drilling programs have allowed the hydrogeologic environment and the geometry of the contaminated groundwater body to be described in detail. A monitoring program was designed and initiated after the drilling program was completed. Interpretation of the data collected during the monitoring program has been used successfully to interpret the dynamics of the groundwater system, impacts of pumping, and migration of contaminants within the groundwater system.

With the results of the field and monitoring programs, groundwater management, abatement, and control programs have been designed and initiated. It appears that carrying out these new programs will control the plume underlying the plant, thereby managing the groundwater problem that exists at the plant.

REFERENCES

Bear, J. 1972. *Dynamics of Fluids in Porous Media*. New York: American Elsevier Publishing Company.

Bear, J. 1979. *Hydraulics of Groundwater*. New York: McGraw-Hill International Book Company.

Braids, O. C., Wilson, G. R., and Miller, D. W. 1977. Effects of industrial hazardous waste disposal on the ground water resource. In *Drinking Water Quality Enhancement Through Source Protection*, pp. 179–207, Robert B. Pojasek (Ed.). Ann Arbor, MI: Ann Arbor Science Publishers, Inc.

Davis, S. N. and De Wiest, R. J. M. 1966. *Hydrogeology*. New York: John Wiley and Sons, Inc.

Deutsch, M. 1963. *Groundwater Contamination and Legal Controls in Michigan*. Washington, DC: U.S. Geological Survey Water Supply Paper 1961.

Driscoll, F. G. 1986. *Groundwater and Wells*. St. Paul, MN: Johnson Division.

Ferris, J. G., Knowles, O. B., Brown, R. H., and Stallman, R. W. 1962. *Theory of Aquifer Tests*. U.S. Geological Survey Water Supply Paper 1536-E.

Fetter, Jr., C. W. 1980. *Applied Hydrogeology*. Columbus, OH: Charles E. Merrill Publishing Company.

Freeze, R. A. and Cherry, J. A. 1979. *Groundwater*. Englewood Cliffs, NJ: Prentice-Hall, Inc.

Freeze, R. A. and Witherspoon, R. A. 1966. Theoretical analysis of regional ground-water flow: Analytical and numerical solutions to the mathematical model. *Water Resources Research,* 2:641–656.

Grisak, G. E., Jackson, R. E., and Pickens, J. F. 1978. Monitoring ground-water quality: The technical difficulties in establishment of water quality monitoring programs. In *American Water Resources Association Symposium Proceedings,* Minneapolis, MN.

Heath, R. C. 1984. *Basic Ground-Water Hydrology*. Washington, DC: U.S. Geological Survey Water-Supply Paper 2220, pg. 84.

Heath, R. C. and Trainer, F. W. 1968. *Introduction to Ground-Water Hydrology*. New York: John Wiley and Sons, Inc.

Javandel, I., Doughty, C., and Tsong, C. F. 1984. *Groundwater Transport: Handbook of Mathematical Models*. EPA-600 LT-84-051, NTIS PB 84-222694.

Johnson, Edward E., Inc. 1966. *Groundwater and Wells,* St. Paul, MN: Johnson Division.

LeGrande, H. E. 1965. Patterns of contaminated zones of water in the ground-water. *Water Resources Research,* 1:83–95.

McDonald, M. and Harbaugh, A. W. 1984. A Modular Three Dimensional Finite Difference Ground-Water Flow Model. Washington, DC: U.S. Geological Survey Open File Report 83-875.

McWhorter, D. B. and Sunada, D. K. 1977. *Ground-Water Hydrology and Hydraulics*. Ft. Collins, Co: Water Resources Publications.

Mercer, J. W. and Faust, C. R. 1981. *Ground-Water Modeling*. Columbus, OH: National Water Well Association.

Miller, D. W. (Ed.). 1977. The Report to Congress, *Waste Disposal Practices and Their Effects on Groundwater*. Office of Water Supply and Office of Solid Waste Management Programs. Washington, DC: U.S. Environmental Protection Agency.

Morris, D. A. and Johnson, A. W. 1967. *Summary of Hydrologic and Physical Properties of Rock and Soil Material as Analyzed by the Hydrologic Laboratory of the U.S.G.S. 1938–1960.* Washington, DC: U.S. Geological Survey Water-Supply Paper, 1839-D.

Pastrovich, T. L. 1979. Protection of groundwater from oil pollution. CONCAWE Report Number 3/79, Water Pollution Special Task Force No. 11. Den Haag, Netherlands.

Perlmutter, N. M. and Lieber, M. 1970. *Dispersal of Plating Wastes and Sewage Contaminants in Groundwater and Surface Water, South Farmingdale–Massapequa Area, Nassau County, New York.* Washington, DC: U.S. Geological Survey Water-Supply Paper 1879-G.

Pricket, T. A. 1979. Groundwater computer models—State of the art. *Ground Water,* **17**, 2, pp. 167–173.

Pricket, T. A. and Lonnquist, C. G. 1971. Selected computer techniques for ground-water resource evaluation. Illinois State Water Survey, Bulletin 55, Urbana, IL.

Roberts, P. and MacKay, D. M. 1986. A natural gradient experiment on solute transport in a sand aquifer, R. S. Kerr Environmental Research Laboratory. Ada, OK: U.S. Environmental Protection Agency.

Schmidt, K. D. 1977. Water quality variations for pumping wells. *Ground Water,* **5**(2): 130–137.

Schwille, F. 1975. Ground-water pollution by mineral oil products. In *Proceedings of the Moscow Symposium* IAHS-Publication No. 103.

Stollar, R. L. 1983a. Guidelines for a ground-water monitoring program. *J. Association of Environmental Professionals,* **9**, 4.

Stollar, R. L. 1983b. Defining and managing a ground-water contamination problem. *J. Association of Environmental Professionals,* **9**, 5.

Stollar, R. L., Pellisier, R., and Studebaker, P. 1983. Management of a ground-water problem. *J. Water Pollution Control Federation.* **55**(11):1393–1403.

Stollar, R. L. and Roux, P. 1975. Earth resistivity surveys—A method for defining ground-water contamination. *Ground Water* **13**(2):145–150.

Todd, D. K. 1959. *Ground-Water Hydrology.* New York: John Wiley & Sons, Inc.

Todd, D. K. 1980. *Ground-Water Hydrology,* 2nd ed. New York: John Wiley & Sons, Inc.

Toth, J. 1962. A theory of groundwater motion in small drainage basins in central Alberta. *J. Geophys. Res.,* **67**:4375–4387.

Toth, J. 1963. A theoretical analysis of groundwater flow in small drainage basins. *J. Geophys. Res.,* **68**:4795–4812.

USEPA. 1977. The Report to Congress, Waste Disposals and their Effects on Groundwater. Washington, DC: Office of Water Supply, Office of Solid Waste Management Programs.

USEPA. 1985. Guidance on feasibility studies under CERCLA. Washington, DC: Office of Solid Waste and Emergency Response.

USEPA. 1987. Data quality objectives for remedial response activities. EPA Contract No. 68-01-6939. Washington, DC: Office of Solid Waste and Emergency Response.

USEPA. 1988. Draft Guidance for Conducting Remedial Investigations and Feasibility Studies under CERCLA, EPA Contract No. 68-01-7090. Washington, DC: Office of Emergency and Remedial Response.

Walton, W. C. 1962. Selected Analytical Methods for Well and Aquifer Evaluation. Illinois State Water Survey Bulletin 49, Urbana, IL.

Walton, W. C. 1970. *Ground-Water Resource Evaluation.* New York: McGraw-Hill Book Company.

Walton, W. C. 1984. *Practical Aspects of Ground-Water Modeling.* Columbus, OH: National Water Well Association.

4

Microbiology of Subsurface Wastes

by Joseph S. Devinny
Associate Professor of Civil and Environmental Engineering
University of Southern California

SOIL MICROBIOLOGY

Only a few years ago it was accepted wisdom that aquifers well below the surface soil layer contained few microorganisms. Oxygen is often not available, no photosynthesis is possible, and concentrations of substrate (food chemicals for microorganisms) are usually low. It was reasonable to suppose that little microbiological activity could be supported, and little was found.

It is now realized that this "wisdom" was largely a result of the lack of adequate methods for detecting and counting microorganisms in soil. Some methods attempted to wash the cells from the soil, and to count them by culturing. Many cells, however, stick firmly to the soil, and were not counted. The microorganisms present often did not grow on the culture media used. Direct microscopic examination missed most of the cells because they were obscured by the soil or because the soil particles and the cells could not be distinguished.

The development of fluorescence microscopy has demonstrated something new. In this technique, the soil sample is stained with a fluorescent dye and examined under UV light. Dye which is on the soil glows, but dye which is absorbed by living cells glows a different color. The cells become brilliantly obvious, and investigators now know that they are numerous. Aquifers commonly contain 1 to 10 million cells per gram of soil, making subsurface waters the second largest microbiological culture in nature, after the oceans (Wilson, 1986).

This abundance explains many of the transformations which chemicals undergo in aquifers. Far from being sterile sand filters, aquifer soils are culture vats teeming with life.

The microbiological activity in aquifers makes the subsurface migration of hazardous wastes more difficult to predict, and so creates some problems. But far more important, it also creates opportunities for waste cleanup which are inexpensive and which carry minimal environmental side effects.

Microbial Metabolism

Microorganisms use organic matter as human beings do: to provide energy for their metabolic processes and as feedstock for the synthesis of necessary structural chemicals. Also like higher organisms, they may break down a few chemicals specifically because they want to eliminate their toxic effects.

In most cases, this biological activity will be the primary cause of chemical transformations of organic matter below the surface. The organisms can speed many chemical reactions which would otherwise occur slowly, and will promote reactions which might never occur in their absence. An understanding of the microbiology of the waste and soil is necessary for an understanding of the fate and transport of the contaminant.

The microorganisms in soil live primarily on the surfaces of the particles. They are less common within particles because oxygen is less available. Of course, many particles in the soil are impenetrable solids, but even the porous grains are not usually filled with active organisms. Microorganisms are also less likely to be found suspended in the water away from soil surfaces, where they could be carried away by the flow. They find an advantage in clinging to the surfaces while the water moves by, so that they are constantly exposed to a stream of passing nutrients and food. Microorganisms suspended in the water must expend energy swimming away from the microscopic regions they have depleted of nutrients and filled with wastes.

Microbially Mediated Reactions.   Microorganisms participate in a bewildering array of chemical reactions. Most, however, can be put in one of two classes. *Gross reactions,* which consume substantial amounts of material and produce equally large volumes of product, are only microbially mediated if they produce energy. The organisms catalyze these reactions in order to utilize the energy. The reactions of interest are inevitably oxidation-reduction (redox) reactions: those in which an electron is transferred from one chemical to another. Redox reactions are also of particular import to hazardous waste and groundwater contamination problems (Chapter 2).

The microorganisms promote a second class of reactions utilizing smaller amounts of material. These *synthetic reactions* produce the highly specialized chemicals necessary for life, such as proteins, carbohydrates, and DNA. The microorganisms promote these reactions for chemical products rather than for energy. Indeed, it is often necessary for the microorganisms to supply energy in order to make the reactions occur. Much smaller amounts of material are processed, and only rarely are the products of direct concern in hazardous waste issues. They may, however, have important secondary effects. For example, the polysaccharides (polymers of sugars) generated by the microorganisms increase the adsorption of contaminants by soil, and its tendency to stick together in clumps.

The fact that the gross reactions are redox reactions allows them to be viewed in a conveniently systematic manner. Each involves the transfer of an electron (or several electrons) from one chemical species to another. The character of both chemicals changes, and other species may be involved.

Electrons are not found in isolation in water solutions (which make up all living organisms). In a redox reaction, the electron is donated by one species and accepted by another. Further, all of the reactions are reversible: the species which accepts an electron in one reaction can give it up in another. As suggested by Stumm and Morgan (1981), a catalogue of reactions may therefore be built by listing the important reactions in which electrons are given up, then reversing each to produce a list of reactions in which electrons are accepted (Table 4-1). In analogy with the common reactions of atmospheric oxygen, reactions which remove electrons from the species of interest are called oxidations. Reactions which add electrons are called reductions.

The reductions are listed in order of the amount of energy they provide to the organisms. The reduction which provides the most energy is at the top of the list.

The energy relationships are also reversible. That is, the energy provided by a reduction is equal to the energy consumed by the corresponding oxidation. When a reduction and an oxidation are combined, the net energy release is the energy of the reduction minus the energy of the oxidation. Because the reactions are listed in order of net energy release, any reduction combined with an oxidation listed below it produces an overall redox reaction which can provide the organism with energy. Combining a reduction with an oxidation above it on the list produces a redox reaction which consumes energy: the organism must supply energy in some way to make the reaction occur.

For example, when we respire, we breathe in oxygen and convert it to water. At the same time, we convert organic matter like carbohydrates to carbon dioxide and water. The combined redox reaction produces the energy we use to power our bod-

Table 4-1. Reductions and Oxidations Important for Microorganisms in Soils and Groundwater.

Reductions, which consume electrons, listed in order of the energy they produce, most energy production first . . .	can be reversed to make a list of . . .	Oxidations, which generate electrons, in order of the energy they consume, most energy consumption first.
Oxygen to water $O_2 \rightarrow H_2O$		Water to oxygen $H_2O \rightarrow O_2$
Nitrate to nitrogen gas $NO_3^= \rightarrow N_2$		Nitrogen gas to Nitrate $N_2 \rightarrow NO_3^=$
Nitrate to nitrite $NO_3^= \rightarrow NO_2^-$		Nitrite to nitrate $NO_2^- \rightarrow NO_3^=$
Nitrite to ammonia $NO_2^- \rightarrow NH_3$		Ammonia to nitrate $NH_3 \rightarrow NO_2^-$
Ferric iron to ferrous iron $Fe^{3+} \rightarrow Fe^{2+}$		Ferrous iron to ferric iron $Fe^{2+} \rightarrow Fe^{3+}$
Carbohydrate to alcohol $CH_2O \rightarrow CH_3OH$		Alcohol to carbohydrate $CH_3OH \rightarrow CH_2O$
Sulfate to sulfide $SO_4^= \rightarrow H_2S$		Sulfide to sulfate $H_2S \rightarrow SO_4^=$
Carbon dioxide to methane $CO_2 \rightarrow CH_4$		Methane to carbon dioxide $CH_4 \rightarrow CO_2$
Nitrogen gas to ammonia $N_2 \rightarrow NH_3$		Ammonia to nitrogen gas $NH_3 \rightarrow N_2$
Hydrogen ion to hydrogen gas $H^+ \rightarrow H_2$		Hydrogen gas to hydrogen ion $H_2 \rightarrow H^+$
Carbon dioxide to organic matter $CO_2 \rightarrow CH_2O$		Organic matter to carbon dioxide $CH_2O \rightarrow CO_2$

ies. Plants can reverse the process, converting carbon dioxide to carbohydrates as they convert water to oxygen. This reverse process, however, consumes large amounts of energy. Plants are able to do it only because they can use the energy of sunlight to power the reaction.

A second principle allows us to not only predict which reactions will occur, but the approximate order in which they will occur. In nonliving chemistry, the amount of energy released by a reaction is not specifically related to the rate at which the reaction proceeds. Kinetics are not predicted by thermodynamics. For biologically mediated reactions, however, there is at least a general relationship. Organisms are faced with the need to process chemicals in order to extract the energy. Because the reactions produce different amounts of energy, processing a gram of one chemical produces more benefit than processing a gram of another. Organisms that are utilizing the more energetic reactions are more efficient, and will likely dominate the ecological competition for food and space. They will grow vigorously at the expense of their neighbors. They will remain the most abundant species until the chemicals which they are using as food are exhausted. Then their decline will allow organisms using the second-most efficient form of metabolism to take over. When the second substrate is used up, a third group of microorganisms will dominate, and so on down the list of redox reactions for which chemicals are available. The result is that biologically catalyzed redox reactions tend to occur in order of the amount of energy they produce.

This is a generality rather than an absolute rule. Often, several of these reactions occur at one time. The presence and abundance of microorganisms is influenced by many factors other than substrate availability. Further, adjacent microenvironments may be in different stages of the process.

Nevertheless, reaction energy is often a valuable guide to understanding why some reactions occur, while others do not. It is almost always true, for example, that aerobic metabolism (utilizing the energy-rich combination of oxygen with organic material) will predominate over anaerobic metabolism (utilizing the less energetic reactions without oxygen) if oxygen is available.

The redox reactions produced by various combinations of reductions and oxidations can be classified. *Respiration,* the "oxidation by oxygen" of organic food to carbon dioxide and water, is the basic metabolism of all multicelled organisms and most single-celled organisms. If oxygen is abundant but no organic matter is available, the reduction of oxygen can be combined with the sulfide, iron, ammonia, or hydrogen oxidations in *chemoautotrophy.* If organic food is abundant, but oxygen is absent, *anaerobic respiration* is accomplished by microorganisms which combine the oxidation of organic matter with the reduction of inorganic species like nitrate, sulfate, and carbon dioxide. Organisms perform *fermentation* by combining the oxidation of one organic species with the reduction of another.

Aerobic Respiration.   The lists in Table 4–1 show that the redox reaction which produces the most energy is the first reduction combined with the last oxidation. The oxidation of carbohydrates and hydrocarbons by oxygen from the atmosphere is the source of fire and many chemical explosions, so we are familiar with the amount of energy involved. Controlled within the cell, it is energy-generating metabolism. In large animals it was named respiration, because the oxygen is breathed in and the carbon dioxide is breathed out. By analogy, the term respiration is used for the same chemical process even when carried out by organisms without lungs.

The tremendous energy value of aerobic respiration has made it the process used by essentially all multicelled organisms. Because of this, the loss of oxygen in a water environment is an ecological disaster: all of the fish and other large organisms die. This can occur where sewage or other effluents containing large amounts of organic material are discharged. The excess organics are decomposed rapidly by respiring microorganisms and the oxygen supply is depleted.

Aerobic respiration can be used to eliminate pollutants. Trickling filters and activated sludge units, mainstays of sewage treatment, are devices designed to supply large amounts of oxygen to a rich culture of microorganisms engaged in converting unwanted organic matter to carbon dioxide.

Biological oxidation may also be a treatment of choice for contaminated groundwater. The respiring organisms must be microorganisms that can live in the pores of the aquifer, and the process is often limited by the lack of oxygen. Nevertheless, for small amounts of material, natural aerobic respiration is often sufficient to eliminate it. Further, if the spill is large, and will obviously deplete the available oxygen in the groundwater, it may still be possible to promote biodegradation by adding oxygen artificially. Oxygenation and the addition of other needed nutrients constitute some of the treatment processes described in later chapters.

Aerobic Chemoautotrophy.   Sometimes, the microorganisms will consume all of the organic matter while oxygen remains abundant. The redox series can be used as a guide to predict the chemoautotrophic reactions which will occur next. An organism is "autotrophic" when it makes its own food; chemoautotrophs are able to do so using inorganic chemicals as a source of energy.

Table 4–2 is constructed combining the reduction of oxygen with each of the remaining oxidations. The resulting redox reactions will be considered in order of total energy (that is, beginning at the bottom of the oxidations list).

In an environment where oxygen is present but organic matter is not, the most energetic redox reaction will be the combination of oxygen reduction and the oxidation of *hydrogen* to water. Because hydrogen is rare in natural environments, this is not widely encountered. But it does occur.

There are also chemoautotrophic organisms which can oxidize *methane* to carbon dioxide (listed next above the oxidation of hydrogen). They are found in abundance where methane is being evolved by other microbiological processes, or where it is seeping from natural gas deposits. They are abundant in soils surrounding gas leaks in natural gas delivery systems.

The methane-utilizing organisms have an importance which arises in an almost incidental way. Some are capable of aerobic degradation of chlorinated solvents such as trichloroethylene and tetrachloroethylene. These chemicals are cleaning fluids used for degreasing parts and dry cleaning clothes. It seems likely that microbial enzymes which are configured to attack methane will also attack the solvents. It will be possible to use these species in the future for spill cleanup.

If neither organic matter nor methane are available, organisms which can oxidize *sulfide* may become dominant. Sulfide is not common in natural aerobic environments, but it does occur. Sulfide-oxidizing bacteria are found in springs, for exam-

Table 4–2. Metabolic Reactions When Oxygen is Abundant.

Reduction of oxygen, which provides the most energy, will be combined with . . .	Oxidation of various substrates beginning with the last, which consumes the least energy, and proceeding upwards, to produce . . .	Redox Reactions of aerobic respiration and chemoautotrophy
Oxygen to water $O_2 \rightarrow H_2O$	Nitrogen gas to Nitrate $N_2 \rightarrow NO_3^=$	9. (Does not occur, kinetically hindered)
	Nitrite to nitrate $NO_2^- \rightarrow NO_3^=$	8. Nitrification
	Ammonia to nitrate $NH_3 \rightarrow NO_2^-$	7. Nitrification
	Ferrous iron to ferric iron $Fe^{2+} \rightarrow Fe^{3+}$	6. Iron oxidation
	Sulfide to sulfate $H_2S \rightarrow SO_{4=}$	5. Sulfide oxidation
	Methane to carbon dioxide $CH_4 \rightarrow CO_2$	4. Methane oxidation
	Ammonia to nitrogen gas $NH_3 \rightarrow N_2$	3. (Does not occur, kinetically hindered)
	Hydrogen gas to hydrogen ion $H_2 \rightarrow H^+$	2. Hydrogen oxidation
	Organic matter to carbon dioxide $CH_2O \rightarrow CO_2$	1. Aerobic respiration

ple, where sulfide generated in deep aquifers reaches the surface. They may also become important where a variety of pollutant processes produce and release sulfide.

Springs which bring anaerobic groundwater into contact with the atmosphere may also support cultures of iron-oxidizing bacteria. Oxidation of *ferrous iron* (Fe^{++}) is the next reaction on the oxidations list (assuming no organic matter is present, so there is no alcohol). It will occur readily when ferrous iron, oxygen, and the appropriate bacteria are present. It can occur in water delivery systems, where the metallic iron of the pipes is first oxidized to the ferrous form and then to the *ferric* (Fe^{3+}) form. This result is undesirable in several ways: the bacteria may cause the water to have an offensive taste and odor, the oxidation contributes to pipe corrosion, and the ferric iron precipitates as ferric hydroxide to form a colloid of rust particles, which stains the water and plumbing fixtures.

Reduced (ferrous) iron is common in groundwater. Many minerals contain reduced iron, which can dissolve. If the groundwater is anaerobic, ferric iron may be reduced biologically. When this water comes to the surface and is exposed to oxygen, the iron-oxidizing bacteria can become abundant. Again rust-colored water and rust deposits will result.

If oxygen is abundant and there is no organic matter, methane, sulfide, or ferrous iron, the presence of *ammonia* (NH_3) may still prompt biological activity. Many species are capable of converting ammonia to nitrite, and then to nitrate. This occurs in sewage treatment plants, where the process is the first step in "nitrification-denitrification," used to remove the nitrogen species which pollute lakes and rivers. It may have a significant influence on subsurface migration of nitrogen: ammonia is adsorbed by soils but nitrate is quite mobile. Ammonia appears in soils as the product of microbial degradation of organic matter or because it is added directly as fertilizer. Conversion to nitrate and subsequent movement of the nitrate has often polluted groundwater. Nitrate is toxic to infants, and its presence has caused the closure of many wells in agricultural areas where fertilizer use is common.

It is reasonable to assume that a final reaction could be combined with the reduction of oxygen (Table 4–2). The oxidation of *nitrogen gas* (N_2) to nitrate by oxygen is thermodynamically favored. Yet we know that nitrogen gas is abundant in the atmosphere, and nitrate is often in short supply. The direct conversion does not occur under ordinary conditions. The unusually strong triple bond in the N_2 molecule produces such a severe hindrance that the reaction cannot be used by organisms. Reactions with nitrogen gas do occur in the activities of certain specialized species. But these species do not convert nitrogen gas directly to nitrate; they reduce it to ammonia.

Anaerobic Respiration.   This discussion of redox reactions has so far assumed the presence of abundant oxygen. This assumption is always valid for systems exposed to the modern atmosphere. When the earth first formed, however, the atmosphere contained no oxygen. The first organisms arose in this environment, composing themselves of the abundant organic matter created by photochemical reactions in the ultraviolet-rich sunlight. This same organic material was the energy-rich substrate they used as food. They were able to thrive and populate the earth for about 2.5 billion years before free oxygen became common.

As the organic matter created by photochemical reactions was depleted, some organisms developed photosynthesis: they used the energy of sunlight to power "reverse aerobic respiration," creating organic materials. They obtained the necessary carbon from atmospheric carbon dioxide, and molecular oxygen (O_2) was released. The oxygen accumulated in the atmosphere, allowing the development of new species of animals who lived by aerobic respiration.

Oxygen is actually toxic to many anaerobic organisms, and competition from the efficient aerobic organisms became fierce. The anaerobes became rare species in environments where oxygen is available. Their descendants remain, however, and are important actors in the biogeochemistry of the earth. They are in tiny microenvironments, such as within clods of earth, and they thrive in large environments separated from the atmosphere, such as swamp muds and buried ocean sediments.

Soils and aquifers below the surface are also separated from direct contact with the atmosphere. Oxygen is replenished by diffusion of the gas, but the process is slow. Water passing into the soil is usually aerated, but oxygen is not very soluble in water. At saturation, water holds only 8 or 9 milligrams of oxygen per liter. If food is available for microorganisms seeking energy, the oxygen will soon be consumed.

Nature creates such food. Dissolved organics are abundant in swamp water and most soils, and may be carried downward by infiltration. Human activity releases biodegradable materials such as petroleum hydrocarbons, solvents, and others which can easily overwhelm the oxygen resources of the subsurface when spills occur. In either case, the biological oxygen demand exhausts the oxygen, anaerobic organisms take over, and anaerobic chemistry determines the fate of the contaminants.

The list of common redox reactions (Table 4–1) can be rewritten on the assumption that oxygen is exhausted and organic matter is abundant. This requires combining the last oxidation with each of the reductions, again in order of energy provided (Table 4–3). The most energetic available redox reaction is the reduction of *nitrate to nitrogen gas.* Denitrification occurs in soils saturated with water, to the dismay of rice farmers who value the nitrate as a plant nutrient. They must replace it with expensive artificial fertilizers. An anaerobic environment is intentionally created in some sewage treatment processes, with the objective of removing nitrate. This prevents nitrate accumulation, which can cause nuisance blooms of algae, in waters where the sewage is being disposed. Converting the nitrate to nitrogen gas, which is released harmlessly to the atmosphere, solves the problem. For subsurface waters, nitrate reduction is another way in which organic material may be decomposed. In a few cases, nitrate has been added to aquifers to promote anaerobic degradation of contaminants. It is an effective oxidizing agent for this application because it is a negative ion, and so is not adsorbed strongly by the soil. It can move readily to the site of contamination. Unlike oxygen itself, it can be supplied in high concentrations because its solubility in water is high.

Under some conditions, the nitrate will all be consumed by the conversion to nitrogen gas. At other times, however, substantial amounts of nitrogen will be converted to *ammonia.* In fermentative nitrate reduction, organisms use nitrate as an electron acceptor by producing ammonia. In assimilatory nitrate reduction, the ob-

Table 4–3. Metabolic Reactions When Organic Matter Is Abundant.

Reduction of various substrates beginning with the first, which provides the most energy, will be combined with . . .	Oxidation of organic matter, which provides the most energy, to produce . . .	Redox Reactions of aerobic respiration and anaerobic respiration.
Oxygen to water $O_2 \rightarrow H_2O$	Organic matter to carbon dioxide $CH_2O \rightarrow CO_2$	1. Aerobic respiration
Nitrate to nitrogen gas $NO_3^= \rightarrow N_2$		2. Denitrification
Nitrate to nitrite $NO_3^= \rightarrow NO_2^-$ or Nitrite to ammonia $NO_2^- \rightarrow NH_3$		3. Assimilatory ammonification or fermentative nitrate reduction
Ferric iron to ferrous iron $Fe^{3+} \rightarrow Fe^{2+}$		4. Iron reduction
Carbohydrate to alcohol $CH_2O \rightarrow CH_3OH$		5. Fermentation
Sulfate to sulfide $SO_4^= \rightarrow H_2S$		6. Sulfide generation
Carbon dioxide to methane $CO_2 \rightarrow CH_4$		7. Methane generation
Nitrogen gas to ammonia $N_2 \rightarrow NH_3$		8. Nitrogen fixation
Hydrogen ion to hydrogen gas $H^+ \rightarrow H_2$		9. Hydrogen generation
Carbon dioxide to organic matter $CO_2 \rightarrow CH_2O$		10. Fermentation

jective of the organism is to produce the organic nitrogen compounds, such as proteins and DNA, which are necessary for life processes. The organisms eventually release the ammonia as a waste product, or may die, so that the ammonia is released as it decomposes.

If the nitrate is all consumed, microbially-mediated redox reactions may continue, using *ferric iron* (Fe^{3+}) to accept the electrons. The ferrous iron ions produced may be important in water quality. Ferrous iron is far more soluble. It will move with the water to be drawn up in wells. When it reaches the treatment plant or the household, the contact with oxygen will convert it back to ferric iron, creating the rust stain problems described previously.

The next reaction in order of energy on the list is the conversion of *carbohydrate to alcohol.* Coupling this reaction with the oxidation of organic matter produces a special kind of anaerobic respiration called fermentation. The best known fermentations are those which produce the alcohol of beer and wine and the carbon dioxide which causes bread to rise.

In many cases, *sulfate* will be the next oxidizing agent utilized by the microorganisms. This is particularly so in seawater environments, where sulfate is present in high concentration. As it is used to oxidize organic matter to carbon dioxide, sulfide is generated. The sulfide has many secondary effects in the local environment. It has a very strong odor, and contributes to the odors of sewers. It is the reason swamps (with their anaerobic mud bottoms) have the reputation for foul odors.

Sulfides are also toxic. In rare cases, biologically generated sulfide has caused deaths in treatment facilities. In natural systems, the toxicity may change the characteristics of the biological community by suppressing some species.

The sulfide also radically alters the chemistry of the local environment. Many metal sulfides have very low solubilities, so the presence of sulfide will immobilize the metals.

When the sulfate is exhausted, further anaerobic microbiological activity may occur using *carbon dioxide* as an electron acceptor. Methane is generated. The substantial problems caused by methane have been described in Chapter 2.

The redox reaction list suggests that oxidation of organic matter may be coupled with the conversion of *nitrogen gas to ammonia.* The process is called nitrogen fixation, and it is very important to the maintenance of natural ecosystems. Nitrogen gas cannot be used as a nitrogen source by plants, but ammonia can. The kinetic hindrance seen for other reactions of nitrogen gas is also seen here, however. The relatively few organisms which can fix nitrogen use an elaborate biochemical process which requires far more energy than the theoretical minimum. Nitrogen fixing organisms are most successful where fixed nitrogen is in short supply. Where ammonia or nitrate are available, non-fixing species, unencumbered by the large energy costs of fixation, will dominate.

The conversion of *hydrogen ion to hydrogen gas* is less commonly observed, but it does occur. Hydrogen gas can be detected in the emanations from bogs and in the digestive tracts of animals.

The final reduction listed in Table 4–3 is the conversion of *carbon dioxide to organic matter.* In the combined redox reaction, organic matter is both created from carbon dioxide and degraded to carbon dioxide. Organic matter, of course, is tremendously diverse in its structure, its reactions, and the energies released or consumed in its reactions. As has been described for fermentation to produce alcohol, it is possible to combine reduction of one kind of organic matter with oxidation of another kind and make redox reactions which produce energy. These reactions, in which one organic species is the electron donor and another is the electron acceptor, do not produce as much energy as the others, but there is enough to make it a way of life for some microorganisms. Several are important in controlling the composition of wastes in aquifers.

This systematic treatment of microbial metabolism has been far from complete. Detailed descriptions were given only to reactions involving the most energetic reduction (Table 4-2) or the most energetic oxidation (Table 4-3). Other combinations are also found. Further, the overall list (Table 4-1) included only a fraction of the oxidations and reductions performed by microorganisms. Many others are known, and even more occur that have not yet been studied.

Cometabolism. Microorganisms utilize gross reactions to generate the energy necessary for life. They carry out synthetic reactions in smaller amounts to produce the building materials of the cells. There is a third type of metabolism, called cometabolism, whose purpose is less clear.

Cometabolic reactions are decomposing reactions: that is, the reactions involve the breakdown of hydrocarbons into small molecules like carbon dioxide, in a way that releases energy. They differ from the gross reactions in that the cells are not capable of utilizing this energy for growth. They must have another chemical present as the primary substrate on which they live. Thus, cells that are kept well fed on a diet of the organic ion, acetate, will also break down methylethyl ketone, a common solvent. But the same organisms will not live on a diet of only methylethyl ketone.

Little is known about why the microorganisms carry on cometabolic reactions. Active organisms, feeding on the primary substrate, synthesize a tremendous variety of enzymes. It may be that some of these, intended for breaking down food chemicals, work on the cosubstrate essentially by chance.

Possibly, the enzymes involved are intended only for detoxifying chemicals naturally present in small amounts, and the organisms have not developed the complete series of enzymes necessary to harness the energy produced.

Whatever the microorganism's purpose for cometabolism, the phenomenon has importance for the cleanup of subsurface hazardous waste. Very often, the hazardous materials are those which the microorganisms cannot use as a primary substrate, but which are decomposible as a cosubstrate. Cleanup by biodegradation will therefore employ the addition of primary substrate. An easily biodegradable and innocuous substrate like acetate is added to the soil or water, and a vigorous culture of microorganisms develops. While they go about the business of consuming the acetate, the bacteria also cometabolize the hazardous waste, and the soil is decontaminated.

Similar substrate additions may be appropriate simply because the hazardous waste is present in concentrations too small to support vigorous biological activity. Again, a culture of organisms developing on the added material will simultaneously degrade the waste.

Growth Kinetics

The Growth Curve.   Microbiologists have long been familiar with the typical pattern of growth for populations of microorganisms. When a seed culture is first exposed to a new environment with abundant food and optimum conditions, the organisms hesitate. There is a *lag phase* or *acclimation period,* during which the organisms do not reproduce by cell division. It is not, however, a period when nothing is happening. Experiments have shown that protein synthesis is occurring during the phase. Presumably, the cells are manufacturing enzymes and the other chemicals necessary to adapt their metabolism precisely to the new conditions. Genes are being activated and inactivated, as necessary. The individual cells often grow larger.

After acclimation has occurred, the population can expand rapidly. During this second phase, the acclimated cells put their energy into reproduction. Generation time is short for many species, often as little as 30 minutes. Each cell divides to produce two cells, so the size of the population doubles with each generation, and the number of organisms rises exponentially. The result is the *exponential phase* or *log phase* of population growth.

Growth soon becomes very rapid, but it cannot proceed without limit. The population becomes dense, and begins to exhaust its food supply and perhaps to poison itself with waste products. The culture enters the *stationary phase,* during which there is little growth. Metabolism slows down, the organisms become efficient at living on the declining resources, and little reproduction occurs.

Eventually substrate concentrations are reduced to levels too low for the organisms to survive. The population begins to die off. The dead cells break up, and become food for the remaining live organisms. Death is rapid at first, then slows later as most of the organisms are gone. This suggests the term *log death phase.* Because the organisms are consuming each other and their own stores of food, it is also called the *endogenous phase.*

These phases of growth in microbial cultures have substantial implications for the subsurface migration of hazardous waste and for the design of remedial actions.

The Lag Phase.   Recent research has determined that the importance of lag phase kinetics is much greater than was appreciated only a few years ago. Some organic chemicals that were thought to be essentially nonbiodegradable have instead been found only to require a very long lag phase. Bouwer and McCarty (1985), using acetate as a substrate, exposed microorganisms on glass beads to low concentrations of dichlorobenzene under aerobic conditions. The 1,2 isomer (with the two chlorine atoms bonded to adjacent carbons on the six-carbon benzene ring) and the 1,4 isomer (with the chlorines at opposite carbons) degraded readily. The 1,3 isomer (chlorines separated by one carbon) was almost entirely resistant for one to two years, until acclimation occurred. Thereafter, the organisms on the beads became quite efficient. The reactor was capable of complete removal of 1,3 dichlorobenzene from the water with only a 20-minute residence time.

Similar results were obtained under anaerobic conditions. 1,1,1-trichloroethane ($CC1_3$-CH_3), tetrachloroethylene ($CC1_2$=$CC1_2$), and chloroform ($CHC1_3$) could be completely degraded by microorganisms in the reactor in two days, but only after a ten-week acclimation period.

It is obvious why these results were not obtained earlier. With no suspicion that such long lag phases occurred, no one had reason to perform such long-term experiments.

The implications for groundwater contamination are considerable. Contaminants released to the environment may travel substantial distances before the indigenous microorganisms are acclimated to degrade them, and begin limiting the spread of the pollution.

The result is hopeful, however, in that it suggests that biodegradation may be a workable solution for cleanup at more sites that was previously thought. While the lag phases are long, in relation to convenient laboratory experiment times, they are not so long compared to the typical cleanup project. It will be possible in many cases to wait a year or two for the microorganisms to adjust in order to complete a cleanup which is effective, inexpensive, and causes minimum disruption.

The Exponential Phase. During the period when the population of organisms is growing rapidly, they degrade the substrate rapidly. Treatment systems are sometimes operated with the microorganism culture in this phase. An abundant supply of waste is provided for food, and the organism culture is washed out at a relatively high rate so that substrate depletion does not occur. This style of operation decomposes large amounts of waste rapidly. It requires less time and a smaller facility. It has the disadvantage, however, that degradation is not complete. Abundant substrate in the reactor becomes high concentrations of waste in the effluent. For most applications, this is not acceptable.

The Stationary and Endogenous Phases. The rapid growth phase ends when the organisms begin to deplete their food resource. Reproduction slows and stops, and the organisms adjust to become efficient scavengers on a limiting resource. A treatment system operated to produce this phase of microbial activity will produce an effluent of of good quality, because the microorganisms are working hard to collect their substrate, which is our waste. When the concentrations get very low, the microorganisms begin to die, and those remaining alive consume the remains of their dead fellows. Because good treatment requires that the microorganisms be removed, this endogenous stage of the growth process is also valuable to the treatment process.

Groundwater cleanup requires all of the growth phases. Early in the process, substrate will be abundant, and exponential growth will occur. Later, the stationary and endogenous phases will complete the removal of the contaminants. If microbiological treatment is attempted, it is important to be aware of the growth phases, and to provide the appropriate amounts of oxygen and nutrients at the appropriate times. This is especially so where treatment is attempted *in situ,* without disturbing the soil. Injections of oxygen and nutrients must occur at the proper times and places.

Growth Limitation

Promotion of microbial activity in aquifers means removing obstacles to microorganism growth and substrate decomposition. There are many obstacles which can arise.

Physical and Chemical Inhibition. Microorganisms are only active within temperature ranges appropriate for each species. While some species are tolerant of lower temperatures than others, overall decomposition rates generally fall as the temperature falls. Below 10°C, biodegradation is very slow. Where soil or groundwater pollution has occurred in soils near the surface, this may be an important limiting factor. Microbiological cleanup of spills is not possible during the winter in north temperate zones, and may never be possible in the arctic, unless the soil is warmed artificially.

The pH also varies in soils, and of course may be substantially altered where wastes have been discharged. Decomposition proceeds best near neutrality, from pH 6 to pH 8. Holden (1986) reports that groundwater contamination by the pesticide aldicarb is worse in regions of low soil pH because biodegradation is inhibited. Contaminant degradation does occur at pH's outside these values, however, and may serve to solve the contamination problem if lower rates are acceptable. More often, it is economically worthwhile to adjust the pH through the addition of inexpensive acids or bases to produce optimum conditions.

Below the water table, water is of course present in abundance. In this regime, there may be "too much" water, in the sense that it prevents access to the atmosphere and needed oxygen. Above the water table, however, it is common that soils are too dry for optimal biological activity. Most bacteria must be submerged in liquid water to be active, and can only move and seek food by swimming. This requires enough water to form films or small droplets among the soil particles. The fungi common in soils are generally more resistant to drying. This comes partly as a result of specific metabolic and biochemical machinery employed by the fungi, but also from their pattern of growth. They are capable of growing long filamentous hyphae. These filaments allow the organisms to seek food and water in pores without the need for swimming. Rather than moving to what they need, they grow to it. Actinomycetes are a class of bacteria with characteristics somewhat similar to the fungi. They are also often filamentous, and more resistant to dry conditions. Accordingly, they are commonly abundant in soils, and may be major participants in waste biodegradation.

Soils. In some cases, the physical nature of the soils may limit the effectiveness of microorganisms. If the soil is tightly packed, and has a low permeability, it may prevent movement of the water, air, and nutrients needed by the microbes. It may further retard activity by preventing organism movement. As the culture develops in one microenvironment in the soil, the organisms must move to others in order that the cleanup be complete. In very tightly packed soils this is impossible. It may be particularly so where growth is vigorous, because thick growths of microorganisms will clog the few small pores which are available.

Interference of transport by microbial clogging has been recognized in the oil industry for many years. In tertiary oil recovery operations, involving the injection of water or aqueous solutions, it is often necessary to sterilize the solutions injected. If this is not done, the pores of the rock near the injection well will clog with growing organisms and the well will be made useless. Both for reasons of oil recovery and for hazardous waste site cleanup, it may be desirable to inject cultures of special organisms. Accordingly, research is being done on ways to improve the transport of bacteria in porous media (Sharma et al., 1985).

For soils of modest organic content, there may be some positive correlation of degradation rate with organic content. Zhong et al. (1986) found higher rates of degradation of the pesticide aldicarb in a soil with more organic matter. Rao et al. (1986) found rates lower in subsoils, where the organic content was lower. The specific phenomena responsible for the correlation have not been established.

Burns (1979) described studies on the interaction of soils and biodegradation. Some kinds of degradation are promoted and some are inhibited, usually for reasons not well understood. Organic molecules may be adsorbed by clay particles, for example, so that they are inaccessible to cells. Natural humic matter may also be complex or adsorb molecules, or provide protective pores. However, both clays and organic matter tend to stabilize pH, promoting microbial activity.

Some degradative reactions are performed outside the cells. Microorganisms excrete exoenzymes, which work in the surrounding solution to break up molecules too large or insoluble to be brought through the cell membrane. Soils may serve as a reservoir for exoenzymes. A subtle interaction is possible: the presence of the exoenzyme in the soil means it will be ready for the episodic appearance of substrate. When degradation begins, the cell will detect the presence of the products, and produce more enzyme. In this way, the cell can respond only when the substrate is present, rather than producing large quantities of all exoenzymes at all times.

Substrate. For some highly toxic chemicals, the minimum concentration necessary for microbial growth may exceed acceptable levels. The organisms will cease activity for lack of food before the degradation is complete. In some cases, it may be that the energy available from the degradation of the chemical is no longer enough to make the search for it worthwhile. McCarty et al. (1984), however, have suggested that more complicated mechanisms may also be involved. Microorganisms in aquifers are predominantly on the surfaces of the soil particles in biofilms. These films are formed of the bodies of microorganisms, exuded polysaccharides, and other chemicals. Mathematical models and experimental measurements indicate that there is a minimum concentration of substrate necessary for maintenance of the films. If the concentration falls below this minimum, the loss of organisms exceeds their production, and the film disintegrates. Biological activity will be drastically reduced.

Substrate concentrations may be too high. This certainly happens when the substrate is toxic. It is well recognized that petroleum hydrocarbons in low concentration in water are degradable. But if high concentrations of the low molecular weight straight-chain hydrocarbons are present, they will disrupt the cell walls of the microorganisms and kill them. There is such a thing as too much food. (This is especially fortunate for us in one way: we don't have to worry about the biodegradation of gasoline in our gas tanks.)

This has an unexpected implication for the investigation of bacteria in oil-polluted soils. Bacteria may flourish in soils containing 10 percent oil, for example, but fail to grow when placed in a laboratory culture medium containing only 1 percent oil. The toxic components of the oil are more soluble in the medium than in the soil pore waters and so have a much greater effect on the organisms. Special techniques are necessary to culture and count the organisms and monitor their progress in the contaminated soils.

The implications of substrate limitation for site cleanup are more obvious. In some cases, it may be necessary to dilute the contaminants to promote biodegradation. If substrate concentrations are too low, it will be necessary to provide a second, nontoxic substrate to keep the organisms growing while decomposition of the contaminant proceeds.

Oxygen. As previously emphasized, aerobic respiration, which converts organic materials to carbon dioxide and water, is generally the fastest form of biodegradation. Indeed, many chemicals seem to be degradable only when oxygen is present. In soils and aquifers, it is often in short supply. In the pores of the vadose zone, air currents are strongly suppressed. While there may be some movement in response to thermal expansion and contraction near the surface, most oxygen transport must occur by molecular diffusion, which is an inherently slow process.

Transport is further inhibited below the water table. The solubility of oxygen in water is low, and diffusion is much slower than in air. Where significant concentrations of substrate are available, the microorganisms will consume oxygen. It is likely that oxygen demand will exceed replenishment, and biodegradation will be limited.

Some chemicals, however, seem to be degradable only when oxygen is absent. Apparently those organisms capable of anaerobic metabolism have unique enzymes that work on these species. Trihalomethanes, chemicals which consist of a carbon atom bonded to one hydrogen and three atoms of chlorine or bromine, are common pollutants. McCarty et al. (1984) reported that three trihalomethanes were degradable only under anaerobic conditions. (The phrase "seem to be degradable only. . . ." has been used here intentionally. The history of biodegradation studies has generally been that what seems impossible very often turns out to be possible when the right organisms, conditions, and timing are found.)

Mineral Nutrients. The first 28 elements in the periodic table, from hydrogen to nickle, are required for life. At least some of the others are also needed. Selenium, for example, is necessary in small amounts.

For many elements, the amounts required are so small that they are rarely limiting in soils. Cobalt, for example, is absolutely required because it is part of the B vitamins. In only a very few cases, however, has cobalt scarcity been shown to be limiting in natural soils. Usually enough is present to satisfy the needs of the organisms.

Usable forms of nitrogen and phosphorus, however, are often in short supply. Relatively larger amounts are needed by organisms. Nitrogen is abundant in the atmosphere as nitrogen gas, but few organisms can utilize this form. For most, ammonium ions, nitrite, nitrate, organic nitrogen, or another available form is required. Phosphorus must be present as some form of phosphate, in which the phosphorous atom is bonded to four oxygens.

Nitrogen and phosphorous are the nutrients commonly used as fertilizers by farmers. Plants require these just as microorganisms do, and grow poorly when they are in short supply.

Because this limitation is common, the addition of nitrogen and phosphorous to systems where biodegradation is to be promoted is common. It is particularly appropriate when the waste to be degraded is low in nitrogen. This will certainly be the case for chlorinated hydrocarbons and other hydrocarbons, which have no nitrogen. It is also true for petroleum, which contains only small amounts of nitrogen.

Projects using biological degradation for site cleanup include frequent measurement of the concentrations of nitrogen and phosphorous, and replenish them as needed.

Adsorption.   Organic molecules sometimes stick to the surfaces of soil particles. The *adsorption* may either aid or inhibit degradation. Promotion of degradation occurs where sorption serves to increase the concentration of the chemical in the immediate microenvironment of the organisms. This is particularly the result expected if a well-developed biological film is present, so that adsorption consists of the transport of the chemical from interstitial water into the film.

Reduction in substrate availability, and therefore limitation in microorganism growth, may occur if adsorption limits access to the molecules. Often, bacteria and fungi attack solid or adsorbed substrates by excreting exoenzymes. The substrate is broken down, and the cell absorbs the soluble products for use in its metabolism. In some cases, however, the enzymes used for decomposition of the chemical act only within the organism. When this is so, a firmly adsorbed molecule, which cannot be taken in, cannot be decomposed. Even where the important enzymes are released, they may not break down the chemical if the site of enzymatic attack is protected by the soil.

If adsorption has occurred within very small pores in the soil, it may be that the microorganisms cannot get near the substrate. This will be the case for organic molecules adsorbed between the layers of clays (Burns, 1979).

Sequential Reactions.   Often, the degradation of a complex organic molecule to carbon dioxide and water is done in many steps, involving several species. It may require a consortium of microorganisms, rather than a single species. Slater and Lovatt (1984) have emphasized the importance of microorganism *communities* in biodegradation, and classified some possible relationships.

In one kind of community, the activities of the first species may provide the nutrients necessary for the second, allowing it to survive. Examples have been found in which each of two species provides a nutrient for the other. Sometimes the reaction that is beneficial to a second organism is cometabolic for the first.

In other communities, an organism may serve by degrading harmful products produced by another. This prevents self-inhibition and allows biodegradation to proceed. Often, the product removed is hydrogen or some other electron carrier. The effect is to maintain redox conditions appropriate for the vigorous function of the degrading organisms.

Organisms may also cooperate in a combined metabolic attack on a difficult substrate. One organism may break crucial chemical bonds in the food molecule. The product molecule becomes the substrate for another organism, which breaks more bonds, and so on.

Investigators have also observed groups of microorganisms which simply attack the same substrate at the same time. Such groups may coexist for long periods of time even under varying conditions. The interactions which prevent competition from eliminating some of the species are not yet understood.

Microbiological communities have developed naturally in this way. Presumably, they have arisen because groups of species are capable of filling some ecological niches more efficiently than a single species. Several species specializing in particular parts of the biodegradation tasks can be more effective than one species doing the whole job. For some substrates, it may be physically impossible for a single species to carry all of the necessary enzymes for complete degradation.

Natural communities, such as those which break down wood fibers, have been studied in nature for some time. The groups of species which work on various artificial contaminants in soil are less known. Presumably, a community is assembled fortuitously from microorganisms which evolution has prepared for other tasks. Because they are decomposing chemicals never seen in nature (or rarely seen in significant concentrations), they may not naturally occur together. It is possible that the limiting factor for the biodegradation of some chemicals may be the absence of one or two crucial species of microorganisms from the community.

This has produced confusion in past efforts to promote biodegradation. Some contaminants pronounced nondegradable have degraded when the appropriate "team" was assembled. Efforts at promoting biodegradation in site cleanups should include mixed inocula from several sources such as sewage sludge and active organic-rich soils to insure that the maximum possible variety of species will be present.

Processes in Soils

Soil Microorganisms.   Representatives of all significant groups of microorganisms are found in soils. Bacteria and fungi are abundant, actinomycetes and protozoa are common, and even photosynthetic algae can be found near the surface. Aerobic and anaerobic species are quite common, even in "aerobic" soils. Apparently there are many microenvironments within common soils that differ significantly from the bulk or average environment around them. Clods, or the smaller agglomerations called peds, can contain anaerobic organisms, even when the large pores of the soil are oxygenated.

Abundances in soils are high. Further, in any given sample, the number of species present is also likely to be high. Soils are thus highly diverse and capable biological reactors.

Grant and Long (1981) have proposed generalizations describing the characteristics of the microbial ecosystem in soils. Most of the bacteria are in the group referred to as Gram-positive, because they are colored by a preparative procedure called Gram staining. The genus *Arthrobacter* is dominant, and *Bacillus* and *Micrococcus* are also common. Gram-negative genera, which do not stain, include *Pseudomonas,* recognized for an ability to decompose a tremendous variety of substrates, and *Flavobacterium.* Species of *Clostridium,* notorious for the production of deadly toxins, are common among the anaerobic organisms. The major actinomycete genus is *Streptomyces.*

The fungi are abundant in soils, commonly comprising more biomass than any other group. Their resistance to dry conditions and their ability to decompose difficult substrates, like wood and other plant fibers, make them very successful. The fungi imperfecti are commonly the most abundant group, but the mycorrhizal species (which grow in close association with the roots of higher plants) may also constitute a substantial part of the biomass.

Soil Formation. Soils are generated primarily by three phenomena. Rocks are broken up by physical forces such as wind abrasion, water erosion, or ice expansion to form sand. Sand particles are chemically quite similar to the parent rock: they are essentially tiny rocks. But over long periods of time, a second process occurs: the sand particles are subject to chemical weathering, which substantially alters their composition. The products are the clays and dissolved ions. The third process is biological activity, which produces organic debris, humic materials, and a host of organic chemicals.

These processes commonly produce layered soils. In geologically recent soils at the surface, the sand and clay have usually been deposited by flowing water. As the focus of erosion and deposition moved elsewhere, plants became established at the surface, and gradually the organic components of the soil were produced. The result is a series of horizons, beginning at the surface with a layer of plant litter. Below this is a layer of soil containing large amounts of organic matter and its decomposition products. Deeper in the soil, organic matter becomes rarer, and finally the deep soils are almost entirely mineral particles. Eventually, at some depth, bedrock is encountered.

The surface soil, rich in organic matter and the nutrients released by microbial decomposition of organic matter, is dark in color. It holds water well. The water and nutrients make plants grow well. Farmers know that good topsoil is necessary for productive farming.

The layered structure of soil has implications for the subsurface migration of hazardous wastes. Interactions with natural organic matter will be important only in the first few feet of most soils. Microbiological activity will also be more vigorous near the surface, where air is available and nutrients are abundant.

In contrast, subsoils may be less active. A lower concentration of organic matter will support fewer organisms and adsorb less contaminant. Ghiorse and Balkwill (1983) examined samples from below the topsoil for microbiological activity. They found lower numbers of organisms, mostly rod-shaped or coccoid bacteria. Most appeared to be *oligotrophs,* adapted for life on low concentrations of organic substrate. White et al. (1983) investigated microorganisms in deep aquifers by analyzing the biochemicals present. The results indicated a moderately dense population of simple bacteria, enriched with anaerobes, and particularly including sulfate reducers.

Biofilms. Microorganisms in water are often more active when solid surfaces are present. The cells exude various adhesives which hold them in place. There are advantages to this style of life. Many chemicals that the organisms can use for food are concentrated near or on the surfaces. Cells attached to surfaces can take advantage of water movement: food will be brought to them, and their waste products will be taken away. Cells suspended in the water must expend energy swimming to accomplish these tasks. This general principle has been used to advantage in traditional sewage treatment for years. Trickling filters utilize fist-sized stones as a surface for microbial growth. Activated sludge systems use a suspension of microscopic particles.

When conditions for growth are good, the cells may become so numerous that they pile up on each other. Their bodies and the materials they exude to hold themselves in place can form a layer. The layer may be thin in human terms, but as thick as tens or hundreds of organisms.

This *biofilm* on the surface of soil particles constitutes a biological system with characteristics beyond those of the individual organisms. The deeper portions of the film, for example, may be anaerobic even in systems where the water at the film surface is well aerated. The activity of the film is governed by the diffusion of oxygen, mineral nutrients, and organic substrate into the film.

It is likely that biofilms are only present where conditions for growth are good. McCarty et al. (1984) developed numerical models indicating that biofilms would only be stable where concentrations of nutrients exceeded a certain threshold. The microscopic examination of subsoils with low nutrient concentrations by Ghiorse and Balkwill (1983) showed organisms in small clumps rather than films.

MICROBIOLOGY AND SUBSURFACE MIGRATION

The influence of microbial ecology on subsurface chemistry is profound. Toxic species may be altered or destroyed, their ultimate environmental fates may be changed, or new toxic substances may be formed. Understanding the subsurface migration of hazardous wastes requires knowledge of subsurface microbiology.

Processes in Landfills

Pohland et al. (1985) have summarized observations on the microbiological processes which occur in landfills. They follow a well-ordered sequence understandable from the preceding discussion of microbial metabolism. Newly deposited municipal waste is largely aerobic. While the interior of various pieces of the garbage may have no oxygen, the spaces between have oxygen remaining from collection and dumping. The material is usually relatively dry, in the sense that water does not flow out of it when it is deposited (the water content is less than field capacity). It is damp enough to support some microorganisms, however, at least from place to place within the waste mass. The microorganisms, presented with abundant organic substrate and oxygen, proceed with aerobic metabolism. Far more organic matter is present than oxygen, however, and when the mass is buried and sealed from the atmosphere, the oxygen is depleted.

As this process occurs, the waste is accumulating water. Some of the material left at the landfill may contain substantial amounts of free water. Sewage sludge, for example, is often codisposed with municipal wastes. Precipitation also contributes water steadily. Eventually, the waste reaches field capacity, and leachate begins to flow downward through the mass. This may contribute to oxygen deprivation, because water retards oxygen diffusion.

The transition to anaerobic waste degradation begins as predicted by the energies available from the possible anaerobic reactions (Table 4–3). The microbes utilize nitrate for oxidation. As the nitrate is consumed, sulfate-reducing organisms become dominant. Because the waste is sealed from the atmosphere, carbon dioxide begins to accumulate in the pore spaces within the waste. Dissolved carbon dioxide is an acid, so the pH of the waste and water mass is reduced. Volatile organic acids, a common product of anaerobic decomposition, begin to accumulate.

In the acid formation stage, the volatile organic acids become abundant. The pH drops sharply. The acid dissolves trace metals in the waste mass. Leachate produced during this phase of landfill life is highly contaminated with the acids, the metals, and an abundance of other chemicals being dissolved from the solid matter.

When the more energetic electron acceptors have been depleted and the microorganisms using carbon dioxide become plentiful, methane is generated. Consumption of the organic acids begins. This allows the pH to increase. The redox potential in the waste reaches its lowest level, but leachate quality begins to improve. Organic acids are no longer so abundant in the leachate, and the rising pH may cause some metals to reprecipitate or to form organic complexes. It is during this stage that the landfill generates large quantities of methane. Off-site migration of the methane can cause explosion hazards, or carry odorous and carcinogenic chemicals.

Final maturation of the landfill occurs when the bulk of the readily decomposible organic matter has been consumed. Microbial activity falls, and nutrient concentrations may fall because the leachate has carried them away. Gas production falls to low levels. Ultimately, oxygen may reinvade the waste mass, which has taken on some of the characteristics of an organic-rich soil. High-molecular-weight decomposition products, similar to the humic substances in soils, will appear. Ironically, leachate quality may worsen because some of these substances will form soluble complexes with trace metals, allowing them to dissolve in the leachate.

These stages of landfill decomposition occur over relatively long times. Anaerobic decomposition is a slow process, and may be further delayed at various times by lack of water, excess acid, or other imbalances. Typically, large landfills will not reach final maturation for forty or fifty years.

The effects of these processes are complicated because not all parts of the landfill are the same age. Large installations may be operated for thirty years or more, so that the first material placed may be in the final stages even as fresh waste is being added. Leachate from parts of the landfill in one stage of the process mixes with leachate from parts in other stages. The results may be hard to predict.

Decomposition of Toxic Substances

It is well established that many of the hazardous substances which are causing concern for groundwater quality are biodegradable. While accomplishment of biodegradation in the field is never assured, success in the laboratory is an indication that microorganisms may be usable for site remediation when the technology is developed.

Products of Decomposition

Microbial decomposition is generally viewed favorably in the context of soil and groundwater contamination. In most cases, the microorganisms break down the complex toxic molecules into smaller, simpler, and nontoxic products. Petroleum, for example, is converted to carbon dioxide and water. The chemicals that microorganisms synthesize are most often benign. Some products of microbial activity, however, can generate problems.

Toxic Substances.   Some breakdown products are a threat to public health. The generation of vinyl chloride, a carcinogenic gas, from the common degreasing agent trichloroethylene, has been described. Microbial degradation of the notorious pesticide DDT produces DDE, a equally toxic chemical. Hydrogen sulfide is a product of anaerobic respiration, and is toxic at high concentrations.

Odors.   Hydrogen sulfide also has a very strong odor, and is a common problem at landfills. The anaerobic conditions which favor its generation also lead to the production of mercaptans, organic sulfides which are noticeable as unpleasant odors to humans even at very low concentrations.

Acids.   Both aerobic and anaerobic degradation produce carbon dioxide. When this gas dissolves in water, it forms carbonic acid, and will substantially lower the pH. Organic acids are also generated. The result is an acidic landfill leachate with a substantially greater capacity for the dissolution and transport of toxic metals. Microbial oxidation of sulfide minerals in coal mine waste piles forms sulfate in the form of sulfuric acid. Mine drainage waters are often highly acidic, and the resulting water pollution problems can be severe.

Production of Liquids, Solids, and Gases

The physical phase of chemicals is a primary determinant controlling their fate in the subsurface environment (Chapter 2). Microbiological activity may, therefore, have special import when the products of metabolic activity have a different phase than the chemical the microorganism is using for food.

Biological Dissolution.   Many of the chemicals that serve as substrate for microbes are poorly soluble. These are usually high-molecular-weight species and polymers. Cellulose and lignin, the degradation-resistant polymers that make up wood and other plant fibers, are obvious examples. Some of the products of partial decomposition are much more soluble. These solutes may have profound effects on leachate quality. Organic acids reduce the pH and dissolve metals. Humic and fulvic acids dissolve metals, form soluble complexes with organic molecules, and color the water brown. Microorganisms may promote dissolution and migration of hydrocarbons through the production of soaplike chemicals. High-molecular-weight polysaccharides released by the microbes may act as a detergent to solubilize oil.

Biological Precipitation.   Other biological transformations have the opposite effect. Carbonates formed by the dissolution of carbon dioxide can precipitate some trace metals. Other species are less soluble when the pH falls. The microorganisms themselves are composed of insoluble parts such as membranes, which will remain as solids when the organisms die.

The accumulation of precipitates has secondary effects on soil quality. The pores may be clogged, increasing the resistance to leachate migration.

Biological Gasification.   Anaerobic degradation of sewage and garbage generates methane and carbon dioxide in substantial quantities. Fires and explosions have occurred near landfills and in sewage treatment plants when sufficient methane has accumulated. California law now requires landfill operators to check for methane generation and install control systems.

The generation of carbon dioxide has also been put to use. Swallow and Gschwend (1983) report measuring carbon dioxide concentrations in the soil as a method of outlining the subsurface plume of a degradable groundwater contaminant.

Biological Influence on Adsorption.   Microorganisms in groundwater are most active on the surfaces of soil particles. Because they occupy space on the surfaces and exude adhesives which coat surfaces, they can substantially change the adsorptive nature of the soil. Inorganic, charged surfaces become organic, uncharged surfaces. If the organisms are sufficiently abundant to form a well-defined biofilm, the surface adsorption processes of the soil are completely replaced by diffusion processes which transport chemicals in and out of the film.

Promotion of Biodegradation

Microbiological decomposition occurs as an intentional or unintentional solution to the problems of subsurface contamination. No doubt, the impact of many contamination episodes has been limited by the capability of soil microorganisms to detoxify organic chemicals. Many small spills have occurred and been cleaned up "naturally" without humans even becoming aware they happened. In other cases, well-designed remedial activities have been needed to promote biodegradation.

On-Site Treatment.   Biodegradation can be accomplished by pumping the contaminated water out of the aquifer and treating it at the surface. (The treatment techniques are those well known for sewage and water treatment, and will not be described here.) The treated water is often reinjected. Groundwater flow patterns must be carefully analyzed to ensure that all of the contaminated water is collected. Such methods work only where the contaminant is readily desorbed from the soil. Where the solubility of the pollutant is very low, or desorption is very slow, most of the contaminant is on the soil rather than in the water. It is necessary to flush the soil many times over to wash out all of the pollutant. The process may not be economically feasible.

It may also be possible to excavate the soil and treat it at the site. In "land treatment,'' the soil is spread out, aerated, kept damp, and treated with nutrients to promote contaminant biodegradation.

***In Situ* Treatment.**   Substantial economic benefits are obtained if it is possible to promote biodegradation without disturbing the soil or the water. This *in situ* treatment requires the provision of an appropriate physical and chemical environment for the microorganisms in the aquifer. The cleanup operation must supply the necessary nutrients, including an oxidant (usually oxygen), mineral nutrients (such as nitrogen and phosphate), and sometimes an organic substrate (such as acetate) to promote cometabolism. Most commonly, these are dissolved in water and pumped into the contaminated aquifer.

There are special problems in getting oxygen to the site of the contamination. Its solubility in water is low: around 8 mg/l. But the consumption of oxygen is high where aerobic microorganisms are active. Providing enough oxygen dissolved in water may require far more water than is acceptable. In some cases, wells have been used as subsurface aerators. This is only possible, however, where aquifer conditions are such that the oxygen will be transported throughout the contaminated area by groundwater movement.

The problems of providing oxygen have led to some innovative solutions. Hydrogen peroxide, which has a high solubility in water, can be pumped into the aquifer in large amounts. It breaks down over a period of time, releasing oxygen. Oxygen may be abandoned entirely by promoting anaerobic respiration. Nitrate has been used as the oxidizing agent. Contaminants can be destroyed by fermentation if the appropriate substrate is provided.

Efforts have also been made to provide prepared cultures of microorganisms. While this seems an obvious move, culture additions have limitations. In most cases, indigenous organisms seem to be capable of effective biodegradation once environmental conditions are optimized. They have, after all, been exposed to the contaminant since it was released. Presumably, those which are abundant are those which are capable of utilizing the pollutant as substrate.

Delivering the microbes to the contaminated aquifer is also difficult. Soils are an effective filter and adsorbent for most cells. Simple injection usually results in clogging of the soils immediately surrounding the well. This is a problem well-recognized in the oil industry.

The introduction of microorganisms that are not ecologically viable may have little lasting effect. They will die out in competition with others already present. The microorganisms in the soil constitute a complex ecosystem, where individual species survive and spread only if they are well adapted to the environment. They must compete with other species and withstand predators. A cultured species will not thrive in the soil ecosystem simply because it is capable of degrading the waste which is present.

It has also been found that cultures of microorganisms rapidly become "lab acclimated." Genetic change in microorganisms occurs quickly because they reproduce so often. The strains which are best adapted to the lab soon dominate. They will not be those which are best adapted to life in the complex soil ecosystem. "Pet microbes" may be as unsuccessful in the wild as a pet poodle joining a wolf pack.

The complex ecological interactions that determine which species will be successful are very poorly understood. It is not possible to predict which species will succeed or why.

Even so, it is possible to imagine conditions under which additions of prepared cultures would be valuable. It is well recognized that microorganisms undergo a "lag phase" growth period when they are suddenly exposed to a new set of environmental conditions. This would be the case for soil which has been dry or anaerobic, and which is suddenly altered to optimize conditions of moisture content and aeration. The lag phase might be avoided if a dense culture of acclimated microorganisms were injected into the soil.

It is also possible to imagine that the species of microorganisms most capable of decomposing the limiting contaminant is not ecologically dominant: that is, it is present but not abundant when competing with other organisms. This is especially likely where the contaminant is present in low concentrations, and therefore does not create a significant ecological advantage for those microorganisms which use it. If so, a dense culture of these organisms, prepared in the laboratory and transferred to the soil, would promote biodegradation. The culture would die back in the food-poor environment, but in the process would scavenge the contaminant from the soil.

The organisms capable of degrading the trace contaminant may be *oligotrophs:* those well adapted to growing slowing and efficiently on very small amounts of food. These might degrade the contaminant, but only over time periods so long that a treatment system would not be practical. Growing the organisms in the laboratory would provide an initial high population density that would remove the contaminant from the soil before the organisms die back.

In either of these situations, it is possible that the repeated addition of dense cultures would maintain a high population in the aquifer, even when such a population would not be self-sustaining.

The effectiveness of biodegradation as a site remediation method depends strongly on the physical and chemical characteristics of the site. It is most easily done where the soils are homogeneous and the pollutant was released from a single point source. Nutrients and oxygen can then be released in the same way and their pattern of dispersion will match that of the pollutant. Rapid groundwater movement complicates the application of the nutrients, and may carry the pollutant away before the relatively slow biodegradation process can be effective. If the pollutant has been discharged in large amounts, as is common for hydrocarbons, it may form a layer of product floating on the water table. These pure hydrocarbons are difficult to decompose because of their toxicity, the lack of water, and because a large mass of material must be destroyed. Remediation is also much more difficult where there is substantial pollution of the vadose zone. It is difficult to provide a uniform supply of water, oxygen, and nutrients to these soils because flow is inherently irregular.

Land Treatment. This book is primarily concerned with contaminants which have been applied to the soil inadvertently or without concern for the effects. The ability of soils to degrade organic matter, however, has led people to intentional utilization of soil microbial ecosystems for the purpose of waste disposal. Wastes have been disposed of in this way for as long as there have been farms. The earliest farmers quickly learned that manures not only disappeared, they promoted crop growth.

Many modern sewage treatment systems employ land treatment as a means of decomposing the organic fraction of wastewater. Because such systems require large amounts of land, they are primarily used for small towns in rural areas. They are quite simple in principle: the sewage (usually after some preliminary treatment) is applied to the soil at appropriate rates and under appropriate conditions so that the microbial community develops, remains healthy, and uses the waste as a substrate.

More recently, land treatment has been used for a variety of industrial wastewaters. Food processing wastes of various kinds are particularly amenable, because they are rich suspensions of readily digestible material, and because food processing is often done in rural areas where land is available. Commonly, these systems are operated with crops or other vegetation in place. The soil conditions are not too different from those in entirely natural systems that receive a large organic input.

Much different systems are used for the degradation of oily wastewaters produced by oil refineries. In these, the load of organic material applied is very high. In order to avoid the development of anaerobic conditions, it is necessary to plow the soil weekly, daily, or even more often. The result is a plot of black, oily land where no plants grow. But the microbiological community becomes a vigorous, dense culture capable of the rapid degradation of oil. Rates of decomposition, in mass of organic matter per unit volume per unit time, reach values comparable to those in activated sludge systems used for sewage treatment (Devinny et al., 1986).

These systems are of interest here because they demonstrate the capability of microbial ecosystems on granular media. Clearly, the microorganisms have the ability to solve many of our subsurface contamination problems if we learn how to control and maximize their activities.

Genetic Engineering. Some contaminants are degraded slowly, or not at all, by common microbes. In other cases, decomposition occurs, but only under conditions that are difficult to obtain in the field. The possibility that genetic engineering might create organisms specifically designed to solve these problems is attractive. Experiments done in the laboratory seem to suggest that such an approach is possible.

It is unlikely, however, that such techniques will be used soon. First, the process of producing such organisms is difficult, time-consuming, and expensive. Secondly, more is required than simply an organism that can degrade the chemical in question. It must be capable of multiplying and thriving in the soil and water environment which is contaminated. Because this environment is rich in predators and competitors, and the mechanisms of such interactions are poorly understood, creating the organism may be difficult. Genetically engineered organisms are likely to be strongly lab-acclimated. At the least, it is necessary that the organism created be capable of living long enough to serve as an enrichment culture, as previously described.

Third, the investigators, regulators, and politicians involved in genetic engineering issues have not yet created a social and political consensus on how to release engineered organisms into the environment. Finally, investigation of the capabilities of indigenous organisms has just begun. It is likely that in the near future, far more progress per dollar spent will result from studies of how to use what is available than from inventing something new.

This is not intended to suggest that genetic engineering has no prospects for environmental cleanup. On the contrary, it is a powerful technique that will inevitably be important in the future. But it should not obscure the obvious and more immediately available potential of traditional methods for the enhancement of biodegradation.

REFERENCES

Bouwer, E. J. and McCarty, P. L. 1985. Utilization rates of trace halogenated organic compounds in acetate-supported biofilms. *Biotechnology and Bioengineering,* 27(11):1564–1571.

Burns, R. G. 1979. Interaction of microorganisms, their substrates, and the products with soil surfaces. In *Adhesion of Microorganisms to Surfaces,* Ellwood, D. C., Melling, J. and Rutter, P. (Eds.), pp. 109–138. New York: Academic Press, 216 pp.

Devinny, J. S., Marshall, T. and Cordery, S. 1986. Assessment of Analytical Techniques for Monitoring Petroleum Waste Land Treatment Systems. In *Proceedings of the 1986 National Conference on Environmental Engineering,* pp. 496–502. Cincinnati, OH: American Society of Civil Engineers.

Ghiorse, W. C. and Balkwill, D. L. 1983. Enumeration and morphological characterization of bacteria indigenous to subsurface environments. In *Developments in Industrial Microbiology,* Chapter 16. Nash, C. H., and Underklofer, L. A. (Eds.), pp. 213–224, Vol. 24 Arlington, VA: Society for Industrial Microbiology.

Grant, W. D. and Long, P. E. 1981. *Environmental Microbiology,* New York: John Wiley and Sons, 215 pp.

Holden, P. W. 1986. *Pesticides and Groundwater Quality: Issues and Problems in Four States.* National Academy Press, Washington, DC, 124 pp.

McCarty, P., Rittman, B. E., and Bouwer, E. J. 1984. Processes affecting chemical transformations in groundwater, In *Groundwater Pollution Microbiology,* Bitton, G. and Gerba, C. (Eds.), pp. 89–115. New York: John Wiley and Sons.

Pohland, F. G., Gould, J. P., Ghosh, S. B. 1985. Management of hazardous wastes by landfill codisposal with municipal refuse. *Hazardous Wastes and Hazardous Materials,* 2(2):143–148.

Rao, P. S. C., Edvardsson, K. S. V., Ou, L. T., Jessup, R. E., Nkidi-Kizza, P., and Hornsby, A. B. 1986. Spatial variation of pesticide sorption and degradation parameters. In *Evaluation of Pesticides in Ground Water,* Garner, W. Y., Honeycutt, R. C., and Nigg, H. N. (Eds.), pp. 100–115. Washington, DC: American Chemical Society, 573 pp.

Sharma, M. M., Chang, Y. I., and Yen, T. F. 1985. Reversible and irreversible surface charge modification of bacteria for facilitating transport through porous media. *Colloids and Surfaces,* 16(2):193–206.

Slater, J. H. and Lovatt, D. 1984. Biodegradation and the significance of Microbial communities. In *Microbial Degradation of Organic Compounds,* Gibson, D. T. (Ed.), pp. 439–485. New York: Marcel Dekker, Inc.

Stumm, W. and Morgan, J. J. 1981. *Aquatic Chemistry.* New York: John Wiley and Sons, 780 pp.

Swallow, J. A. and Gschwend, P. M. 1983. Volatilization of organic compounds from unconfined aquifers, In *Proceedings of the Third National Symposium on Aquifer Restoration and Groundwater Monitoring,* pp. 327–333. Dublin, OH: National Water Well Association.

White, D. C., Smith, G. A., Gehron, M. J., Parker, J. H., Findlay, R. H., Martz, R. F., and Fredrickson, H. L. 1983. The groundwater aquifer microbiota: biomass, community structure, and nutritional status. In *Developments in Industrial Microbiology,* Chapter 15, Nash III, C. H., and Underklofer, L. A. (Eds.), pp. 201–211. Arlington, VA: Society for Industrial Microbiology.

Wilson, J. T. 1986. Aerobic Transformations of Halogenated Aliphatic Compounds. Seminar, Stanford University, June 17, 1986.

Zhong, W. Z., Lemley, A. T., and Wagenet, R. J. 1986. Quantifying pesticide adsorption and degradation during transport through soil to groundwater. In *Evaluation of Pesticides in Ground Water,* Garner, W. Y., Honeycutt, R. C., and Nigg, H. N. (Eds.), Washington, DC: American Chemical Society, 573 pp.

5

Chemical and Physical Alteration of Wastes and Leachates

by Joseph S. Devinny
Associate Professor of Civil and Environmental Engineering
University of Southern California

Between the time that a waste is generated and the time that it creates problems as a groundwater contaminant, its composition may be substantially altered. Chapter 4 described changes that might occur as a result of microbiological activity. The soils, water, and other wastes that the fluid contacts as it migrates will also have profound effects, often producing a leachate with entirely unexpected potential for groundwater and soil contamination.

DILUTION

Mixing Before Disposal

Some changes begin almost as soon as the waste is generated. Solvents used to clean parts are spilled on the floor and mixed with water, soap, and other spilled wastes. Wastes from several processes are drained into common holding tanks. When tanks are washed, the wash water will contain small amounts of the material that the tank held, mixed with large volumes of water and small amounts of soap or other cleaner.

For the purposes of wise disposal and environmental protection, this initial dilution is usually undesirable. A small volume of waste is converted into a large volume, so that hauling, piping, or any other waste-handling process becomes proportionately more expensive and difficult.

Where the dilution is by water containing another waste, problems may be compounded. Often the second pollutant is not treatable by the same methods, so that two kinds of treatment become necessary. Further, the compounds may interfere in such a way that one or both treatment processes are unusable. Even analytical methods, used just for determining concentrations of the contaminants, may be made more difficult by the mixing of two kinds of materials. For highly complex mixtures, such as partially degraded crude oil in soil, analytical techniques capable of fully characterizing the mix may not exist, and each material may require a substantial research project to determine the methods which come closest to telling what is present.

Mixing During Disposal

Codisposal. Many disposal systems mix wastes from several sources. This is often a matter of convenience or short-term economic benefit. A single system will cost less than two systems, especially if the ultimate fate of the waste is simple landfilling. In the past, it has also been done intentionally with the intent of diluting waste. Acids were mixed with bases to neutralize both, producing a noncorrosive waste. Acidic or caustic wastes were mixed with municipal wastes to take advantage of the neutralizing capacity of the garbage. It was also presumed that adsorption of metals on solids and biodegradation of organic chemicals would provide some protection.

Studies of codisposal practices in landfills confirmed some of these assumptions. Work by Walsh et al. (1984) on experimental waste cells showed little effect on the overall properties of the generated leachate when various industrial wastes were added to municipal solid waste. In some cases, specific metals present in the industrial waste did appear in elevated concentrations in the leachate, but the amounts were not large, even when the industrial waste fraction was substantial.

More recent developments, however, have reduced the attractiveness of codisposal to landfills. In most cases, the industrial wastes are formally classified as hazardous wastes, and codisposal is not permitted. Further, the leachate from municipal waste itself has been recognized as a hazardous waste that can produce monumental soil and groundwater contamination problems. The fact that codisposal makes it no worse is little comfort: substantial *improvement* is a necessary goal.

The newer emphasis on waste treatment and recycling, especially on-site processes, also works against the mixing of wastes. It is more difficult to design adequate treatment processes for a complex mixture of materials than for individual wastes. Water contaminated with small amounts of oil, for example, can be treated effectively using a biological system such as activated sludge. If wastewater containing chlorinated solvents is added, the system may fail. Such solvents do not biodegrade rapidly, so the effluent from the combined waste treatment process would not meet standards. It is much more effective to treat the oil biologically, and deal with the chlorinated solvents separately.

Even dilution with water may be very undesirable. If a liter of water containing tetrachloroethylene (dry cleaning fluid) is diluted with a liter of uncontaminated water, the effect is to make two liters of hazardous waste from one. The problems and costs are doubled. Efforts are now being made in many industries to avoid the incidental dilution that comes during washing procedures and other processes which involve mixing potential waste materials.

Mixing with Soil. The soil itself can serve as a diluting agent. A waste that is dumped on land, either intentionally or accidentally, will be present at a reduced concentration per unit volume. The soil may also act as an agent which causes much greater dilution in percolating water. Contaminants may adsorb to the surfaces of the particles, then be gradually released to large volumes of water passing through.

Again, such dilution may be desirable or undesirable. Certainly in some cases, it has had the effect of reducing concentrations of chemicals to levels which are no longer considered hazardous. In other cases, however, the effect has just been to create a much larger volume of hazardous waste. In many cleanup efforts, the greatest expense is the cost of excavation, transportation, and disposal of these large volumes of contaminated soils.

Dilution in Landfills. Many landfills remain open to percolation by precipitation. Water which falls on the surface raises the water content above the field capacity, and drainage begins. As the water passes through the waste material, it picks up solutes and particles, and carries them along. The water has become a contaminated leachate.

The concentration of pollutants in the leachate produced will depend on many factors, but an important one is the amount of water. Walsh et al. (1984) did studies on confined columns of municipal waste and found that this dilution factor was perhaps the most important control on leachate strength. Small amounts of water produced a concentrated leachate, while large amounts produced a more dilute solution. This was true for a variety of leachate characteristics, including organic content, metals content, and the concentrations of other inorganic species.

This result may have some significance for approaches to leachate management. A common response to the production of contaminated leachate is to use caps and drains to limit water input to the landfill. But this may reduce leachate volume only in proportion to an increase in leachate strength. The same amount of contaminants may be transported to the groundwater. Throughput control may only be effective if water input can be reduced to zero.

Disposal Time and Place. The details of how and when wastes are discharged can be of fundamental importance in controlling the physical and chemical processes that occur. Acid wastes may solubilize trace metal wastes that have been previously dumped at the same site. However, if the acid is discharged first, and neutralized by reaction with the soil, trace metals in subsequent discharges may be firmly adsorbed. It is common that concentrations of contaminants in soils vary sharply over short distances. Especially for small spills, analyses commonly show concentrations sharply different ten feet away or two feet down. In such conditions, it may be difficult to predict interactions because of the uncertainty as to how much mixing has occurred for the various wastes.

Mixing in Groundwater

In the typical case of groundwater contamination, a waste stream from the surface descends to join groundwater flowing more horizontally. The characteristics of the result will depend on the rates of flow of the waste stream and groundwater. If the waste flow is large with respect to the groundwater flow, the dilution ratio may be low. The descending waste will spread in all directions. A "mound" will be created in the water table, altering the hydraulic gradient which carries the waste away.

In contrast, if the waste stream is small with respect to the groundwater flow, dilution ratios will be large. Only a small portion of the waste will enter each parcel of groundwater as it passes. The waste will have relatively little effect on the water table and the flow regime, and the waste plume will be highly asymmetrical, stretched out in the direction of groundwater flow like a smoke plume on a windy day. The downstream groundwater will be contaminated, while the upstream water will not.

The most extreme cases are seen in the relatively rare "karst" terrains. In these areas, the rock is limestone, which dissolves quickly in water (in geological time). Seepage tends to form underground rivulets, where more dissolution occurs, eventually forming underground streams and rivers. Water movement occurs as flow within subsurface caves (which sometimes collapse to form sinkholes). Toxic contaminants entering such a flow regime will be carried rapidly downstream, and may not be detected in surrounding rocks and soils. Groundwater monitoring for such geological strata will require specially designed systems. It is necessary to drill wells which intercept the underground streams (Quinlan and Ewers, 1986).

Dispersion. A body of water flowing through porous media (like soils) does not have all of its molecules moving at the same speed. Some water passes quickly down the middle of the pores, while some lingers in spaces between soil particles. If the water is compared to a group of people, the proper analogy is not with a marching regiment of soldiers, in which each individual maintains a fixed relationship with the others. Rather, the water moves like a crowd at a shopping mall, with some individuals moving quickly down the center which others stop to look in the windows. School children who begin as a small, homogeneous group are soon dispersed among the other shoppers.

In the same way, contaminants which enter moving water do not travel with a rigidly defined front. Some of the contaminant moves with the fast water and is diluted, creating a region of low concentration ahead of the mass. This process spreads and dilutes the contaminant. Spreading will occur in the direction of flow (longitudinal dispersion) and perpendicular to the direction of flow (transverse dispersion).

Johnson et al. (1985) present evidence for the phenomenon on a large scale. They investigated the movement of contaminants in an area of fractured rock. Just as on the microscopic scale, irregular macroscopic flow produces longitudinal dispersion. An effective model of contaminant movement was made by assuming that the water was in two fractions, one mobile and one immobile. Because contaminants contained in the immobile fraction are retarded with respect to those in the mobile fraction, dispersion occurs.

Waste Interactions

Wastes are often discharged to the land as mixtures. Landfills receive every imaginable sort of material in a lumpy, irregular mix, including hazardous and nonhazardous species side by side. Spills may include materials from more than one tank or car. Where the site has been the scene of casual or illegal dumping over a period of time, it is common that many different species will have been discharged. Chemical reactions within the mix may give it characteristics different from those of any of its components, complicating the problems of subsurface migration and site cleanup.

Acids and Bases.   Many materials are hazardous because of their low or high pH. Where these are disposed of together (if the disposer survives the violent reaction), neutralization will occur. Natural soils display a degree of acid- or base-neutralizing capacity. Because of this, the pH of the materials will generally change as they migrate. In some cases, this characteristic behavior has been used to reduce the degree of hazard. Acids have been intentionally mixed with large amounts of garbage or other bulky waste in order to provide a degree of neutralization. Under controlled conditions in a treatment facility, and where the compositions of the wastes are well known, it is possible to mix acids and bases and eliminate the hazards of both.

Acids and Cyanide.   More serious problems occur where acid wastes come in contact with metal plating wastes containing cyanide ions. The reaction product is hydrogen cyanide, a highly poisonous gas which is a threat to anyone nearby.

Acids and Metals.   Copper, zinc, lead, and most of the other heavy metals are more soluble in acids than in neutral or basic solutions. Where the two are disposed of together, production of a metal-rich leachate becomes likely.

Solvents.   Many contaminants have only limited solubility in water, and will not migrate readily. They remain adsorbed on soil particles. This can be changed, however, if they are codisposed with a solvent. Pure solvent may be present, either migrating downward through the soil or floating on top of the groundwater. This pure chemical will be capable of carrying large amounts of contaminant with it. Even small amounts of solvent dissolved in water, however, can significantly increase the solubility of many organic chemicals. For example, the aromatic chemicals in petroleum, such as the carcinogenic polynuclear aromatic hydrocarbons, are much more mobile when solvents are present.

There are also some indications that organic solvents will compete with metals for adsorption sites on the soil. This may cause increased mobilization of the metals.

DISSOLUTION AND PRECIPITATION

Control of Dissolution

Water dissolves solids with exceptional facility. Trace metals, many kinds of toxic salts, acids, bases, and many toxic organics will contaminate water on contact. The slow passage of water through soil, and the large surface area of the soil particles, provides ideal conditions for dissolution.

Sometimes, the mechanism is as simple as rain falling on a pile of potentially hazardous material. It produces acid drainage when it falls on mining waste. Similarly, rain falling on wastes in a landfill can generate hazardous leachate.

Local topography may worsen the situation by collecting large amounts of water from surrounding areas. Run-on, or flow into the facility, brings this additional water into contact with the waste. This is a common occurrence for carelessly designed landfills, often built in canyons or old quarries, where water collects.

If the landfill is in a deep hole and the water table is shallow, the wastes at the bottom may actually be immersed in the aquifer. These will be readily dissolved.

Even where the water table is deep, perched water may produce saturated flow into a waste mass. In one case, water from the septic tanks of homes surrounding a landfill encountered clay layers and collected above them. Although this was far above the groundwater, saturated zones developed and the water migrated laterally into the waste mass.

Many wastes that are classified as solid contain considerable amounts of water. Processed sewage sludge is an example. The pressure of overlying layers of waste may squeeze the water out of the sludge (Lu et al., 1985).

Microbiological activity often produces water. The amount is small, but it may be enough to make an already damp waste generate a moving leachate.

The readiness with which a solid waste dissolves is often the key characteristic which determines whether it is a threat to groundwater quality. U.S. Environmental Protection Agency standards for the identification of hazardous wastes include a precisely defined "Extraction Procedure" to determine what species will dissolve when the the waste contacts water (Chapter 2). The phenomena that govern dissolution are complex and imperfectly known. It is also difficult to anticipate the exact environmental conditions that will be encountered by the waste.

Physical Control.   Wastes of similar chemical composition may produce quite different leachates because of differences in physical factors. Particle size is important, because smaller particles have a higher surface-to-volume ratio. Dissolution occurs at the surface of the particle, so small particles are likely to release more contaminants more rapidly than large particles.

Soil permeability will indirectly affect dissolution because it controls the movement of water. Dissolution may be much slower in impermeable soils simply because the water cannot pass through.

Temperature influences many aspects of the dissolution process. Warm water speeds the reactions that occur at the surface of the particle. A commonly accepted generalization is that the rate of a chemical reaction doubles for each 10°C rise in temperature. The effect of temperature on the natural leaching of minerals has been recognized: certain highly leached clays, like bauxite, are more commonly found in tropical areas. It is likely that temperature is an important parameter in the leaching of solid waste.

Temperature will also exert an indirect influence through its control over microbiological activity. When the waste is organic, or when it is codisposed with organic wastes, microbiological activity will be important in dissolution processes. Like chemical dissolution, microbiological decomposition occurs at rates which rise with temperature.

The depth of the waste body is often referred to as column length, in analogy to laboratory experiments made with confined columns of material. It will have an influence on the dissolution of chemicals because a greater column length means a longer time of contact between the water and the waste. However, if there are channels in the waste, created by large objects, rodent burrows, or the decomposition of tree roots, the water will flow rapidly down these channels and have little contact with the material. Correspondingly less will dissolve.

Compaction and compression of the waste mass reduces its porosity and may close the channels. Often the waste is intentionally compressed during the disposal operation for this purpose and to increase the capacity of the landfill. For wastes containing some water, however, compaction during filling or compaction under the mass of subsequent waste loads may force water out of the material, generating leachate.

The characteristics of the waste will change with age. As weathering and microbiological decomposition proceed, the waste will generally (but not always) become less dangerous. In a sanitary landfill operation, different parts of the fill will have various ages, and different microbial populations (Chapter 4). Some fills proceed over 20 or 30 years. As new wastes are deposited on the old, leachate generated from above will move downward to interact with the old waste. A complex set of multiple interactions is possible. Acid generated by the new waste may promote further leaching of the old, for example, or the old waste may adsorb chemicals leaching from the new.

Chemical Control.   The chemical processes occurring in disposed waste are numerous, complex, and poorly understood. The chemistry of the burial environment will change each waste, and each waste will act upon the others with which it is buried.

The chemical composition of a waste exerts an obvious influence over what species may dissolve in the groundwater. Wastes containing metals generate leachate contaminated with trace metals, and organic wastes generate leachate containing solvents. It is surprising, however, how little correlation there is beyond this. Environmental factors may promote rapid dissolution of a contained species, or completely suppress it.

The pH of interstitial waters within the waste or contaminated soil controls, and is controlled by, interactions with the solids. Many solids produce excess hydrogen or hydroxide ions on interaction with water. Commonly, the products of microbial degradation are acids. In turn, acid waters are more effective at dissolving metals. Basic solutions are more likely to precipitate the metals as hydroxides.

The redox potential has important effects on dissolution, because elements in different oxidation states have much different solubilities. Iron or manganese, in their reduced forms, are relatively soluble, but the oxidized forms precipitate. Thus, an anaerobic waste mass is much more likely to produce a leachate containing these metals.

All chemical reactions occur at finite rates. Where those rates are modest or low, the time of contact will be an important factor determining the concentrations of pollutants in groundwater. Concentrations under these conditions are said to be under "kinetic" control, meaning that predictions or models must be based on the rates of reactions and the time of contact. Dissolution does not continue forever, however. Eventually, the concentrations of pollutant in the water will reach equilibrium with the solid. After this condition develops, no further change will occur, and the concentration is said to be under "thermodynamic" or "equilibrium" control, and the solution is "saturated." In this case, modeling the phenomena will depend on equilibrium constants. The groundwater chemist commonly encounters both kinetic and equilibrium conditions.

The chemical characteristics of soils are complex and varied. Different soil layers may be present over and under a waste mass, and often soils will be included in it. These soils may be almost inert sand, with little chemical reactivity and the large pores which reduce contact time, or they may be clays, with high ion exchange capacities, strong adsorption characteristics, and small pores which produce long contact times. Humus in organic-rich soils will also exhibit a wide range of reactivities. Dissolution of disposed wastes will thus be strongly affected by included or associated soils.

Precipitation

Where the concentrations of chemicals in the interstitial waters exceed equilibrium values, precipitation will occur. The chemicals will form solid particles in suspension, or a layer of solid particles on available surfaces. At first consideration, this seems unlikely: if the species were present in concentrations in excess of saturation, it would have precipitated before the waste was discharged. However, there are several mechanisms by which precipitation can occur within soils and aquifers. Elements may be soluble under the conditions existing in discharged wastewaters, but not so when conditions are altered. When the water reaches the soil, changes in environmental conditions, such as pH or temperature, or changes in speciation of the element, may drastically lower the equilibrium concentration, producing an oversaturated solution. Precipitation will begin, converting the element to a solid form much less likely to migrate with the groundwater.

Neutralization.   For many species, the saturation concentration is a strong function of pH. A change in pH may reduce the saturation concentration, so that the waste becomes supersaturated, even though the concentration of the chemical of interest has not changed. Many trace metals, for example, are more soluble in water of low pH. If acids containing the metals are discharged to the soil and neutralized, the metals may precipitate. Clogging problems have resulted from similar processes where crushed limestone has been used for leachate collection systems in landfills. Acidic leachates partially dissolved the calcium carbonate in the rock. Eventually, the acid was neutralized, and the calcium carbonate reprecipitated. The resulting cementation clogged the leachate collection system (Ghassemi, 1986).

Temperature.   Often, saturation solubilities are a function of temperature. If the waste cools when it is discharged, the solubilities of some species will be reduced, and precipitation may occur.

Chemical Reaction.   A discharged waste mixed with soils or other waste may undergo chemical reactions which produce new species of low solubility. There are many simple inorganic recombinations of this kind. If ferrous chloride, for example, encounters sulfide in groundwaters, it will likely precipitate as ferrous sulfide, which has a very low solubility. Redox reactions often have the same effects. Manganous ion (Mn^{2+}) is common in groundwaters, and is moderately soluble. If it is oxidized to manganese oxide, however, it has a low solubility and will precipitate.

Effects on Groundwater Quality

The most obvious effect of precipitation is that it removes the species involved from the water. Thus, precipitation of the metals present in plating wastes may hold the metals relatively near the surface even as the water moves deeper. Wagner and Steele (1985) was able to detect the precipitation or dissolution of calcium in groundwater by comparing the concentrations of the species present with those in rain and dry deposition at the study site.

Secondary effects may also occur. The precipitating chemical may form a colloid, or suspension of tiny particles, which eventually coagulates or is filtered out of the water by the soil. Some of these colloids have highly active adsorptive surfaces. When they form, they will adsorb many other chemicals, both organic and inorganic. When they are removed, the adsorbed species are removed with them.

Effects on Soils

Precipitation will also change the quality of the soils in which it occurs. Shuckrow et al. (1982) said that precipitates may clog the pores of soils, reducing porosity and retarding the flow of water or leachate.

Precipitation tends to occlude the particle surface. Pohland et al. (1985) noted that the metals in sewage sludge can be encapsulated, and thus prevented from migration, by the formation of a sulfide or carbonate layer.

Precipitation may also tend to harden soils. The solids formed at the surfaces of the particles will tend to cement the particles together. In natural systems, this produces "desert pavement," a hard layer on the surface of desert soils (which is disrupted by off-road vehicles). Over longer periods of time, it contributes to the formation of sedimentary rocks, such as sandstone.

FILTRATION AND ADSORPTION

Filtration

The most easily visualized effect a soil may have on an infiltrating fluid is filtration. Particles in the fluid which are larger than the pores in the soil will be removed just as spaghetti is removed from water in a colander. This may be an important factor where suspensions of relatively large particles have been discharged.

The term filtration is often used to include physical removal phenomena which are different in microscopic detail. As the fluid passes through the soil pores, it is held for a time in millions of tiny chambers which act as sedimentation basins. Particles which are slightly more dense than the water, as is the case for most, will sink. Although small particles sink very slowly, the small size of the soil pores means that they must sink only short distances before contacting the soil particle below. Depending on the surface chemistry of the particle and the rate of fluid flow, many of the particles will stick.

Often, the factor limiting the rate of coagulation for these small particles is the frequency of collisions with other particles. Movement through soil pores produces a great deal of irregular flow. The water flowing through the middle of the pore flows more rapidly than that near the walls, so that adjacent portions are sliding past each other. (This kind of movement is called "shear." The tremendous amount of shear which is generated in the multitudinous minute pores creates the resistance that makes water flow slowly in porous media.) Particles suspended in the faster-moving water will strike particles suspended in adjacent slower-moving water. Under appropriate conditions, these collisions will result in coagulations: the particles will stick together. The one larger particle which results will sink faster than the two small particles which formed it, so once again, removal of the suspended material will be facilitated.

Overall, filtration effects are substantial. As was emphasized in the second chapter, particulate matter is generally not transported for significant distances in porous media.

Adsorption

Adsorption is the process by which molecules of dissolved chemicals in a fluid attach to solid surfaces. Concern for subsurface migration of hazardous waste almost always involves adsorption of metals or organic species dissolved in water. Adsorption is of great importance in determining the seriousness of soil and groundwater contamination incidents. If the contaminant in question is tightly adsorbed on soil particles, it will be retained quite near the site of the discharge even as the water moves on, or as rainfall washes through the soil. The amount of soil contaminated will be small, groundwater contamination will not occur, and cleanup by excavation will be simple. Polychlorinated biphenyls (PCB's) and heavy petroleum products have this characteristic. In contrast, cotaminants which are not adsorbed readily may create much more serious problems. Trichloroethylene (TCE), tetrachloroethylene (PCE), and trichloroethane (TCA) move readily with the water and contaminate large aquifers. At some sites in southern California, large water supply aquifers have been contaminated by TCE over long periods of time. TCE dumped on the ground over many years has gradually migrated downwards, ultimately ruining water supplies hundreds of feet deep.

Principles of Adsorption

Adsorption is a complex phenomenon, not fully undestood. Several different mechanisms commonly cause molecules to leave solution and stick to the surfaces of soil particles.

Chemical Bond Formation.   Molecules may be held at surfaces by the formation of covalent chemical bonds with the solid. These bonds are the ones that hold molecules together, and are often quite strong. Breaking a covalent bond may require an energy as high as 100 kcal/mole. Where this is the case, the solute molecule essentially becomes part of the surface, and the adsorption may be effectively irreversible.

Ion Exchange. The particles that make up soils often have ions held at their surfaces by electrostatic attraction. The identity of the ions depends on the history of the soil and the parent material from which it was made. When a solution is passing through the soil that contains other ions, exchange is possible. The dissolved ion will replace the adsorbed species if the solid has a greater affinity for it. The released ion will appear in the solution.

The charged sites on soil arise in several ways. Weathering reactions produce clays, which are complex oxides of silicon, aluminum, and other materials. At the edges of the clay particles, the oxides have unsatisfied bonds, producing a negatively charged surface. Soil organic material may have carboxyl groups at the surface, capable of exchanging the hydrogen ion for metals.

The affinity of the charged sites for ions in solution depends on several factors. Smaller ions are generally more tightly held. Their small radius allows them to approach the charge on the surface more closely. Ions with a greater charge are adsorbed more strongly. The multiply charged ion becomes a "package" in which two, three, or four units of charge are held together despite their mutual repulsion. Thus, they can neutralize charges at the particle surface without the tendency to drive each other off the surface. In a few cases, soil components adsorb specific ions with a high selectivity. This occurs if the adsorbant site on the surface is restrained by a special geometry that is suitable for only a certain ion.

The relative affinities of the soils for ions can be overcome by concentration. That is, the ion that is individually less tightly adsorbed may occupy all of the sites, if that species is present in solution at high concentration. Exchange may occur because an ion in high concentration replaces an ion of low concentration. When an acid is passing through a waste, for example, the hydrogen ions may displace metal ions. If the acid comes from acid rainfall, and the particles are soil in a natural ecosystem, the result may be lower soil fertility. If the acid is generated by microbiological activity in a landfill, and passes through contaminated soil, the result may be a leachate with high concentrations of toxic metals.

Most ion exchange sites on soils are negative, and thus adsorb positive ions. It has long been recognized, for example, that soils tend to adsorb ammonium ion (NH^+_4) far more effectively than nitrate ion (NO^-_3). Nitrate may be less desirable as a fertilizer because it is more easily washed out of the soil. It is also more likely to become a groundwater pollutant.

Hydrophobic Bonding. Molecules may exhibit polarity; one end carries a small positive charge, while the other carries a small negative charge. Molecules which are nonpolar generally have a low solubility in water, which is strongly polar. Although water is obviously liquid, it retains some structure because of the strong polar interactions between its molecules. They form long chains and complex networks in which the positively charged hydrogens stick to the negatively charged oxygens on adjacent molecules. These networks are constantly forming and breaking up as they are battered by the thermal motion of the molecules.

Dissolution of nonpolar molecules requires "inserting" them in this structure, and breaking the association between water molecules. Breaking the association in turn requires energy. In the reverse process, energy is released when the nonpolar molecules leave solution. When nonpolar surfaces are available, therefore, there is a strong tendency for the nonpolar molecules to leave solution and stick to the surfaces. This "hydrophobic" bonding is an important factor controlling the fate of many organic pollutants in soils.

The strength of hydrophobic bonding depends on the degree to which the molecule is "rejected" by water. This can be measured by a simple experiment: water, octanol, and the pollutant can be put in a flask and allowed to reach equilibrium. Octanol is an alcohol with eight carbon atoms, and is relatively nonpolar. Octanol and water do not mix, but form separate phases. Hydrophobic pollutants will be concentrated in the nonpolar octanol phase. The ratio between the concentration in the octanol and the concentration in the water is the octanol-water partition coefficient for the pollutant. It will be high for hydrophobic contaminants, and low for hydrophilic contaminants.

The situation in the flask is analogous to that in the soil. The pollutant is free to pass from the water into the organic matter which is part of the soil. Because hydrophobic bonding is so important for organic pollutants, the degree of adsorption can often be reliably predicted from the organic content of the soil and the octanol-water partition coefficient for the contaminant.

Van der Waals Forces.   Molecules consist of "skeletons" defined by the nuclei of the constituent atoms, and surrounded by clouds of electrons. The electron clouds are not rigid: they can be distorted by electrical forces. When this occurs, the balance of charges in the molecule is changed. As the electrons shift toward one end of the molecule, they carry the negative charge with them and leave the unneutralized positive charge behind. The molecule gains a temporary induced polarity. For example, as a negative ion approaches a surface, it may induce a dipole by repelling the electrons. It will then be attracted to the remaining positive charge at the surface, and be adsorbed.

The electron cloud may also oscillate back and forth, producing alternating dipoles. When such a molecule nears a surface, the molecular dipole can induce an alternating dipole of the opposite sign in the electrons of the surface. The oscillating, synchronous dipoles attract each other because the corresponding "poles" are of the opposite charge on the surface and on the molecule.

The result of these phenomena is a "Van der Waals" attraction between many kinds of molecules. It is responsible for a substantial fraction of the force which creates surface tension in water, for example. In soils, it may the dominant force causing adsorption of solutes.

Multiple Mechanisms.   Quite often, adsorption of a given type of molecule to a given surface results from two or more of these mechanisms working at the same time. Complex organic chemicals, for example, may be nonpolar over much of their molecular surface, and yet have polar groups. The nonpolar region may engage in hydrophobic bonding while an ionized group is involved in ion exchange. Where reactions occur, they will commonly be adding to other forces.

Isotherms. Irrespective of the specific nature of the adsorption mechanism, it is possible to experimentally determine the amount of adsorption expected for a given soil and a given concentration of groundwater contaminant. Experiments are performed by mixing solutions of various concentrations with soil, allowing time for the establishment of equilibrium, and measuring the amount of contaminant left in solution. The results are commonly presented as a plot of the logarithm of the amount of chemical adsorbed versus the concentration in solution. These plots are termed "isotherms" because the experiments are performed at constant temperature.

Researchers have developed many theoretical models of adsorption phenomena. The two most commonly used are the Langmuir model and the Freundlich model (Figure 5–1). The Langmuir model assumes a limited number of surface sites available for interaction with the solute, each of which is either occupied or not occupied at a given solute concentration. It is assumed that the total number of sites is constant. The chemical reaction of adsorption can then be represented as:

$$a + S = S-a$$

where a is the adsorbate, S represents the unoccupied surface site, and S-a represents the adsorbate bound to the surface site. As for all chemical reactions, the relationships among the concentrations of the species at equilibrium is:

$$K_1 = \frac{[S-a]}{[a]\,[S]}$$

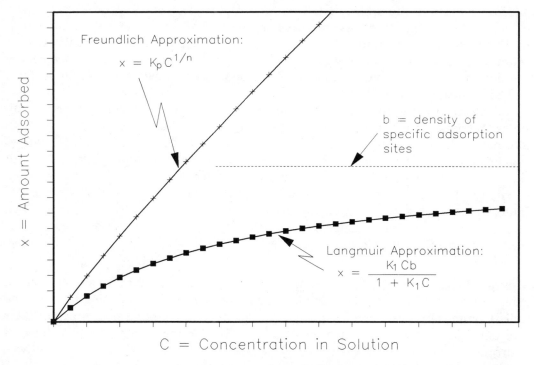

Figure 5–1. Comparison of the Freundlich and Langmuir Isotherms. The Langmuir approximation assumes an adsorption limit, resulting from a finite number of specific adsorption sites.

where the brackets represent concentrations: [S-a] is the concentration of occupied sites, [a] is the concentration of the adsorbate in solution, and [S] is the concentration of unoccupied sites. K_1 is the equilibrium constant for the adsorption reaction. Solving for the concentration of occupied sites produces:

$$[S - a] = K_1 [a] \frac{\{[S] + [S-a]\}}{\{1 + K_1 [a]\}}$$

More conventional notation uses [S-a] = x, the amount of adsorbate held, [a] = C, the concentration of adsorbate in solution, and [S-a] + [S] = b, the total number of sites (and therefore the maximum amount of adsorbate that can be held). The equation becomes

$$x = \frac{K_1 Cb}{\{1 + K_1 C\}}$$

The model is appropriate for highly specific adsorption mechanisms, such as chemical bonding of one type with a limited number of surface functional groups. It predicts a rapid rise in the amount adsorbed as the concentration in solution increases from low values, then gradual approach to the ultimate value, b, as concentrations are high.

The model may be modified by assuming an infinite number of adsorption sites (or at least a number of sites large with respect to the number of solute molecules). This is appropriate where the adsorption phenomena are nonspecific and the adsorbate molecules can pile up on the surface and on each other, essentially indefinitely. It is also true in dilute solutions, where the amount of adsorbate is small. In either case, the number of unoccupied sites can be taken as a constant. Beginning again with the relationship for concentrations at equilibrium:

$$K_1 = \frac{[S-a]}{[a] [S]}$$

Because [S] is constant, $K_1[S]$ is also constant, so we can define:

$$K_p = K_1[S]$$

and, substituting in the first equation, we get:

$$K_p = \frac{[S-a]}{[a]}$$

and solving again for [S-a]:

$$[S-a] = K_p[a]$$

or, again converting to more conventional notation:

$$x = K_p C$$

This final result is a form of the Freundlich isotherm. It is an appealingly simple formula, directly relating the amount of chemical adsorbed to the concentration in solution. Experimental results, however, often show some deviation from this straightforward model. Many phenomena may be the cause, but the results can often be well described with a curve of the form:

$$x = K_p C^{1/n}$$

This is the more complete form of the Freundlich isotherm. The exponent $1/n$ is essentially a "fudge factor," which allows correction for deviation from the simple model. The value of n is determined experimentally.

The Freundlich equation is widely used to describe and predict adsorption of organic chemicals from groundwater, and often the value of n is taken to be 1. That is, the simple form of the model is adequate.

K_p is called the "partition coefficient." It is the ratio of the concentration of adsorbed contaminant to the concentration of dissolved contaminant. It reflects the affinity of the solute for the soil. The affinity is in turn controlled by the various physical or chemical adsorption phenomena previously described. Of obvious importance is the total amount of surface available. Twice as much surface will adsorb twice as much solute. Soils vary tremendously in the amount of surface area that they present to groundwater for adsorptive reactions: they may easily be as low as $10 \text{ m}^2/\text{g}$ or as high as $300 \text{ m}^2/\text{g}$. The surface-to-volume ratio varies inversely with particle radius, so small particle soils will be able to adsorb more. Available surface will also be high where the particles are porous.

Kinetics. For many applications of adsorption technology, the rate of adsorption is important. Both the Langmuir and Freundlich models assume thermodynamic equilibrium. If there has not been sufficient time for the establishment of equilibrium, both will overestimate adsorption.

Groundwater problems occur over relatively long periods of time. Groundwater movement is slow compared to other phenomena. Further, the body of water is very finely divided within the pores of the soil, so that molecules of solute may readily migrate to the surfaces. It has often been assumed that groundwaters are at equilibrium for adsorption.

However, this assumption is not valid for all cases. Schwarzenbach and Westall (1981) showed in the laboratory that kinetic effects were significant in sand where water flow velocities exceeded 10^{-2}cm/sec. The water was moving fast enough that equilibrium was not attained before the contaminant moved along.

Pignatello et al. (1987) have discovered a phenomenon with important implications for soil cleanup. When contaminants have been in porous soils for long periods of time, they apparently diffuse to chemically remote pores within the particles. These have small openings, so that diffusion into the pores is slow. Diffusion back out is equally hindered. Thus, a contaminated soil may exude small amounts of pollutants for years or even decades after the large pores have been cleaned. The physical, rather than chemical, nature of the bonding of this residual material is demonstrated by its release when the soil is milled.

Adsorptive Capacity. For all adsorbant-solute combinations, there is some ultimate adsorptive capacity. If the adsorbant is described by the Langmuir model, there are a limited number of sites. When all of these are occupied, no more adsorption can occur. In the case of Freundlich adsorption, the concentration of the solution in the soil rises as more and more adsorbate is held, until it essentially equals the concentration of disposed waste. (In many cases, the Freundlich approximation is inaccurate at high concentrations.)

When soil reaches its adsorptive capacity because of the disposal of large amounts of contaminant, then further amounts of contaminant will pass through the soil unaffected, and pollute regions beyond. This condition is called *breakthrough*. After breakthrough occurs for a given volume of soil, pollution will spread much more rapidly.

Soils will reach their adsorptive capacity in the region near the source of the spill where concentrations are high. The size of the region will depend on the amount of chemical discharged and the adsorptive capacity of the soil. Beyond this region, concentrations will begin to decline as the water passes into soil still capable of adsorption.

Desorption. In most cases, adsorption is not permanent. Molecules which have become attached to the surfaces of soil particles when the concentration of the solute was high return to solution when the concentration drops. An equilibrium relationship, perhaps described by the Freundlich isotherm, is maintained.

This suggests a method by which contaminated aquifers can be cleaned up. If the water containing the solute is pumped out and replaced with clean water, some of the contaminant will be desorbed as equilibrium is established. Because some of the contaminant has been removed, concentrations in the water and the amounts adsorbed on the soils will be reduced. If that water is replaced by more clean water, further reductions will occur. Successive replacements can eventually reduce the contaminant concentrations to acceptable levels. Soil and aquifer "flushing" is a workable remediation alternative in some cases.

Desorption by flushing will work well where the amount of contaminant adsorbed is moderate in comparison to the solubility of the contaminant in water. In this situation, water passing through the soil will carry away substantial amounts of contaminant and the cleaning will be done in a short time. In some cases, however, there may be large amounts of relatively insoluble chemicals on the soil. Each volume of water which passes through will remove only a small fraction of the chemical present. The amount of water to be pumped through the aquifer will be very large, and flushing may not be a workable alternative for aquifer restoration.

Processes reflecting the same phenomena occur wherever groundwater movement is carrying a mass of pollutants with it. As the first traces arrive at a given portion of soil, adsorption begins (Figure 5-2). Equilibrium may be established with the concentrations in the water, but subsequent parcels of water have higher concentrations of pollutant, so adsorption continues. After the bulk of the contaminant has passed, however, concentrations begin to fall. Because this means the solute concentrations are below equilibrium, desorption begins. As ever cleaner water passes the soil, the adsorbed materials are gradually depleted until the soil is clean. Each portion of soil therefore adsorbs pollutant, holds it for a period of time, then releases it again into the groundwater flow. The overall result is that movement of the pollutant is retarded with respect to the movement of the groundwater.

The process is quantified by the "retardation factor," which is the ratio between

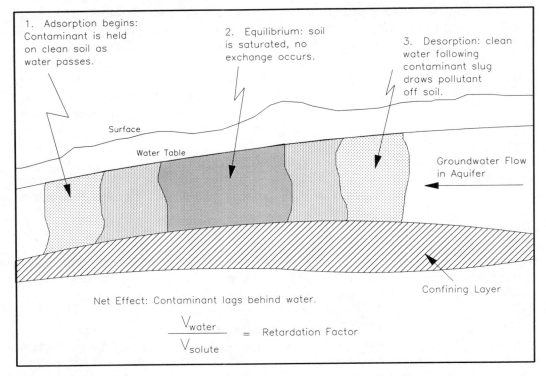

1. Adsorption begins: Contaminant is held on clean soil as water passes.

2. Equilibrium: soil is saturated, no exchange occurs.

3. Desorption: clean water following contaminant slug draws pollutant off soil.

Surface

Water Table

Groundwater Flow in Aquifer

Confining Layer

Net Effect: Contaminant lags behind water.

$$\frac{V_{water}}{V_{solute}} = \text{Retardation Factor}$$

Figure 5–2. Contaminant retardation. Pollutants which adsorb to soils will travel more slowly than the water.

the rate of water movement and the rate of contaminant movement. For strongly adsorbed pollutants, retardation factors may be quite high. For those which are not adsorbed, retardation factors are near 1.

The velocity of the contaminant is the average of the velocity of the dissolved material and the velocity of the adsorbed material (Figure 5–3).

$$V_{solute} = \frac{V_{water} \times \{amount\ dissolved\} + V_{adsorbed} \times \{amount\ adsorbed\}}{\{amount\ dissolved\ +\ amount\ adsorbed\}}$$

where

V_{solute} = the overall velocity of the contaminant,

W_{water} = velocity of the water, and

$V_{adsorbed}$ = velocity of the adsorbed fraction.

Dissolved molecules move with the water, at essentially the same speed. Adsorbed molecules do not move at all, so the velocity of the adsorbed fraction is zero.

$$V_{solute} = \frac{V_{water} \times \{amount\ dissolved\}}{\{amount\ dissolved\ +\ amount\ adsorbed\}}$$

or

$$\frac{V_{water}}{V_{solute}} = \frac{\{amount\ dissolved\ +\ amount\ adsorbed\}}{\{amount\ dissolved\}} = retardation\ factor$$

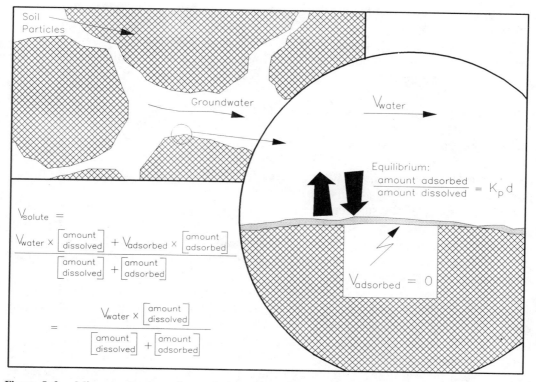

Figure 5–3. Microscopic view of retardation. The relative velocities of water and contaminant can be calculated from adsorption equilibria.

From the simple form of the Freundlich equation, we have :

Amount adsorbed (per cm³ soil) = CK_pd, and

Amount dissolved (per cm³) soil) = $C\theta$

where:

C = the concentration of solute in the water,

K_p = the partition coefficient,

d = the soil density, and

θ = soil porosity, the volume fraction of the soil between the solid particles so:

$$\textit{Retardation Factor} = \frac{C\theta + CK_pd}{C\theta} = 1 + \frac{K_pd}{\theta}$$

Thus the retardation factor depends on the partition coefficient in a simple and important way. The partition coefficient is commonly equal to 5, or 30, or even 100. When it is high, the retardation factor will also be high, and the contaminants will lag far behind the movement of the groundwater.

If data are only available at a single point, the effects of retardation may be confused with those of attenuation. That is, it may be impossible to determine whether the pollutant is held back by adsorption or destroyed by biological or chemical reactions. However, Sutton and Baker (1985) showed that attenuation and retardation can be separated if it is possible to compare both peak heights and peak positions with those for a conservative tracer. In experiments in a sandy aquifer, butyric acid, phenol, p-chlorophenol, and dimethyl phthalate were attenuated, but not retarded. Anaerobic degradation was presumably destroying them.

Soil Type.   The partition coefficient is influenced by many variables, including the amount of surface area per unit volume of soil, and the chemical characteristics of the soil particles.

Soils made of small particles will present the largest surface area for adsorption. Available surface area can also be influenced by particle shape. Spherical particles present the least possible surface for a given mass of material. Clay particles are often in the form of tiny flat platelets, which will have a greater surface area. Porous particles may have very large total surface areas (the porosity of activated carbon particles accounts for its great adsorptive capacity).

Access to some of the surface area may be prevented. Where particles are packed tightly together, they may occlude some surface area. If pores are extremely small and the solute molecules large, it may be impossible for the molecules to enter. The particles may be covered with precipitates or biogenic chemicals which prevent certain adsorptive reactions.

Not all surfaces are equal. Clay platelets, for example, are layers of atoms bonded to each other and to atoms in adjacent layers. At the edges of the plates, the atoms do not have a complete set of neighbors. The edges have different chemical characteristics than on the faces. Ion exchange with the edges is much more common.

Organic molecules are best adsorbed by organic matter. There is a strong correlation between partition coefficients and the organic content of the soil (Johnson et al., 1985). Soils containing more organic material offer more hydrophobic surfaces for the attachment of organic pollutants. In some cases, contaminants may be capable of dissolving in semisolid or liquid organic materials in the soil, so the content of organic matter will reflect the volume of the organic phase available. Correlation of partition coefficients measured for various soils with their organic matter content shows a strong positive relationship.

The organic matter present is primarily dead plant material and its decomposition products, called humic material. Khan (1975) describes humic substances as "amorphous, dark colored, hydrophilic, acidic, partly aromatic, chemically complex organic substances that range in molecular weights from a few hundred to a few thousand." They are are produced by partial microbial breakdown of plant material like leaves, roots, and twigs. The chemical nature of humic substances, particularly their random structure and aromatic groups, suggest the biological polymer lignin as the parent material. Few details are known, however, about the processes producing humic substances. Even the composition of humic substances is only crudely described, because they constitute a very complex mixture of chemicals which cannot be precisely characterized.

Polysaccharides, released by bacteria, may also be present in substantial amounts. All of the organics present show strong stratification with depth. Commonly, the organic content in topsoil is high. Deeper in the soil column, the mineral phases dominate. Thus, a small contaminant spill might be readily adsorbed by the topsoil, when a spill only moderately larger would penetrate to the subsoil and travel much farther before complete adsorption.

Organic chemicals will absorb strongly to the surfaces of mineral soil particles if the soil is very dry (Vlasaraj and Thibodeaux, 1988). This does not occur in wet soil because the polar water molecules compete for the polar mineral surface sites effectively, driving the organics off. But when the water content drops below that necessary to form a monolayer on the soil, the organics can bond to the bare mineral. This may be an important factor controlling the effectiveness of air stripping as a remediation method (Devinny, 1989).

Interstitial Fluid Composition.   The solution which carries hazardous materials in subsurface migration varies substantially in composition. In the case of massive releases of petroleum products or solvents, the product may be present as the nearly pure phase. Adsorption of organics under these conditions will be totally different. Hydrophobic bonding, which is a primary factor for most adsorption of organics, will not occur. Many chemicals which are strongly bound when the interstitial fluid is water will be released in large amounts into hydrocarbon liquids.

Where the interstitial fluid is water, the presence of various solutes will effect adsorption. Subsurface waters differ in concentrations of salts, ranging from nearly pure to more saline than seawater. In general, higher salt concentrations will further reduce the solubility of nonpolar organics, promoting more adsorption. Salts in high concentration may compete in ion exchange reactions, releasing toxic trace metals.

It is common to find groundwaters polluted with more than one organic material. The presence of a slightly soluble solvent may in turn decrease the adsorption of a second pollutant. Two mechanisms are possible: the solvent may increase the solubility of the second pollutant, or it may be competing for adsorption sites on the soil.

Fu and Luthy, (1986) showed that naphthol, naphthalene, quinoline, and 3,5-dichloroaniline had lower partition coefficients in water with low concentrations of methanol or acetone added. The values of K_p declined semilogarithmically with increasing solvent concentration.

Adsorption of Organic Chemicals on Soils

The prominence of organic chemicals among soil and groundwater pollutants has focussed particular attention on their adsorption. Partition coefficients for organic chemicals display a tremendous range of values. At one extreme, the low-molecular-weight chlorinated hydrocarbon solvents such as TCE are highly mobile, contaminating aquifers far from the point of discharge. At the other, there are some high-molecular-weight petroleum hydrocarbons, like tars and asphalts, which are so strongly adsorbed and so insoluble that they have defied the development of a procedure to dissolve them so that they can be characterized in the laboratory.

Nevertheless, experience suggests some generalization. In most cases, adsorption of the organics is well approximated by the Freundlich isotherm in which the exponent is taken as one. That is:

$$x = K_p C.$$

The equation says the amount of chemical adsorbed is proportional to the concentration in solution and the partition coefficient. Schwarzenback and Westall (1981) further investigated factors controlling the value of K_p. The sorption is dominantly hydrophobic, reflecting the transfer of organics from the aqueous solution to small amounts of organic material in the soil. K_p is proportional to the size of the organic phase, measured as the soil organic fraction. It also reflects the hydrophobicity of each contaminant, which is approximately indicated by the octanol-water partition coefficient of the contaminant. Thus, K_p can be estimated within a factor of two, where the soil organic fraction and the octanol-water partition coefficient have been measured.

This simple relationship breaks down for soils which contain little organic matter. Adsorption is then controlled by several different, relatively weak interactions, and contaminants are likely to be mobile.

Fu and Luthy (1986) emphasized this relationship. They also found that secondary solvents in the water could cause swelling of the soil organic material, increasing its capacity for adsorption. The organic material swells rather than dissolving because it consists of cross-linked humin or kerogen (the stable end products of biological degradation of plant matter).

Rao et al. (1986) showed that concentrations of the pesticide aldicarb in soils could be correlated with soil organic carbon concentrations even for quite small-scale variability (tens to hundreds of meters).

Roberts (1986) measured retardation factors for chlorinated hydrocarbons injected with treated wastewater. They found values of 3 for chloroform and 30 for chlorobenzene. In a carefully controlled experiment with small pulses of chlorinated hydrocarbons, retardation factors from 2 to 9 were measured.

Adsorption may be of particular importance in soils where flow is slow, as it is in tightly packed clays. Because of this restriction of groundwater movement, tight clays are often chosen for use in caps and slurry walls for the purpose of confining hazardous waste in landfills or at spill sites (see Chapter 9). It has only recently been pointed out, however, that in regimes of very slow flow, the movement of some solutes is no longer controlled by the movement of the water. Molecular diffusion, which inevitably carries solutes from areas of high concentration to areas of low concentration when there is a hydraulic connection, becomes the dominant mode of transport. While such transport is slow compared to movement in typical aquifers, it may be sufficiently rapid to defeat the purposes of the clay barriers. Gray and Weber (1984) calculated that concentrations outside a three-foot-thick clay wall could rise to 1 percent of those inside the wall in less than ten years. This would occur even where there was actually slow flow into the landfill as a result of a negative hydraulic gradient. While such movement is slow in human terms, it may well be enough to allow escape of pollutants and contamination of soils and groundwater.

This effect was seen in some field examples for cations, and so seems a well-established phenomenon. Further calculation, however, predicted that it will not be seen for species which are adsorbed to an appreciable extent on the clay in the slurry wall. When a partition coefficient of 5 was assumed, calculations indicated breakthrough would not occur for more than 100 years. It may thus be reasonable to include some organic content in clays used for slurry walls to prevent significant diffusional transport. It also follows that careful quality control for slurry walls will be necessary. The time to breakthrough for the solutes is proportional to the square of the thickness of the wall. Thin spots resulting from inaccurate trenching during wall installation will therefore have disproportionate effects on contaminant escape.

The discussion of soil adsorption of organics emphasizes hydrophobic adsorption, which is the dominant mechanism for most. Organic ions also occur, however, and may participate in other interactions. Faust (1975) noted that cationic herbicides are strongly held in soils, as would be expected from experience with inorganic cations. Khan (1975) found that the herbicides paraquat and diquat, though they are aromatic hydrocarbons, are adsorbed in soils primarily through ion exchange. These species are doubly charged positive ions in solution, and can replace hydrogen ions on soil particulates. This was confirmed in experiments in which hydrogen ions were released from soils when the herbicides were added.

Adsorption of Trace Metals on Soils

For trace metals, ion exchange is the dominant form of adsorption. The process has substantial effects on the quality of soils and soil water, and thus on the subsurface migration of trace metal ions. Many metals are strongly held, reducing their migration and the extent of contamination. They may be released, however, by exchange with ions in solutions applied at some later time.

Leonard et al. (1984) found that selenium is adsorbed in significant amounts by the amorphous hydroxides found in a natural limestone. They predicted that this effect would be sufficient to protect water supplies in a limestone aquifer from selenium leached from disposed fly ash.

Bailey et al. (1986) showed that the adsorption of trace metals is reduced by the presence of some organic solvents. Hexane, benzene, and phenol all diminished adsorption of cadmium, copper, nickel, and zinc. Presumably, the organic species were competing for adsorption sites with the metals.

CHEMICAL REACTIONS

While biological activity is the major agent for the breakdown of contaminants in the soil, there are some species that are susceptible to chemical reactions in the absence of life. The complexity of the soil environment and the variety of chemical pollutants ensure a large number of possible chemical reactions. Only a fraction of these are known and only a fraction of those known can be described here.

Acid-Base Reactions

The polarity of water molecules makes them particularly effective at dissolving protons. Because the protons pass easily into water, the solution serves well to transport protons from acids to bases. Acid-base reactions in water are therefore commonly rapid and acid-base systems are usually at equilibrium in soils and groundwater.

Such reactions are common and important for subsurface migration of hazardous wastes. Where acidic or basic solutions are discharged, they will react rapidly with the soil constituents. If the soil has a high neutralizing capacity (alkalinity or acidity), the influence of the discharge will be limited.

Hydrolysis

The addition of a water molecule to an organic molecule is a common reaction. Often both split, so that the products of this hydrolysis are two parts of the original organic molecule, one with an added hydrogen atom and another with the hydroxide. These hydrolysis reactions can be important steps in the degradation of organic contaminants.

Jones (1986) reported chemical hydrolysis of aldicarb. Dierberg and Given (1986) showed that aldicarb and two of its degradation products degrade rapidly in soils even when the soil has been sterilized. The reaction is likely hydrolysis of the carbamate ester bond. In Florida soils with neutral pH (6.8–7.8), the half-life for the materials was only 9–24 days. Interestingly, degradation of the aldicarb was catalyzed by the soil. Degradation rates in groundwater samples with the soil removed were much lower. This was not true of the sulfoxide or sulfone degradation products, so a specific catalytic reaction was apparently responsible. This chemical degradation is influenced by the nature of the groundwater and soil system.

On Long Island, where the groundwater is more acid (pH 4.2–5.9), the hydrolysis of aldicarb occurred much more slowly. There, the half-life of the chemical was 2–3 years. This is of substantial importance: a two to three year half-life will allow contamination of many water supplies which would not be touched in 9–24 days.

Ethylene dibromide, a gas commonly used for soil fumigation, is hydrolyzed in soil to produce ethylene glycol and hydrogen and bromide ions (Weintraub et al., 1986). The process was more rapid at higher temperatures, but was not affected by a pH between 4 and 9. The hydrolysis rate constant was independent of concentration in pure water, but was not in typical groundwater samples. This suggested groundwater constituents were influencing reaction rates.

Redox Reactions

In contrast to acid-base reactions, redox reactions are commonly kinetically hindered. Electrons are not readily dissolved in water, so the transfer from reductant to oxidant is slow. For this reason, redox reactions are most commonly catalyzed by biological activity. Biological redox was treated earlier in this chapter.

Generalizations similar to those developed for pH are nevertheless possible. The availability of electrons is controlled mainly by the presence of oxygen. Molecular oxygen is a powerful oxidant, and reacts with electrons to form oxides. Where oxygen is abundant, as it is in the atmosphere, electrons are in short supply and the oxidized forms of the elements will predominate. Where oxygen has been depleted, electrons are readily available and the reduced forms of the elements will be common. The changes in form associated with the redox reactions have far-reaching effects. Carbon as solvents or chlorinated hydrocarbons is in a relatively reduced form. If it is oxidized by oxygen under the influence of biological catalysis, it becomes carbon dioxide. Carbon dioxide is not toxic, but it is acidic and may reduce the pH of the interstitial solution. Metals will also be affected.

Complexation

Where metals and soluble organic species are present, complexes between the two may form. These often have chemical characteristics quite different than the metal ions. Complexation may increase the solubility of the metals and promote leaching that would otherwise not occur. In some cases, however, the complexes may be adsorbed by soils, thus preventing the movement of the metals.

Two organic molecules may also form complexes, again with characteristics different from either of the two reacting species. Of particular concern are complexes formed with humic acid. The natural abundance of humic acid means that it is available to participate in many reactions. Laboratory work indicates that it may react with PCB's, increasing their solubility and even interfering with their adsorption on activated carbon.

Photochemical Degradation

Reactions which occur as a result of sunlight activation of molecules are obviously not important below the surface of the soil. They may nevertheless have some significance for the subsurface migration of hazardous waste, because they can alter a spilled or discharged waste in the time before it percolates.

Oil spills may be particularly affected because the oil clings strongly to soil particles and may stay near the surface. Some of the very numerous organic chemicals in oil will break when they absorb ultraviolet radiation to form free radicals, molecules with unsatisfied chemical bonds. These highly energetic species will react with other chemicals and oxygen to form peroxy radicals and eventually oxygenated organic chemicals such as alcohols, aldehydes, and ketones (Bossert and Bartha, 1984). The ultimate effects may be desirable or undesirable: these species are more mobile, but they are also more rapidly biodegraded.

Humic and fulvic acids may act as photosensitizers. They absorb the energy from the light, then pass it on to other organic species which may be degraded.

VOLATILIZATION

Volatilization is a likely fate for contaminants discharged to soils. If the spill remains at or near the surface for a long time, a significant fraction may evaporate. This will occur, for example, if the soil has a very low permeability. This will affect the composition of the remaining material. In petroleum spills, the light fraction is lost most rapidly. This is the most toxic material, so evaporation may convert an oil which resists biodegradation to one which decomposes more readily.

As soon as the material has percolated below the surface, contact with the atmosphere is substantially reduced. Evaporation is limited because the vapors must move by diffusion through the soil pores to reach the atmosphere. Concentrations of the vapors near the mass of the contaminant reach saturation in the soil atmosphere, so that further evaporation is suppressed.

Reports of volatilization of oil vary. Some investigators have found up to 40 percent of the oil evaporating, while other see only 0.1 percent (Bossert and Bartha, 1984). These differences may well be real, indicating the substantial variability in environmental control of volatilization.

Gasoline contains lighter hydrocarbons, and evaporates at much higher rates than crude oil. In some cases, this may contribute to spill cleanup. If the contaminated soil is exposed to the atmosphere, much of the gasoline will evaporate, while the remainder is biodegraded. This has become a widely used cleanup method in cases of leaking underground storage tanks where the leak is of modest size. Air pollution regulations, however, will limit volatile releases in the future.

The volatility of gasoline may also contribute by making the contamination more easily traceable. Many spills have been followed without expensive drilling by inserting probes to collect vapors (Marrin, 1985). The distribution of vapors detectable a few feet from the surface mimics the distribution of gasoline contamination in the soils below. The correlation is best, however, when the gasoline is present as pure product floating on the groundwater, so that vapor concentrations are high. Where the gasoline is only adsorbed on the soil, or dissolved in the water, vapors will be released much more slowly. Biodegradation in the soils above the contaminated soil may be sufficient to destroy the vapors before they reach the surface and are detected.

Volatilization is promoted by fluctuations in the water table (Lappala and Thompson, 1985). The rising water gives the contaminants the opportunity to adsorb to soil above. When the water falls again, the soil is exposed to the air and the contaminants evaporate. Movement of vapors often constitutes a significant hazard. Toxic and carcinogenic chemicals may migrate to homes or other buildings. Gasoline vapors have sometimes caused violent explosions, with loss of life.

Piwoni et al. (1986) conducted laboratory experiments to determine the fates of organic contaminants included in treated wastewater disposed by rapid infiltration. A substantial fraction of the contaminants was released to the atmosphere, either during the application period or by evaporation from the soil between doses. Release was greatest for the volatile chlorinated solvents. 94 percent of the tetrachloroethylene and 89 percent of the 1,1,1-trichloroethane were passed to the atmosphere, and values for chloroform, 1,1-dichloroethane, trichloroethylene, and 1,2-dibromo-3-chloropropane were all above 50 percent. Small, but significant, fractions of the substituted benzenes were also evaporated. An estimated 89 percent of the 1,2,4-trichlorobenzene was transferred to the atmosphere, but the fractions for chlorobenzene, 1,2 dichlorobenzene, and toluene ranged from 14 percent to 21 percent. Phenol and some of its chlorinated derivatives were also tested, but their primary fate was biodegradation rather than evaporation.

REFERENCES

Bailey, S. J., O'Shaughnessy, J. C., and Blanc, F. C. 1986. Effects of organic compounds on the adsorption and desorption of metals. In *Environmental Engineering, Proceedings of the 1986 Specialty Conference,* Cawley, W. A., and Morand, J. M. (Eds.), pp. 334–340. New York: American Society of Civil Engineers.

Bossert, I. and Bartha, R. 1984. The fate of petroleum in soil ecosystems. In *Petroleum Microbiology,* Atlas, R. M. (Ed.), Chapter Ten. New York: MacMillan Publishing Co. 692 pp.

Devinny, J. S. 1989. Soil water content and air stripping. In *Proceedings National Conference on Environmental Engineering,* Austin, TX: American Society of Civil Engineers, July 10–12.

Dierbert, F. E. and Given, C. J. 1986. Aldicarb studies in ground waters from Florida citrus groves and their relation to ground-water protection. *Ground Water* 24(1):16–22.

Faust, S. D. 1975. Chemical mechanisms affecting the fate of organic pollutants in natural aquatic environments. In *Fate of Pollutants in the Air and Water Environments, Part 2. Chemical and Biological Fate of Pollutants in the Environment, Vol. 8.* Suffett, I. H. (Ed.), New York: John Wiley and Sons.

Fu, J. K. and Luthy, R. G. 1986. Effect of organic solvent on sorption of aromatic solutes onto soils. *J. Env. Eng.* 112(2):346–366.

Ghassemi, M. 1986. Leachate collection systems. *J. Env. Eng.,*112(3):613–617.

Gray, D. H. and Weber, W. J. 1984. Diffusional transport of hazardous waste leachate across clay barriers. In *Proceedings Seventh Annual Madison Waste Conference, Municipal and Hazardous Waste,* pp. 373–389. Madison, WI: University of Wisconsin.

Johnson, R. L., Brillante, S. M., Isabelle, L. M., Houch, J. D, and Pankow, J. F. 1985. Migration of chlorophenolic compounds at the chemical waste disposal site at Alkali Lake, Oregon—2. Contaminant distributions, transport, and retardation. *Ground Water* 23(5):652–666.

Jones, R. L. 1986. Field, laboratory, and modeling studies on the degradation and transport of aldicarb residues in soil and ground water. In *Evaluation of Pesticides in Ground Water,* Garner, W. Y., Honeycutt, R. C., and Nigg, H. N. (Eds.), Washington, DC: American Chemical Society. 573 pp.

Khan, S. U. 1975. Interaction of humic substances with herbicides in soil and aquatic environments. In *Fate of Pollutants in the Air and Water Environments, Part 2. Chemical and Biological Fate of Pollutants in the Environment, Vol. 8.* Suffet, I. H. (Ed.), New York: John Wiley and Sons.

Lappala, E. G. and Thompson, G. M. 1984. Detection of groundwater contamination by shallow soil gas sampling in the vadose zone theory and applications. In *Management of Uncontrolled Hazardous Waste Sites, Proceedings of the Fifth National Conference,* pp. 20–28. Silver Springs, MD: Hazardous Materials Control Research Institute.

Leonard, D., Unites, K. F., and Kebe, J. O. 1984. Fly ash disposal in a limestone quarry: hydrochemical considerations. In *Proceedings Seventh Annual Madison Waste Conference, Municipal and Hazardous Waste,* pp. 198–218. Madison, WI: University of Wisconsin.

Lu, J. C. S., Eichenberger, B., and Stearns, R. J. 1985. *Leachate from Municipal Landfills, Production and Management.* Park Ridge, NJ: Noyes Publications. 453 pp.

Mahmood, R. J. and Sims, R. C. 1986. Mobility of organics in land treatment systems. *J. Env. Eng.,* 112(2)236–245.

Marrin, D. L. 1985. Delineation of gasoline hydrocarbons in groundwater by soil gas analysis. In *Proceedings of the Hazardous Materials Conference/West,* pp. 112–119. Wheaton, IL: Tower Conference Management Company.

Pignatello, J. J., Sawhney, B. L., and Frink, C. R. 1987. EDB: Persistence in soil, *Science* 236(4808):898.

Piwoni, M. D., Wilson, J. T., Walters, D. M., Wilson, B. H., and Enfield, C. G. 1986. Behavior of organic pollutants during rapid-infiltration of wastewater into soil: I. Processes, definition, and characterization using a microcosm. *Hazardous Waste and Hazardous Materials* 3(1):43–55.

Pohland, F. G., Gould, J. P., and Ghosh, S. B. 1985. Management of hazardous wastes by landfill codisposal with municipal refuse. *Hazardous Waste and Hazardous Materials,* 2(2):143–148.

Quinlan, J. F., and Ewers, R. O. 1986. Data from most monitoring wells in limestone terranes are irrelevant to pollutant detection but reliable monitoring can be attained: Why and how, In *Environmental Engineering, Proceedings of the 1986 Specialty Conference,* Cawley, W. A., and Morand, J. M. (Eds.), pp. 438–444. New York: American Society of Civil Engineers.

Rao, P. S. C., Edvardsson, K. S. V., Ou, L. T., Jessup, R. E., Nkidi-Kizza, P., and Hornsby, A. B. 1986. Spatial variation of pesticide sorption and degradation parameters. In *Evaluation of Pesticides in Ground Water,* Garner, W. Y., Honeycutt, R. C., and Nigg, H. N. (Eds.), pp. 100–115. Washington, D.C.: American Chemical Society. 573 pp.

Rinaldo-Lee, M. B., Hagarman, J. A., and Diefendorf, A. F. 1984. Siting of a metals industry landfill on abandoned soda ash waste beds, In *Proceedings Hazardous and Industrial Waste Management and Testing: Third Symposium,* ASTM STP 851, Jackson, L. P., Rohlik, A R., and Conway, R. A. (Eds.), pp. 171–192. Philadelphia: American Society for Testing and Materials.

Roberts, P., 1986. Seminar at Stanford University, June 16, Palo Alto, CA.

Shuckrow, A. J., Pajak, A. P., and Touhill, C. J. 1982. *Hazardous Waste Leachate Management Manual,* Park Ridge, NJ: Noyes Data Corporation. 379 pp.

Schwarzenbach, R. P. and Westall, J. 1981. Transport of nonpolar organic compounds from surface water to groundwater. Laboratory sorption studies. *Env. Sci. and Technol.* 15(11):1360–1367.

Sutton, P. A. and Baker, J. F. 1985. Migration and attenuation of selected organics in a sandy aquifer—A natural gradient experiment. *Ground Water* 23(1):10–16.

Valsaraj, K. T., and Thibodeaux, L. J. 1988. Equilibrium adsorption of chemical vapors on surface soils, landfills and landfarms—A review. *Journal of Hazardous Materials,* 19:79–99.

Wagner, G. H. and Steele, K. F. 1985. Use of rain and dry deposition compositions for interpreting ground-water chemistry. *Ground Water* **23**(5):611–616.

Walsh, J., Vogt, G., Kinman, R., and Rickabaugh, J. 1984. Long-term impacts of industrial waste co-disposal in landfill simulators. In *Proceedings Seventh Annual Madison Waste Conference, Municipal and Industrial Waste,* pp. 166–167. Madison, WI: University of Wisconsin.

Weintraub, R. A., Jex, G. W., and Moye, H. A. 1986. Chemical and microbial degradation of 1,2-dibromoethane (EDB) in Florida ground water, soil, and sludge. Garner, W. Y., Honeycutt, R. C., and Nigg, H. N. (Eds.), In *Evaluation of Pesticides in Ground Water,* pp. 294–310. Washington, DC: American Chemical Society, 573 pp.

Young, P. J., Baldwin, G., and Wilson, D. C., 1984. Attenuation of heavy metals within municipal waste landfill sites, In *Proceedings Hazardous and Industrial Waste Management and Testing: Third Symposium,* ASTM STP 851, Jackson, L. P., Rohlik, A. R., and Conway, R. A. (Eds.), pp. 193–212. Philadelphia: American Society for Testing and Materials.

6

Groundwater Monitoring

Robert L. Stollar
Principal Hydrogeologist
R. L. Stollar & Associates, Inc.

INTRODUCTION

During the past few years, the technology associated with groundwater monitoring and sampling has developed rapidly. This development is due to federal and state regulations, public awareness, recognition of groundwater contamination as a major environmental concern, and the physical and chemical complexities associated with groundwater contamination.

Although technological advances have been made in the past few years, groundwater monitoring is still an imprecise science. This is due to the enormous number of variables, many of which have to be estimated, which affect not only the drilling and well installation process but also, in the end, the interpretation of the data. To increase our understanding of the groundwater flow and contaminant migration process, and to have a high degree of confidence in the results, it is important that consistent and highly defensible approaches to groundwater monitoring be developed.

A preliminary understanding of the hydrogeologic and chemical systems, the chemical alteration of the sample that can occur due to drilling and sample collection methods and materials, and the chemical changes that can occur due to the storage of samples, is essential before wells are located and monitoring can be initiated. Understanding that these problems occur is important to determining the appropriate procedures, methods, equipment, and materials necessary for constructing a monitoring well, collecting water-quality samples, and ultimately implementing a sound groundwater quality monitoring system. This chapter is devoted to the description of the various methods, equipment, and materials available, their appropriate use, and the advantages and disadvantages of each.

DESIGNING A GROUNDWATER MONITORING PROGRAM

As mentioned above, it is important to have a conceptual understanding of the groundwater flow characteristics at a site, and knowledge of the physical and chemical characteristics of potential contaminants and their interaction, before the monitoring program is designed. Generally, contaminants entering the groundwater system are transported by the processes of advection and dispersion. Advection is the process in which contaminants are carried along with the moving groundwater; dispersion causes contaminants to spread in directions of decreasing concentration. Therefore, the contaminants can move horizontally or vertically, depending on the dispersion, comparative densities, and natural flow patterns. They tend to travel as well-defined slugs or plumes, but can be reduced in concentration with time and distance (attenuated) by such mechanisms as adsorption, ion exchange, dispersion, and decay. The rate of attenuation or retardation is a function of the type of contaminant and of the local hydrogeologic framework. It is apparent that there are a large number of factors to be considered when designing a monitoring program. An outline of pertinent information to be evaluated is listed below:

1. Regional Hydrogeologic System

   - Identify hydrogeologic units
   - Characterize units with respect to hydraulic conductivity and type of media (porous or fractured)
   - Determine regional groundwater flow system (recharge and discharge areas, direction, velocity)
   - Climate (precipitation and evaporation).

2. Characteristics of Waste and Potential Contaminants

   - Form of contaminant (solid or liquid)
   - Type of contaminant (organic or inorganic)
   - Concentration of contaminant
   - Degree of hazard: published limits, guidelines, or statistics
   - Mobility of contaminants
   - Behavior or fate of contaminants in the hydrogeologic system.

3. Site Characteristics

   - Geologic environment (landforms, such as landslides, basin drainages, alluvial fans, etc.)
   - Geomorphic environment
   - Unsaturated zone

        recharge
        soil properties
        fate of contaminants
        depth to water

- Saturated zone

    geometry
    aquifer characteristics
    flow directions
    velocity
    recharge and discharge
    background water quality
    quantity of water
    fate of contaminants

After assumptions are made about these characteristics, or the above data are evaluated, a monitoring network can be designed and installed. The degree of difficulty in completing and installing the system is dependent on the hydrogeologic complexity of the site, the drilling methods, well construction and materials, sampling methods, sample storage, and their relationship to the specific constituent being analyzed. The above are discussed in the remaining sections of this chapter.

In order to obtain much of the data needed to design a monitoring program, it is generally necessary to construct groundwater monitoring wells. The purpose of constructing a groundwater monitoring well is to:

- Collect geologic data
- Collect groundwater data (i.e., depth to water table, aquifer characteristics, etc.)
- Collect chemical data on soil and water
- Provide for long-term monitoring capabilities

The well is completed after a borehole has been drilled. This usually involves the placement of casing and well screens, backfilling the annular space between the borehole and the casing with the appropriate gravel pack, bentonite seal, and grout or other sealant. An example of a typical well is shown in Figure 6-1. After the well is constructed, it is then developed. However, wells in hard-rock formations and some consolidated formations can be left as open holes.

Although some wells may already exist in the study area, they usually were not drilled and constructed for the purpose of groundwater monitoring, and thus could bias the data necessary for the monitoring program. Some boring programs may be divided into two phases, a reconnaissance phase and a monitoring well construction phase. The first phase requires drilling less costly, small-diameter boreholes to collect geologic data and assess the specific needs of an area. The second phase can then utilize the reconnaissance data to determine the optimum locations of the more elaborate and expensive boreholes and groundwater monitoring wells.

It should be emphasized that the results of the monitoring program will be biased by the site characteristics, drilling method, well construction methods and material, and well development. The major sources of bias during these operations include:

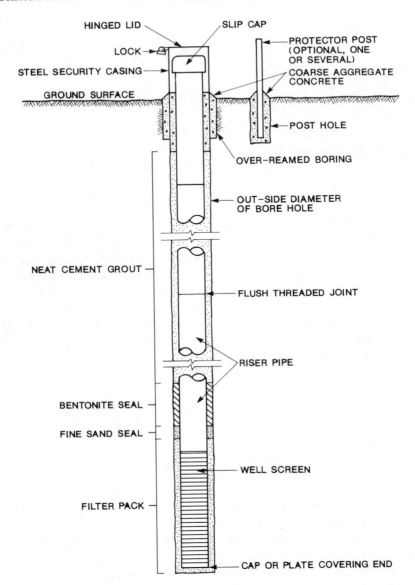

Figure 6-1. Typical monitoring well—Completed above grade.

- Physical disturbance (mixing) during drilling
- Contamination caused by drilling operations
- Contamination caused by materials used in well construction and sampling
- Adsorption onto materials contacted by sample.

For example, if the contaminant is a gasoline product, the results of a monitoring program using the mud-rotary drilling method and materials such as glued PVC casing probably would not represent actual subsurface conditions at this site. This is due to the potential contamination of samples with organic materials such as grease, PVC glue, and drilling-mud additives (unless pure bentonite is used), adsorption and desorption of chemicals on the PVC casing, and in the drilling-mud residue clogging pore space (Stollar and Hume, 1983).

In addition to the construction of the monitoring system, extreme care must be taken during sample collection, storage, and analysis. Removing water from a well has the potential, in itself, to introduce error. The method of sampling is dependent on the constituent of interest. For example, the sampling method can cause error due to contamination of the sample, degassing, atmospheric invasion, volatilization, adsorption, or desorption. The sampling method is also dependent on the hydraulic conductivity of the formation, the volume of water to be flushed from the casing, and the accessibility of the site. Typical sampling methods include bailers, syringes, suction-lift, positive displacement, gas-lift and gas-drive methods, and jet pumps.

It is apparent that there are a large number of factors to be considered when designing a monitoring program and in evaluating groundwater quality data. Some of the factors are interactive, constituent specific, and site specific. For example, sampling methods can vary from being acceptable to unacceptable, depending on the depth to water, the volume of sample required, and the parameters to be analyzed, and some sampling methods can give erroneous results for heavy metals, if iron and manganese are present in the groundwater environment.

As mentioned earlier, the drilling methods, well construction and materials, sampling methods, and sample storage can affect the results of the monitoring program. These problems are related to the specific constituent being analyzed, and therefore are complicated. Presently, much research is being carried out to solve these problems. Excellent discussions of, and solutions to, some of these problems are found in Gillham et al. (1982), Unwin (1982), Gibb et al. (1981), Miller (1982), Knirsch and Blose (1983), National Water Well Association and Plastic Pipe Institute (NWWA and PPI) (1980), Schmidt (1981), U.S. Environmental Protection Agency (USEPA) (1977), and Stollar et al. (1983).

In summary, a number of conditions must be satisfied before a groundwater monitoring system can be successful. These include:

- Hydrogeologic characterization of the site to determine the proper locations (horizontal and vertical) and appropriate number of sample points
- Appropriate monitoring device configuration, construction, and emplacement techniques to insure representative groundwater samples
- Sampling equipment and methods chosen to insure samples representative of aquifer conditions
- Techniques of sample pretreatment, handling, and analysis designed to assure integrity of the sample and validity of the final results
- Data must be interpreted to present an understanding of the groundwater system to aid in decision making and possible remedial action

To achieve the best monitoring program results with the least amount of bias, the variables associated with the well construction process are discussed below. Generally, the well is drilled by a process that will be the most efficient, allow for the best data collection, and establish a viable monitoring point. The well basically consists of the borehole excavated during the drilling process, the well screen, casing, sand or gravel pack outside the screened interval, the bentonite and grout seals, and the well completion appurtenances. These are all discussed in this chapter.

WELL DRILLING AND SOIL SAMPLING

The following section reviews well drilling and soil sampling techniques. More details related to these methods are described in the cited references.

Drilling Methods for Monitoring Wells

The drilling methods summarized in this chapter include auger, mud rotary, reverse circulation rotary drilling, dual-wall reverse circulation rotary, cable tool, driven wells, jetted wells, and jet percussion methods. Table 6–1 lists the advantages and disadvantages of each of these methods. During the drilling process, a boring log and well construction summary form should be completed (Figures 6–2 and 6–3). Further detailed discussions of the various drilling techniques are given by Todd (1980), U.S. Department of the Interior, Water and Power Resources Services (WPR) (1981), Driscoll (1986), USEPA (1977), and USEPA (1987).

Table 6–1. Drilling Methods for Monitoring Wells.

TYPE	ADVANTAGES	DISADVANTAGES
Hollow-stem auger	• No drilling fluid or water is necessary, minimizing contamination problems and the need for a water source. • Formation waters can be sampled during drilling by using a screened lead auger or advancing a well point ahead of the augers. • A core sample can be collected through the hollow stem. • Natural gamma-ray logging can be done inside the in-place augers. • Hole caving before well construction is complete can be overcome by placing screen and casing within the hollow stem before the augers are removed. • Fast. • High mobility rigs can reach most sites. • Equipment generally readily available throughout the U.S. • Usually less expensive than rotary or cable tool drilling. • Depth to water table can usually be determined while drilling.	• Can be used only in unconsoilidated materials. • Limited penetration; maximum normal depth is between 100 and 150 feet, depending on soil type. Works best to depths of 75 and 80 feet. • Soil samples may not be completely accurate, depending upon how they are taken. • Difficult to control heaving sand; cannot keep hole open below the water table in heaving sand without a knock-off plate on lead auger.

Table 6-1.   (Continued)

TYPE	ADVANTAGES	DISADVANTAGES
Mud rotary	• Can be used in both unconsolidated and consolidated formations, but generally not in cavernous rocks. • Capable of drilling to any depth depending on rig size, usually to 500-foot depth, limited by borehole diameter, mudpump capacity, and ability to maintain circulation for most portable rigs. • Coring devices for detailed sampling easy to use. • Can run a complete suite of geophysical logs in the open hole that is mud filled. • Casing not required during drilling. • Flexibility in well construction. • Fast. • Smaller rigs can reach most sites. • Equipment generally readily available throughout the U.S. • Relatively inexpensive.	• Drilling fluid is required, hence a water source is needed. The water and mud is placed in borehole, thereby adding fluids to the hole. • Contaminants may be circulated with the fluid. • In fractured, cavernous, and very coarse material, drilling fluid may be lost and therefore can not be circulated. • The fluid mixes with the formation water and invades the formation and is sometimes very difficult to remove. • Bentonite fluids may absorb metals and may interfere with some other parameters. • Depth to water table is very difficult to determine during drilling process. • Organic fluids may interfere with bacterial analysis and/or organic-related parameters. • Soil samples circulated through the mud difficult to correlate to actual depths. • Well must be developed to remove all drilling fluids and mud cakes from borehole to allow water samples to be representative of formation.
Air rotary	• No drilling fluid is used, minimizing contamination problems. • Can be used in both unconsolidated and consolidated formations; best suited for consolidated rock. • Capable of drilling to any depth, limited by the borehole diamter, compressor capacity, and the ability to maintain circulation. • Soil sampling ranges from excellent in hard, dry formations to nothing when circulation is lost in formations with cavities. • Formation water is blown out of the hole along with cuttings making it possible to determine when the first water-bearing zone is encountered.	• Casing is required to keep the hole open when drilling in soft, caving formation below water table. • Contaminants may be added to borehole if air filters not operating properly. • When more than one water-bearing zone is encountered and hydrostatic pressures are different, flow between zones occurs between the time drilling is completed and the hole can be properly cased and grouted off. • May not be economical for small jobs. • Water samples collected during drilling are aerated. • Requires a minimum 6-inch diameter hole. • May be hazardous when drilling at sites where the contamination is associated with volatile compounds.

(continued)

Table 6-1.   (Continued)

TYPE	ADVANTAGES	DISADVANTAGES
Dual-wall reverse circulation rotary method	• Fast set up and drilling, can usually drill between 40 to 80 feet per hour. • Estimates of aquifer yield can be made easily at many depths in the formation. • Fast penetration rates are possible in coarse alluvial deposits or broken or fissured rock. • Problems of lost circulation are either eliminated or reduced drastically. • Washout zones are reduced or eliminated. • Depth to water table can be determined during drilling process. • Wells can be constructed through the dual tube and thereby reduce the potential for hole collapse during well construction. • Continuous representative formation and water samples can be obtained.	• Air and oil become contaminated with grease if air filter is not working properly. • System is limited to holes greater than 9 to 10 inches in diameter. • System is limited to depths of approximately 1200 to 1400 feet in alluvial deposits (works best to 600 feet and generally up to 2000 feet in hard rocks). • Well-trained drilling crews are needed. • Equipment has limited availability. • Initial cost of drilling rig and equipment is high, and therefore drilling costs are high.
Cable tool	• Only small amounts of drilling fluid (generally water with no additives) are required. • Can be used in both unconsolidated and consolidated formations; well-suited for caving, large gravel-type formations with large cavities above the water table; can drill through boulders and fractured, fissured, broken, or cavernous rocks, which are beyond the capabilities of other types of equipment. • Wide depth range. • Water and lithologic samples can be collected easily, hydraulic conductivity tests can also be accomplished in different water-bearing zones.	• Relatively larger diameters are required (minimum 4-inch casing). • Heavy steel drive pipe must be used and could be subject to corrosion under adverse contaminant characteristics. • Cannot run a complete suite of geophysical well logs because of casing placement. • Drilling is very slow.
Driven wells	• Water samples can be collected at closely-spaced intervals during drilling. • Can expect a good seal between casing and formation only if drilling through loose, well sorted material and formation collapses around well.	• Depth limitations. Applicable to shallow wells primarily less than 50 feet. • No soil samples can be collected, the only information on subsurface material is from penetration rate. • Only certain types of pumping equipment can be used. • Drive point screen may become clogged with clay, if driven through a clay unit. • Can be used only in unconsolidated sediments. • Usually limited to well casings 2.5 inches in diameter.

Table 6–1. (Continued)

TYPE	ADVANTAGES	DISADVANTAGES
Jetted well	• Inexpensive. Light equipment. Drilling contractor not required. • Excellent for shallow boreholes in unconsolidated sediments. • Can obtain vertically-spaced groundwater sample if drive point is forced ahead of borehole and pumped. • Drilling equipment can reach almost any site.	• Slow, especially at depth. • Maximum depth of 100–150 feet. • Cannot penetrate boulders or wash up coarse gravel. • Can be used only in unconsolidated sediment. • Wash water can dilute formation water, which must be taken into account in vertical sampling. • Interpretation of geology from wash samples requires skill. • Can set only short sections of screen without difficulty.
Jet percussion	• Relatively inexpensive. • Simple equipment and operation. • Can obtain a reliable formation water sample at completed depth.	• Slow. • Use of water during drilling can dilute formation water and cause cross-contamination. • No formation water samples can be taken during drilling. • Poor soil samples due to the fact that fines are washed out of sample. • Can only be used on unconsolidated sediments or weathered rock.

Source: After USEPA, 1977; Johnson, 1984; and Driscoll, 1986.

Auger Method. Wells constructed by the auger method can either be constructed by hand-operated tools or power-driven earth augers (Figure 6–4). Hand augers operate with cutting blades on the end of a piece of pipe that bore into the ground with a rotary motion. The auger is removed when the blades are filled with loose dirt. This technique is only used at very shallow depths and in unconsolidated material.

Power-driven earth augers can be divided into three types. These are bucket augers, solid-stem augers, and hollow-stem augers. Drilling with bucket augers, also called rotary bucket drilling, is usually used for drilling large-diameter holes. This technique has a rotating bucket, equipped with auger-type cutting blades, attached to the end of a kelly bar that is rotated by large ring gear.

Solid-stem augers are generally only used to drill though a stable formation. This technique is comprised of either a single flight or continuous flight of augers. When drilling, an auger bit or cutter head is attached to the leading auger flight. Cuttings are brought to the surface of the hole by the flights which act like a screw conveyer.

The hollow-stem auger method, which is the auger method most commonly used in monitoring well drilling when soil sampling is required, uses continuous flights that operate in a manner similar to the solid-stem auger method. The hollow-stem auger consists of a section of steel tube with a spiral flight to which is attached a

R. L. STOLLAR & ASSOCIATES, INC.
FIELD LOG OF BORING

SITE TYPE	SITE ID
BORE	

SHEET ____ OF ____

PROJECT NAME AND LOCATION	PROJECT NUMBER	ELEVATION AND DATUM	
DRILLING COMPANY	DRILLER	DATE AND TIME STARTED	DATE AND TIME COMPLETED
DRILLING EQUIPMENT: METHOD		COMPLETION DEPTH	TOTAL NO. OF SAMPLES
SIZE AND TYPE OF BIT		NO. OF SAMPLES: BULK SS	DRIVE LABORATORY
DRILLING FLUID		WATER LEVEL: FIRST	AFTER ____ HOURS
SAMPLER HAMMER		HYDROGEOLOGIST/DATE	CHECKED BY/DATE
TYPE DRIVING WT. DROP			

DEPTH/FEET	SAMPLES				DESCRIPTION	USCS SYMBOL	ESTIMATED PERCENT OF			MOISTURE	CONSISTENCY	COLOR	COMMENTS
	TYPE AND NUMBER	INTERVAL	RECOVERY	BLOW COUNT			GR	SA	FI				
0													
1													
2													
3													
4													
5													
6													
7													
8													
9													
10													
11													
12													
13													
14													
15													

Figure 6-2. Field log of boring—Example form.

cutter head at the bottom. A plug is sometimes inserted into, or attached to, the lead auger to prevent soil from entering the inside of the hollow center. Adapters at the top of the drill stem and auger flight are designed to allow the auger to advance with the plug in place. Spiral flights of hollow-stem augers are pushed into the ground while being rotated. The spiral action of the augers brings the cuttings to the surface (USEPA, 1977). As the hole is drilled, additional lengths of hollow-stem flights and center stem are added. The major difference between hollow-stem augers and solid-stem augers is that the hollow flights allow drill rods, sampling tubes, well points, and construction materials to pass through their center. The cen-

R. L. STOLLAR & ASSOCIATES, INC.

WELL CONSTRUCTION SUMMARY

SITE TYPE

WELL	

PROJECT ____

PERSONNEL ____

LOCATION OR COORDS. ____ ELEVATION: GROUND LEVEL ____

TOP OF CASING ____

DRILLING SUMMARY

TOTAL DEPTH ____

BOREHOLE DIAMETER ____

DRILLER ____

RIG ____

BIT(S) ____

DRILLING FLUID ____

SURFACE CASING ____

WELL DESIGN

BASIS: GEOLOGIC LOG
GEOPHYSICAL LOG

CASING STRING (S): C = CASING S = SCREEN

CASING C1 ____

C2 ____

C3 ____

C4 ____

SCREEN S1 ____

S2 ____

S3 ____

S4 ____

FILTER MATERIAL ____

CEMENT ____

OTHER ____

CONSTRUCTION TIME LOG

TASK	START		FINISH	
	DATE	TIME	DATE	TIME
DRILLING:				
GEOPHYSICAL LOGGING:				
SCREEN PLACEMENT:				
CEMENTING:				
DEVELOPMENT:				
OTHER::				

WELL DEVELOPMENT

COMMENTS

FOPM23 / OCT 1987

Figure 6–3. Well construction summary form.

Figure 6–4. Auger drill rig—Mobile drill model B-46L22 (photo courtesy of Mobil Drilling Co., Inc.).

ter stem and plug may be removed at any time during the drilling to permit disturbed, undisturbed, or core sampling below the bottom of the cutter head by using the hollow-stem auger flights as casing (Figure 6–5).

Hollow-stem augers can also be used as a temporary casing to prevent caving and sloughing of the borehole wall during well construction. While the hollow flights are still in place, the well casing and screen can easily be lowered into place, centered, and then the sand or gravel pack, seal, and grout can be placed around the casing in the annular space of the borehole. The flights of hollow-stem augers are usually pulled out as the borehole is being grouted.

Power augers are best used when drilling through clay, silt, sand, and gravel (less than 5 cm in diameter). The usual diameter of holes drilled range between 15 to 90 cm (about 6 to 36 in.) and the usual maximum depth is 25 m (about 82 ft). The typical monitoring well diameter is between 2 and 4 in. Table 6–1 lists the advantages and disadvantages of using a hollow-stem auger.

Mud Rotary Drilling. Mud rotary drilling (also referred to as direct circulation rotary (WPR, 1981), direct rotary (Driscoll, 1986), hydraulic rotary (USEPA, 1977), and rotary method (Todd, 1980) operates by rotating a hollow bit through which

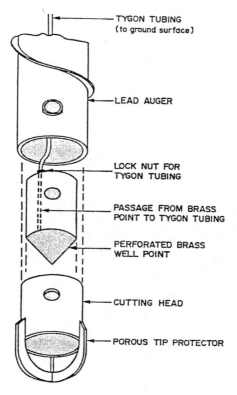

Figure 6–5. Continuous flight hollow-stem augers showing sampling tube passing through center (after Anderson, 1977).

a drilling mud or a mixture of clay and water is forced (Figure 6–6). The mud is pumped from the mud pit through the drill rods and bit and flows upward through the annular space between the borehole wall and drill rods back to the mud pit. The bit breaks up the geologic material and the mud carries these cuttings up through the annular space, between the wall of the boring and drill pipe, to the surface. The mud is discharged into a settling basin (also called the mud pit) where cuttings settle out, and the mud is recirculated down the hole through the drill pipe. The mud serves several functions besides removing cuttings from the borehole. The mud also lubricates and cools the bit. The weight of the mud provides the hydrostatic pressure to hold the hole open. Finally, a thin layer of mud is deposited on the interior of the hole which prevents loss of fluid into the formation (WPR, 1981 and Speedstar Division, Koehring Co., no date). The mud cake and hydrostatic pressure also prevent cross-contamination from occurring after the mud cake is in place.

A sample of the drilling fluid (mud) should be collected. The sample should be analyzed or retained for analyses for the constituents of interest.

The mud rotary drill string may be comprised of up to four components. These include the bit, a drill collar or stabilizer, the drill pipe and, for table-driven machines, a Kelly, or for rotary drilling machines, a top head driver.

There are two general types of bits used in mud rotary drilling. These are the drag bit and the roller cone bit (including the all-purpose tricone bit). For enlarging holes, a reamer or under-reamer is used. For a detailed discussion of drill string compo-

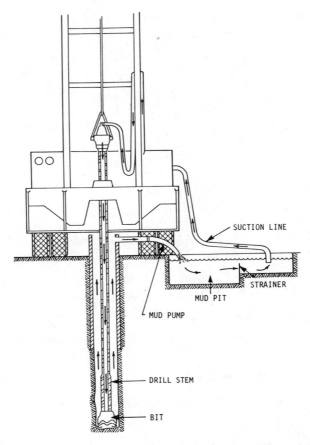

Figure 6–6. Mud rotary drilling method (after Speedstar Div., no date).

nents and drill bits, see Driscoll (1986). The typical mud rotary drill rig consists of a derrick or mast, a table or tophead drive for rotating the drill string, a pump for circulating the drill mud, drawworks, and the engine.

Mud rotary rigs are best suited for drilling through silt, sand, gravel less than 2 cm in size, and soft to hard consolidated rock (Todd, 1980). This drilling method is best suited for drilling to maximum depths of 450 m (about 1500 ft) with diameters of 8 to 45 cm (about 3 to 18 in.). In very coarse materials, such as cobbles and boulders, in highly fractured materials, and in cavernous materials, circulation of the drilling fluid may not occur, that is, the fluid may be discharged to and lost to the formation. Therefore, the mud rotary technique may not be suited for these formations. Table 6–1 provides a list of advantages and disadvantages for using this technique. As mentioned, drilling with this technique causes filter cakes to form on the borehole wall. Therefore, the monitoring well needs to be well developed to remove all filter cake material and mud from the well in order for water samples collected from the well to be representative of the formation.

Air Rotary Drilling. The air-rotary drilling method, also called direct air rotary, operates in a manner similar to mud rotary except that compressed air is circulated down through the drill string to cool the bit and to blow the cuttings to the surface. The cuttings then accumulate around the borehole.

Drilling equipment used in air rotary drilling is essentially the same as the mud rotary method. The major differences are that an air compressor replaces the mud pump, and fluid channels in the bit are uniform in diameter rather than jets (WPR, 1981). There are two key problems related to this method. First, if the air filters in the compressor are not working properly, lubricating oil can contaminate the well. Second, when using air to remove the cuttings and fluids from the hole, it is more difficult to control the flow at the well head. Therefore, if the aquifer is highly contaminated there may be safety problems.

When drilling with the air rotary method in very porous unconsolidated formations, it is often necessary to take steps to insure the integrity of the boring side wall and/or prevent the loss of air into the formation. The use of a drilling foam can help remove cuttings from the boring, reduce the loss of air into the formation, and provide some side wall support. Using foam can also affect the quality of the water sample, and therefore should be used with care. If the formation is too coarse or too unstable, it may be necessary to drive casing just behind the bit. Several methods are commonly used for this technique. The most generally available method utilizes a casing driver (usually pneumatic) to drive welded or threaded steel casing. A drive shoe is usually welded to the leading edge of the casing. A second method, the ODEX system, requires special bits and casing, but does not usually require a driver, because the special ODEX bit can overream the hole as it drills, so that the casing often falls into the hole under its own weight.

Two variations to air rotary drilling include the ability to shift to a mud pump when needed and the use of a down-hole air-hammer. Air rotary drilling can successfully be used to drill in semiconsolidated and consolidated materials. Because of this, some air rotary rigs are equipped with a mud pump to drill through completely unconsolidated materials. The down-hole air-hammer method, also referred to as rotary-percussion method or downhole pneumatic hammer, uses a pneumatic piston-hammer drill at the end of the drill pipe which rapidly strikes the rock while the drill pipe is slowly rotated. This technique is extremely effective in penetrating dense, resistant materials. A significant problem with the down-hole rotary percussion method occurs when drilling below the water table. Because water is essentially incompressible, its presence in the boring tends to absorb a large percentage of the energy from the impact of the hammer and reduces the cutting efficiency. If sufficient air pressure and volume can be provided to keep the boring free of liquid, this problem can be avoided.

Air rotary drilling techniques are best suited for penetrating silt, sand, and gravel (less than 5 cm) in a soft to hard consolidated rock (Todd, 1980). This method is ideally used to drill holes with diameters between 30 to 50 cm (12 to 20 in.) and maximum depths up to 600 m (about 2000 ft). Table 6-1 provides a list of advantages and disadvantages for using this technique.

Reverse Circulation Rotary Drilling.   Reverse circulation rotary drilling operates in a manner that is similar to the direct rotary rig. The differences between these two drilling methods is that in reverse circulation, liquid is pumped up through the drill pipe rather than down through it. This method is restricted to the use of liquid as a circulating fluid, because the hydrostatic pressure of the liquid is required to support the borehole wall.

Equipment used by reverse circulation drilling is similar to the direct circulation techniques, except that the equipment is larger. Larger compressors or mud pumps are required because boreholes are larger in diameter. Only table drives are used in reverse circulation because of the larger borehole diameter and the torque required to turn the bit.

The method minimizes disturbance to the walls of the borehole because of the higher head in the hole. It also provides a more rapid removal of cuttings, since the area of the drill rods is smaller than the annulus, thereby having higher upward velocity. A variation to reverse circulation drilling is called inverse drilling, which converts a top head drive, direct rotary machine into a reverse circulation rig (Driscoll, 1986). This technique uses a side discharge swivel assembly and drill pipe with built-in air channels. Compressed air is injected through an injection stem into air channels mounted outside the drill pipe and then into the drilling fluid as it moves up inside the pipe. The drilling fluid and cuttings are assisted to the surface with air.

Reverse circulation is best used when penetrating through silt, sand, gravel, and cobbles. Usual hole diameters required for this technique range between 40 to 120 cm (about 16 to 50 in.), with a maximum depth of 60 m (about 200 ft).

Dual-Wall Reverse Circulation Rotary Method. The dual-wall reverse circulation rotary method, or dual-wall method, is similar to reverse circulation rotary drilling (Figure 6–7). This technique uses flush-jointed, double-walled pipe in which the drilling fluid (air or liquid) circulates by reverse circulation. Drilling fluid flow is contained between the two walls of the dual-wall pipe, and it only comes in contact with the borehole wall near the bit. This is unlike reverse circulation rotary in that the drilling fluid runs down the outside of the drill pipe. Table 6–1 provides a list of advantages and disadvantages to this technique.

Cable Tool Drilling. The cable tool method of drilling, also known as the percussion, churn, or standard method, functions by repeatedly lifting and dropping a string of tools suspended on a cable. A drill bit, attached to the end of the tool string, crushes consolidated material into fragments and loosens uncemented material so that the geologic formation can be penetrated. A well casing is driven to the bottom of the hole either ahead of the bit or to the level of the bit. Well casing material is generally comprised of steel or wrought iron. The crushed and loosened material accumulates at the bottom of the borehole, inside the driven casing, and mixes with water, which accumulates from the formation or is added on an as-needed basis, to form a slurry (USEPA, 1977). Accumulated slurry is removed with a bailer or a sand pump once the drill bit's impact is lessened.

Five components comprise a full string of cable tool drilling equipment. From top to bottom, these are a drill line, a swivel socket, drilling jars, drill stem, and a drill bit. A detailed discussion on these components is given by Driscoll (1986). The drill rig used for the cable tool method consists of a trailer- or truck-mounted assembly that has a mast, a multiline hoist, a walking beam, and a motor.

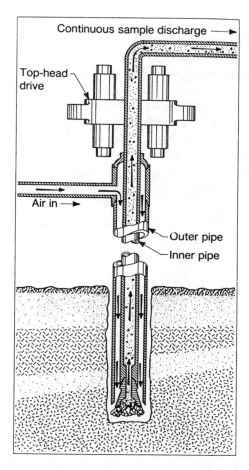

Figure 6-7. Dual-wall reverse circulation method of drilling (after Driscoll, 1986).

Cable tool drilling is best suited for penetrating unconsolidated formations and medium-hard to hard consolidated rock. This drilling method can be used to drill through fractured, fissured, broken, or cavernous rocks and boulders, which are often beyond the drilling capabilities of other techniques. Cable tool rigs are generally limited to drilling depths of less than 2000 ft (600 m) with diameters of 4 to 30 in. (10 to 75 cm) (WPR, 1981). This drilling technique can be used for most applications, except where the required casing may interfere with water-quality analysis.

In studies where detailed geologic, hydrologic, or contaminant data are needed, the casing must be driven ahead of the drill bit. Information regarding water-bearing zones can be obtained easily. Hydraulic conductivity, water quality, and lithologic data can be obtained from the different zones. Contaminated materials can be closely controlled through bailing and containment.

Wells can be constructed inside the casing driven during drilling. This casing needs to be either pulled back to a height above the well screen or entirely removed. Care needs to be given to the interactions between the casing being driven during drilling and the constituents in the water. Table 6-1 provides a list of advantages and disadvantages for this drilling method.

Driven Wells.   A driven well, or drive point well, consists of a series of connected pipe driven by repeated impacts into the ground with a sledgehammer, drive weight, mechanical vibrator, or pneumatic hammer to below the water table. At the end of these pipes is a drive point, which consists of a screened cylindrical section protected at its end by a steel cone (Figure 6–8). Driven wells are usually of a small diameter, 3 to 10 cm (about 1 to 4 in.), with most penetrating to a depth less than 15 m (about 50 ft). These wells are usually limited to unconsolidated material devoid of large gravel and rock. These wells can be constructed within a short time at minimum costs. However, very little control is obtained during drilling and therefore this method is not recommended when important geologic data is to be evaluated. Table 6–1 provides a list of advantages and disadvantages to this technique.

Jetted Well.   The jetted well or wash boring is drilled partly by the chopping and twisting action of a chisel-shaped bit and partly by the jetting action of water pumped through the drill rod and out of the bit (Figure 6–9). The high-velocity water stream and twisting action of the bit wash the soil aside, allowing the well

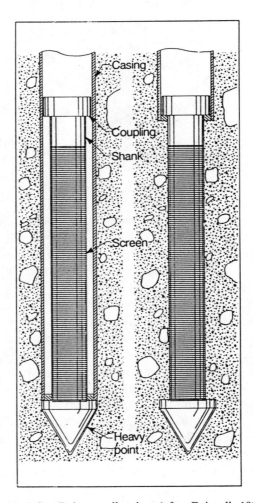

Figure 6–8.   Driven well points (after Driscoll, 1986).

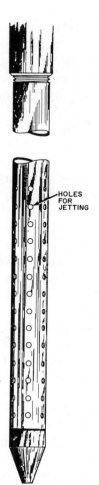

HOLES
FOR
JETTING

Figure 6–9. Jet-drive point (after Departments of the Army and Air Force, 1975).

casing to settle into the hole by its own weight. Water circulating in the annular space between the drill rod and casing carry the cuttings to the surface.

A variation of jetted wells is jet percussion drilling. The major difference between this drilling technique and jetted wells is that the casing is driven into the borehole with a drive weight rather than being allowed to advance by its own weight.

Jetted wells are best adapted to small diameter holes 3 to 10 cm (about 1 to 4 in.), and to depths not greater than 15 m (about 50 ft). Due to the injection of water, this method changes the water chemistry around the borehole. It also has similar disadvantages as the driven well. This technique is best suited for drilling through unconsolidated sediment. Table 6–1 provides a list of advantages and disadvantages for this method.

Soil Sampling Techniques

Sampling and identification of subsurface materials involves many different procedures. These are influenced by hydrogeology, geology, types of contaminants, geographical conditions, purpose of investigation, the type of drilling equipment used, and by the background, training, and experience of the field investigators. Soil samples for hazardous waste studies are required to characterize hydrogeologic conditions, soil contamination conditions, and potential transport pathways to groundwater. During site investigations, soil samples are collected for lithologic interpretation, field headspace analysis, hydrodynamic and solute transport analysis, and/or laboratory chemical analysis. Soil samples are also collected for analyses of grain size and gradation relationships of granular materials to aid in proper well design. For hazardous waste studies and chemical analysis, it is important that soil sampling devises obtain relatively chemically undisturbed soil samples. The drilling method chosen and the corresponding sampling device should be capable of yielding the necessary quantity and quality of samples. Soil samples may be collected by using hand tools or during drilling operations, using drive samplers, core barrels, or from cuttings generated during drilling. Commonly used varieties of hand tube-type samplers include soil sampling tubes and Veihmeyer tubes (also called King tubes). Drive samplers, such as thin-walled tube samplers, split-spoon samplers, and piston samplers are the most useful in obtaining soil samples for hazardous waste investigations. All of these devices are tube type samplers. In addition, diamond core conventional barrels and wire line core barrels, which have been used for many years in the oil industry, are useful in obtaining a continuous sample record for lithologic descriptions. These are also useful for hazardous waste studies, if such an elaborate borehole is required. Due to advancing technologies, many variations to these sampling techniques and devices are currently on the market. Many of these sampling devices are described in the American Society for Testing and Materials (ASTM) *Book of Standards for Geotechnical Sampling*. However, the techniques used in hazardous waste investigations do not strictly adhere to ASTM specifications, since these standards are generally not mandatory for sampling soil for hazardous-waste studies. This includes the data collected from such a device; such as blow counts. Only a general description of the sampling techniques and devices is discussed here. For specific ASTM standards, please consult the ASTM manual.

Regardless of the sampling method used, all sampling equipment should be decontaminated between each sampling event, following procedures described in the subsection on Decontamination of Drilling and Soil Sampling Equipment on page 214. Each hole should be carefully logged with regard to depth and material as the samples are obtained. Samples of unconsolidated material taken by drive sampling or coring should be described as either homogeneous, layered, stratified, and so forth, and the material's nature, thickness, and color should be recorded. Highly disturbed samples of unconsolidated material, such as cuttings obtained with a cable tool, rotary, reverse circulation rig, or grab samples from spoil piles, usually represent a mixture of the materials in the interval sampled. These samples should be examined and described carefully. The disturbed soil samples are useful only as supplemental information or confirmation of geology and of the presence of chemical contaminants. For uniformity, all unconsolidated materials should be logged in accordance with "Description of Soils—Visual Manual Procedure" (ASTM D2488-69), which is based on the Unified Soil Classification System. Drill core samples of consolidated rock should be identified regarding the type of rock, color, cementation, fractures and other similar characteristics. A detailed boring log for each boring should be prepared, using a field log of boring similar to that in Figure 6–2. For surface probes or grab samples, the description should be recorded and logged in a field notebook. All sample containers should be labeled by writing on them with an indelible marker or by affixing a label. The information on the sample tube should include the project number, project name, date of sampling, boring number, sample number, zone of sampling, and any other information that the field engineer or geologist feels is pertinent. After samples are labeled, an accompanying Chain of Custody Record form should be filled out, following the procedures detailed in the subsection on Sampling Procedures on page 248. After labeling, the sample should be stored in an ice chest at 4°C until it can be transferred to the laboratory.

Hand-Operated Soil Samplers.

Soil Probes. Soil-sampling probes or tubes, as described by Everett and Wilson (1977), consist of a hardened cutting tip, a cut-away barrel, and a threaded adaptor. The tube is attached with an adaptor to sections of rods to attain the requisite sampling depth. A tee handle used for turning is attached to the uppermost segment.

The cut-away barrel is designed to facilitate examining soil layering and to allow for the easy removal of soil samples. Generally, the tubes are constructed from high-strength alloy steel (Clements Associates Inc., 1983); often, stainless steel. The sampler is available in three common lengths, 12 inches, 15 inches, and 18 inches. Two modified versions of the tip are available for sampling in either wet or dry soils. Depending on the type of cutting edge, the tube samplers obtain samples varying in diameter from $\frac{11}{16}$ inches to $\frac{3}{4}$ inches.

Extension rods are manufactured from lightweight, durable metals. Extensions are available in a variety of lengths, depending on the manufacturer. Markings on the extensions facilitate determining sample depths.

Sampling with these units requires forcing the tube by hand in vertical increments into the soil. When the tube is filled at each depth the handle is twisted and the assembly is then pulled to the surface. Commercial units are available with attachments which allow foot pressure to be applied in order to force the sampler into the ground.

Veihmeyer Tube. In contrast to the soil probe, the Veihmeyer tube consists of a long, solid tube which is driven (by hand) to the required sampling depth. Components of the Veihmeyer tube consist of a bevelled tip which is threaded into the body tube. The upper end of the cylinder is threaded into a drive head. A weighted drive hammer fits into the tube to facilitate driving the sampler into the soil. Slots in the hammer head fit into ears on the drive head. Pulling or jerking up on the hammer forces the sampler out of the cavity.

The components of this sampler are constructed from hardened metal. The tube is generally marked in even, depth-wise increments (Everett and Wilson, 1977).

Drive Samplers.

In some circumstances, it may be desirable to obtain a relatively "undisturbed" sample. The sampling tubes described in the previous sections may be suitable for some hazardous waste studies, but have a limit to their depth of extraction. An alternative method is to use the so-called drive samplers, which may be hand- or power-driven. Drive samples and/or drill cuttings should be inspected at specified intervals and at significant changes in lithologies. In general, the number of samples and the sampling intervals should be decided upon prior to the commencement of field operations, and should be based on data collected during preliminary investigations. However, where major changes in lithology and/or where obvious contamination occurs, the geologist should be prepared to collect additional samples. Therefore, additional numbers of sampling tubes and other related equipment should be on-site.

Split-Spoon Sampler. One of the most frequently used samplers, which facilitates examining and removing the sample, is the split-tube sampler (refer to ASTM D1586), also called the split-spoon sampler, or split-barrel sampler. In this unit, a thick-walled steel tube is split longitudinally. During sampling, the two halves are placed together with a hardened drive shoe (cutting tip) which is threaded onto the lower end of the tube. The drive head assembly, including a waste barrel and a check ball valve, is screwed onto the other end (Everett and Wilson, 1977). When a boring is advanced to the point that a sample is to be taken, drill tools are removed and the sampler is lowered into the hole on the bottom of the drill rods.

The sampler is then driven into the ground. For geotechnical purposes, the sampler is driven 18 inches into the ground, as described in the standard penetration test (ASTM D1586) and the number of blows necessary to drive the sampler every six inches with a 140 pound hammer are counted. The standard-size split-spoon sampler has an inside diameter (ID) of $1\frac{3}{8}$ to $1\frac{1}{2}$ inches. When soil samples are taken for chemical analysis, it may be desirable to use a 2 or $2\frac{1}{2}$ inch ID sampler, which provides a larger volume of material, but cannot be used to calculate many of the physical properties as stated in the ASTM test method. With this sampler, recovery is best with cohesive soils. Split-spoon samplers are used with retainers or sand traps on the end of the spoon to collect the more cohesionless soil samples. Cohesionless soil samples below the water table, or in very soft soils, are difficult to collect with these samplers.

Some split-spoon samplers are available with a thin-walled split shell or rings (about 2 inches in length), or a split barrel which houses thin-walled sleeves (refer to ASTM D3550), often called a ring-lined barrel sampler. The rings or liners are commonly made of stainless steel, brass, or Teflon, and should be cleaned prior to each use. After the sampler is withdrawn from the borehole, it is opened and the rings can be set out. If possible, the ring to be submitted for chemical analysis should be one from the middle and completely full with soil. Each end should be covered with Teflon tape and secured by fitting a tight cap over the ends and then taping the cap onto the ring with duct tape. These types of sampling rings are useful for collecting samples for volatile organic chemicals. The soil in the other rings can be used for visual inspection and description, for testing physical properties, and can be placed in a glass jar for headspace analyses.

Additional samples, for nonvolatile analyses, can be placed in appropriately sized decontaminated sealable jars and labeled. All samples are kept out of direct sunlight and stored at about 4°C until shipped to the laboratory for analyses. The split-spoon sampler and rings must be decontaminated prior to using them for any sampling event.

All samples should be carefully labeled, including top and bottom (relative to location in boring), project name, number, boring number, date, time, depth at top and bottom of sample interval, recovery, number of sample, and any obvious zones of contamination. For samples for physical analysis, the blow counts should also be entered on the label. After samples are labeled, an accompanying Chain-of-Custody Record form should be filled out, following procedures detailed in the subsection on Sampling Procedures on page 258. After labeling, the sample should be stored in an ice chest at 4°C until it can be transferred to the laboratory. The storing and shipping of samples collected for physical analyses is described in ASTM D1586.

Thin-Walled Tube Sampler.   There are different types of thin-wall samplers, such as the "Shelby tubes," "Z" tubes, and "UD" (undisturbed) tubes. The sampling requirements for the geotechnical use of this sampler are described in the ASTM manual under method ASTM D1587. These tube samplers are commonly used with many types of drilling rigs, but samples can, at times, be obtained with hand-driven equipment. A common variety of this sampler consists of a thin-walled seamless steel tube, with a beveled cutting tip, and a head unit threaded to fit a standard drill rod. The head contains a ball and check valve for releasing air from the cylinder during sampling, which helps to hold the sample in the tube when it is being withdrawn from the ground. The tubes are constructed of steel about 1 millimeter (mm) thick (for tubes 2 inches in diameter) or 3 mm thick (for tubes 5 inches in diameter). Thin-walled tube samples are obtained by several different methods, including pushed tube, Pitcher sampler, Denison sampler, and Piston sampler methods. Choosing the most appropriate method requires planning and that the field personnel use good judgment (Acker, 1974). Since the purpose of thin-walled tube sampling is to obtain the highest quality undisturbed samples possible, special care should be taken in all sampling, handling, packaging, and shipping of these samples.

In obtaining pushed-tube samples (one method for a thin-walled tube sampler), the tube is advanced by hydraulically pushing in one continuous movement with the drill rig. The hydraulic pressure should be recorded. At the end of the designated push interval, and before lifting the sample, the tube is twisted to break the seal at the bottom of the sample.

Upon recovery of a thin-walled tube, the actual length of sample should be measured and recorded (excluding slough or cuttings). At least $\frac{1}{2}$ inch of soil should be cleaned from each end of the tube, and the ends of the soil sample squared off. Usually the top of the sample will contain cuttings or slough. These should be removed before sealing. The soil that has been cleaned from the tube can be used for a visual classification of the sample. The resulting space at each end of the tube is filled with melted sealing material, such as approved wax, or with expandable packers. Previously decontaminated Teflon or stainless steel plugs are also used. The ends of the tube can then be cut off or be closed with tight-fitting metal or plastic caps, and the seam between the cap and tube is wrapped with tape. Finally, the ends can be dipped in hot wax, completely covering the tape to ensure sealing. The tube is marked top and bottom, so that the orientation of the soil sample is known.

The sample containers are labeled carefully. The labeling is discussed above in the sub-subsection on Drive Samplers.

Samples from the tubes are used for physical testing, such as hydraulic conductivity, soil moisture, and other geotechnical properties, as well as for chemical analyses. The samples are extruded from the tubes in the laboratory.

Care during shipping of the samples is important for both physical and chemical analysis. The tubes should be checked for any volatilization of contaminants with organic vapor analyzers, even if no visual signs of contamination were evident. All federal, state, and local laws and regulations related to shipping of hazardous materials should be followed. Preparation for shipping samples for chemical analysis is discussed in the subsection on Soil Sampling Techniques on page 187.

Samples being shipped for physical analysis should be carefully padded and wedged to prevent movement and to minimize vibration. The boxes should be marked "Fragile" and "Keep From Heat and Freezing." All packaging of tubes should be supervised by the field geologist. If the field geologist believes that the samples have been disturbed during shipment, the project manager and soils laboratory should be notified in writing.

Core Barrel Samplers.

Conventional Core Barrel Samplers.   Another technique for obtaining relatively undisturbed soil samples is by coring (refer to ASTM D2113). This method is more commonly used with a rotary-type drill rig, but can be used with other drilling techniques. The objective of the core boring is to obtain samples of geologic material in a core barrel that is rotated. The sampler is termed the core barrel. There are both single- and double-tube core barrels. In the latter type, the inner tube retains the core and does not rotate with the outer tube. The rotating tubes of core barrels usually have drill bits at their cutting ends or are attached to the drilling bits or augers. The cutting surface makes an annular opening in the geologic material and the core barrel gradually slides down into this opening. Therefore, the core barrel contains the rock or soil material cut away from the host material. The core barrel is then lifted to the ground surface either by removing the drill string or by a wire line. The core is kept in the barrel during the lifting process either by relieving pressure above the sample, for example by a ball check valve or piston, or by actually supporting the core from below with attachments such as a core catcher or core spring.

Wire Line Core Barrel Sampler.   Wire line sampling devices can be used with rotary-type or auger drilling methods. With this method, continuous cores can be taken as the bit or auger advances. A thin-walled sample tube and special latching mechanism are placed near the bit or within the lowest hollow stem auger. With this type, the latching arrangements permit the tube to remain stationary while the bit and drill stem or auger rotate. When the sample tube is full, it is pulled to the surface by a wire line drum hoist and can be exchanged for an empty sampler.

Wireline coring has some advantages over conventional coring. One of these is that the whole string of tools does not have to be brought to the surface each time the core barrel is emptied. Thus coring at depth is much faster and easier. The disadvantage is that the bit or auger remains in the hole and the condition of the bit is not observable.

Wire Line Piston Core Sampler. One of the field problems that has been recognized in groundwater contamination studies is to successfully collect relatively undisturbed core samples from below the water table in deposits of cohesionless sand and gravel. Conventional coring techniques that involve thin-walled or split-spoon samplers are generally not suitable because of the poor recovery of saturated sands and gravels due to the cohesionless nature of the material. As the sample is lifted to land surface, the sand and gravel fall from the sampling tube. A core barrel referred to as the Waterloo Cohesionless Aquifer core barrel, and described by Zapico (1987), was developed to provide a technique for collecting samples of considerable length within a liner and with minimal disturbance. The device is able to collect samples of cohesionless material from below the water table.

Again, these core samples can be used for both defining physical properties, and for chemical analyses of the material. Therefore, storing and shipping procedures are the same as discussed above.

The Waterloo coring device can be used in conjunction with conventional hollow-stem augers, but can also be operated efficiently with a cable tool and rotary-drilled boreholes. The rotary rig must be equipped with a hammer. The core barrel consists of an exterior barrel, an interior sample tube, a hardened-steel drive shoe, a piston, the wire line, a drill-rod adapter sub and a drive head. The inner liner is made of inexpensive aluminum or plastic tubing, and the exterior steel housing protects the liner when the core barrel is driven into the aquifer. The core barrel, which is approximately 1.6 m (5.6 feet) long, is attached and lowered to the coring depth inside the hollow-stem augers or other casings. It is advanced ahead of the lead auger by hammering at the surface on drill rods that are attached to the core barrel. After the sampler has been driven 1.5 m (5 feet), the drill rods are detached and a wire line is used to hoist the core barrel, with the sample contained in the aluminum or plastic liner, to the surface. A vacuum developed by the piston during the coring operation provides for good recovery of both the sediment and aquifer fluids contained in the sediment. In the field, the sample tubes can be easily split along their length for on-site inspection, or they can be capped with the pore water fluids inside and transported to the laboratory. The cores are 5 cm (2 in.) in diameter by 1.5 m (5 ft) long. Core acquisition is obtainable to depths of 35 m (115 ft), with a recovery rate of greater than 90 percent. A large-diameter (12.7 cm [5 in.]) version has also been used successfully (Zapico, 1987). This sampler minimizes the use of drill rig time and also retains the original pore water within the sand or gravel contained inside the liner. Retention of the pore water is a feature that offers advantages in studies of groundwater chemistry and microbiology.

Cuttings and Wash Samples. During the use of any of the drilling methods, soil or rock fragment samples are lifted to the surface through the drilling process. These samples, called cuttings or wash samples, are disturbed and aerated. Cuttings or wash samples are usually collected as the boring is advanced. These samples are used to visually inspect the material for lithologic descriptions, and are generally not used to determine either physical or chemical properties of the geologic material. The material can be handled and packaged as outlined in the section dealing with split-spoon samples. An estimate of the depth from which the sample was obtained should be recorded on the log sheet. Cutting samples are usually taken every five feet. Samples should be labeled in the manner outlined above.

MONITORING WELL COMPLETION

The above material discussed the methods of drilling the borehole. Once the borehole is completed, the well is constructed. The following sections discuss the completion of the well, well casings and screens, filter packs, grouting, and well development.

Material Selection for Groundwater Monitoring Well Casings and Screens

Advances in ground-water monitoring technology have led to a number of materials being used to construct monitoring well casings. These materials include mild steel, stainless steel, polyvinyl chloride (PVC), Teflon,[1] polypropylene, and kynar. Several factors must be considered in order to determine the best material for the environmental conditions expected at a site. These factors include well depth and diameter, construction techniques, material strength, groundwater corrosiveness, microbiological activity, sorptive/desorptive properties of the chemical species in question, and material cost. This section will discuss the advantages and disadvantages of different screen and casing materials for each of these factors.

The depth and diameter of a monitoring well determines the amount of stress and temperature that the casing will be exposed to. Deeper wells with wide diameters would require casings to be resistant to stress exerted by the weight of the pipe string, unconsolidated sediment, and an increase in the temperature with increasing depth. Table 6–2 lists the advantages and disadvantages of each material, in response to stress and temperature.

The groundwater chemistry and the interaction with the material being selected are other factors that must be examined before selecting the material for constructing a monitoring well (casing and screen) or material for the sampling tools or sampling methods. Table 6–3 lists the advantages and disadvantages of materials when exposed to corrosive groundwater (groundwater that can cause corrosion by galvanic and electrochemical effects). The monitoring well casing should not act as a catalyst for chemical reactions or leach constituents into or adsorb contaminants from the groundwater.

Major ions in the groundwater such as calcium, magnesium, chloride, and sulfate do not create a serious threat to the integrity of a groundwater sample. However, when trace metals or organics are present, great care must be taken in the selection of materials that will come in contact with the sample. The interaction potential of the groundwater with the material being selected must be reviewed (Gillham et al., 1982).

Plastics are acceptable for sampling inorganic constituents, and are resistant to most naturally occurring compounds and to corrosion by galvanic and electrochemical effects (National Water Well Association (NWWA) and Plastic Pipe Institute (PPI), 1980). However, many plastics are highly susceptible to attack by organic

---

[1] A registered trademark of DuPont, Inc.

Table 6-2. Advantages and Disadvantages of Different Monitoring Well Construction Materials for Stress and Temperature.

MATERIAL	ADVANTAGES	DISADVANTAGES
1 Mild steel	Strong, rigid; not temperature sensitive.	Heavier than 3, 4, and 5.
2 Stainless steel	High strength; not very temperature sensitive.	Heavier than 3, 4, and 5.
3 PVC	Lightweight.	Weaker, less rigid, and more temperature sensitive than materials 1 and 2.
4 Teflon	Lightweight.	Tensile strength and wear resistance low when compared to materials 3 and 5.
5 Polypropylene	Lightweight.	Weaker, less rigid, and more temperature sensitive than materials 1 and 2.

Source: Adapted from Johnson, 1984 and Driscoll, 1986.

compounds, and thus should be chosen carefully when sampling for potential organics (Gillham et al., 1982). Sources of contamination by some plastics include: leaching of accumulated surface deposits, plasticizers used in the manufacturing process, and organics and inorganics from adhesives, solvation, which causes a release of organic constituents to the groundwater, as well as the removal of constituents from the groundwater, and diffusion through the plastic (Gillham, et al., 1982).

Metals can contaminate the groundwater samples in a variety of ways. Metallic

Table 6-3. Advantages and Disadvantages of Different Monitoring Well Construction Materials for Groundwater Corrosiveness.

MATERIAL	ADVANTAGES	DISADVANTAGES
Mild steel	———	May react with and leach some constituents into groundwater.
Stainless steel	Excellent resistance to corrosion and oxidation.	May corrode and leach some chromium in very acidic water. May act as a catalyst in some organic reactions.
PVC	Excellent chemical resistance to weak alkalies, alcohols, aliphatic hydrocarbons,[1] and oils. Good chemical resistance to strong mineral acids, concentrating oxidizing acids and strong alkalies.	May adsorb some constituents from groundwater. Poor chemical resistance to ketones, esters, and aromatic hydrocarbons.[2] May leach plasticizers.
Polypropylene	Excellent chemical resistance to mineral acids. Good to excellent chemical resistance to alkalies, alcohols, ketones, and esters. Good chemical resistance to oils.	May react and leach some constituents into groundwater.
Teflon[3]	Outstanding resistance to chemical attack. Insoluble in all organics except a few exotic fluorinated solvents.	
Kynar	Resistant to most chemicals and solvents.	Poor resistance to ketones, acetones.

[1]Aliphatic hydrocarbons such as: propane, butane, hexane, octane.
[2]Aromatic hydrocarbons such as: xylene, benzene, toluene, ethylbenzene.
[3]Teflon is a registered trademark of DuPont, Inc.
Source: Adapted from Johnson, 1984 and Driscoll, 1986.

materials can: leach heavy metals to the water since they are somewhat soluble; release by-products of the deterioration process, such as oxidation and corrosion; and also serve as an adsorbing surface for organic and inorganic constituents of the groundwater (Gillham et al., 1982). Lubricants used on metallic moving parts during fabrication and storage, or in the assembly process, are also a potential source of contamination unless thorough cleaning of all metal parts is performed.

In instances where more than one casing material is acceptable, the short- and long-term costs will be a deciding factor. Low-priced material such as mild steel, PVC, and polypropylene can be used in most cases. Kynar is considered medium-priced, while Teflon and stainless steel are the most expensive. The expense of higher-priced material, however, may be recouped with time due to lower long-term maintenance or replacement costs. Table 6-4 is a summary of the advantages and disadvantages to the different casing and screen materials, for correlation and reference.

Table 6-4. Well Casing and Screen Materials.

TYPE	ADVANTAGES	DISADVANTAGES
PVC (Polyvinyl chloride)	• Lightweight. • Excellent chemical resistance to weak alkalies, alcohols, aliphatic hydrocarbons, and oils • Good chemical resistance to strong mineral acids, concentrated oxidizing acids, and strong alkalies. • Slotted casing and screens are readily available. • Low priced.[1] $1.47 per ft. for 2-inch diameter schedule 40 pipe, flush threaded joint. $3.13 per ft. for 2-inch diameter schedule 80 pipe, flush threaded joint. $2.30 per ft. for 2-inch diameter schedule 40 slotted pipe, flush threaded joint. $3.44 per ft. for 2-inch diameter schedule 80 slotted pipe, flush threaded joint. $15.13 per ft. for 2-inch diameter wire-wound continuous slot screen.	• Weaker, less rigid, and more temperature sensitive than metallic materials. • May adsorb some constituents from groundwater • May react with and leach some constituents into groundwater. • Poor chemical resistance to ketones, esters, and aromatic hydrocarbons.
Polypropylene	• Lightweight. • Excellent chemical resistance to mineral acids. • Good to excellent chemical resistance to alkalies, alcohols, ketones, and esters. • Good chemical resistance to oils. • Fair chemical resistance to concentrated oxidizing acids, aliphatic hydrocarbons, and aromatic hydrocarbons.	• Weaker, less rigid, and more temperature sensitive than metallic materials. • May react with and leach some constituents into groundwater. • Poor machinability—it cannot be slotted because it melts rather than cuts.

Table 6–4.   (Continued)

TYPE	ADVANTAGES	DISADVANTAGES
	• Low priced.[1] $3.10 per ft. for 2-inch diameter schedule 80 pipe, flush threaded joint. $4.43 per ft. for 2-inch diameter schedule 80 slotted pipe, flush threaded joint.	
Teflon[2]	• Lightweight. • High-impact strength. • Outstanding resistance to chemical attack, insoluble in all organics except fluorinated solvents.	• Tensile strength and wear resistance low in comparison to other engineering plastics. • Expensive.[1] $25.00 per ft. for 2-inch diameter schedule 40 pipe, flush threaded joint. $32.80 per ft. for 2-inch diameter schedule 40 slotted pipe, flush threaded joint.
Kynar	• Greater strength than Teflon. • Resistant to most chemicals and solvents.	• Not readily available. • Poor chemical resistance to ketones, acetone. • Moderately expensive: lower-priced than Teflon.[3] $15.00 per ft. for 2-inch diameter schedule 40 slotted pipe, plain square ends. $10.00 per ft. for 2-inch diameter schedule 40 pipe, plain square ends.
Mild steel	• Strong, rigid, temperature sensitivity not a problem. • Readily available. • Variably priced.[1] $1.62 per ft. for 2-inch diameter schedule 40 pipe, flush threaded joint. $34.10 per ft. for 2-inch diameter wire-wound continuous-slot screen.	• Heavier than the plastics. • May react with and leach some constituents into groundwater. • Not as chemically resistant as stainless steel.
Stainless steel	• High strength at great range of temperatures. • Excellent resistance to corrosion and oxidation. • Readily available. • Moderately expensive.[1] $10.00 per ft. for 2-inch diameter special monitoring pipe. $32.80 per ft. for 2-inch diameter wire-wound continuous-slot screen.	• Heavier than plastics. • May corrode and leach some chromium in very acidic waters. • May act as a catalyst in some organic reactions.

[1]Prices are list prices and may vary somewhat depending upon manufacturer and are based on 1989 prices.
[2]Teflon is a registered trademark of DuPont, Inc. (Trade name of one of the fluorocarbons that are fluorinated and polymerized derivatives of ethylene or propylene.)
[3]Prices are based on 1984 prices.
Source: After Johnson, 1984 and Driscoll, 1986.

Well Screen Types

Well screens for monitoring purposes are placed in the well opposite the interval of the water-bearing zone or aquifer that is to be sampled. Certain criteria, however, should be considered prior to choosing a particular type of screen or screen material. Criteria for production or extraction wells, as described by Todd (1980) and Driscoll (1986), are usually different than criteria for selecting screens for monitoring wells.

The screen criteria chosen should prohibit entry of aquifer material, allow water to enter the well freely, and provide structural support for both the column load of the well and the surrounding loose formation material (collapse strength). In order to obtain the optimum well design, the diameter of the well, the slot size of the screen, screen material type, and the screen design (Figure 6–10) should all be evaluated with respect to the objective of sampling, such as well yield and contaminants of interest.

The diameter of the well can be varied some, but should be based on the desired well yield. Diameters of wells range in size from $1\frac{1}{2}$ to 60 in. The smaller diameter screens generally cost less to purchase, but well efficiency is also less. Before the

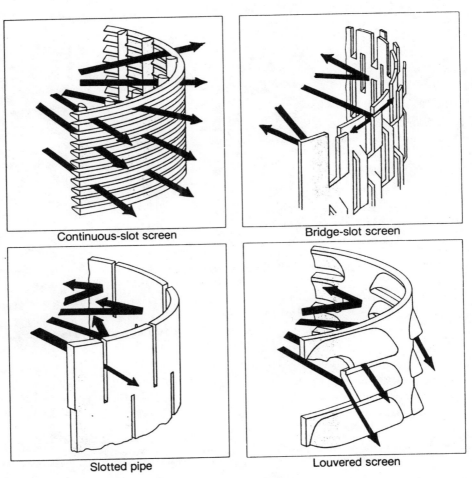

Continuous-slot screen   Bridge-slot screen

Slotted pipe   Louvered screen

Figure 6–10. Well screen slot designs (after Driscoll, 1986).

final diameter is selected, the sampling device and pumping system should also be decided upon.

The slot size, or the size of the screen opening, is another factor to consider. Production wells or extraction wells should base slot size on the grain-size analysis performed on samples obtained from the aquifer. These samples are usually taken during the drilling of a pilot borehole. If an artificial gravel pack is used, the gravel or sand size is also based on the grain-size analysis and the well screen slot size is then based on the gravel pack size and the grain-size analysis (Driscoll, 1986 and Todd, 1980). Monitoring wells usually need to be designed without the benefit of drilling pilot boreholes. Usually, there is not enough time for samples to be analyzed to determine grain size. Thus, geologic and aquifer information is acquired or estimated, preferably from existing wells on the site, borehole logs in the area, which are obtained from state and local agencies, or geologic reports of the region predicting the depositional environments. These sources provide general characteristics of the aquifer, which should enable adequate preselection of screen slot size, and thus the gravel pack. Formation grain sizes that are predominantly fine-grained are compatible with a 10-slot screen (or 0.010 in. slot); in fine-grained sand aquifers, a 20-slot screen would be appropriate; and in coarse-grained aquifers, such as gravel with sand, a 30-slot screen would be appropriate. Variations may be required under unforseen field conditions that may warrant a different screen slot size or, preferably, a modification of the artificial gravel pack. In these situations, since well construction occurs promptly after borehole drilling, modifications of the artificial gravel pack can usually be altered to less than ideal, yet still allow the well to function as a monitoring well. Ideally, the slot size should retain 40 to 50 percent of the aquifer material (Driscoll, 1986) or 95 percent of an artificial gravel pack. For best well efficiency (maximum water production with minimum side-effect), the percentage of open area in the screen should be the same as, or greater than, the average porosity of the aquifer material.

In large production wells, drawdown due to head loss is a function of the total open area of the screen; the smaller the open area, the greater the drawdown for a certain yield (Todd, 1980). That is, given a specific amount of water production, the smaller the screen open area, the less efficient the well, or the lower the water level surrounding the well and, ultimately, the water level in the well is lowered. Although screens with a small amount of open area may be less expensive to purchase, they usually have higher operational expenses due to the higher cost of lifting water to the surface. The larger the slot opening, the lower the entrance velocity of water, thus there is minimum head loss, which minimizes drawdown due to head loss in the well at a given pumping rate. The reduced velocity also reduces the corrosive ability of water to attack the screen openings. Slot sizes vary with screen type, and the screen material, or they can be custom sized, including multiple slot sizes on one section of screen. The latter type is made for nonhomogeneous formations according to individual stratum.

The screen strength should also be considered when selecting a screen type (Figure 6–11). Certain screen types offer little structural support, especially with increasing depths or increasing slot size. It follows then that excessive screen strength (above optimum design) reduces the open area of the screen.

The screen lengths for production wells are chosen based on the thickness of the

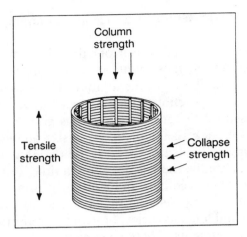

Figure 6–11. Screen tensile, column, and collapse strengths (after Driscoll, 1986).

aquifer, available drawdown, and stratification of the aquifer (Driscoll, 1986). Screen lengths for monitoring wells are chosen by the area of the aquifer that is of interest. The total screen length typical for monitoring wells usually ranges between 5 and 20 ft. Most screens are manufactured in lengths of 5, 10, and 20 ft.

The screen type can be selected based on estimates of the optimum well design. Some screen types are only made from certain materials, which may limit the selection if the construction material does not meet the desired needs. Well screen materials are discussed in the previous section of this chapter. Again, in completing monitoring wells, as soon as the drilling is complete, the screen and casing are placed in the borehole. Therefore, all casing and screen material need to be at the site before drilling is complete. This means that the well is designed before it is drilled and the sizing of screens and gravel packs are estimated. The following discussion provides information on the different well screens available and the advantages and disadvantages of each (Table 6–5). There are five different types of well screen used for monitoring wells; field slotted pipe, factory slotted perforated pipe, wire-wound perforated pipe, manufactured louvre-type or bridge-slot screen, and wire-wound continuous-slot screens.

Field-Slotted Pipe. Field-slotted well screen can be made by hand with a Mills knife or similar tool (WPR, 1981), sawing, machining, or torching. The screens made by hand generally perforate the casing in place and thus have many disadvantages (Table 6–5) which make them undesirable for use in monitoring wells. The screens made by the latter three methods can have satisfactory slots if properly sized and entrance velocity limits can be met (USDI, 1981). However, slots are susceptible to corrosion, have rough edges with slag remnants adhering to openings, and the maximum obtainable open area is low, 12 percent (USDI, 1981), all of which increases maintenance costs. Wells with these screens are also more difficult to develop than continuous-slot or louvre screens.

Factory-Slotted Perforated Pipe. Manufactured or factory-slotted perforated pipe is made by punching or stamping slots in the steel or plastic casing (Figure 6–10).

Table 6–5. Groundwater Monitoring Well Screen Types.

TYPE	ADVANTAGES	DISADVANTAGES
Field slotted pipe	• Readily available. • Very inexpensive.	• Very low amount of open area ($<12\%$) making development difficult to impossible; groundwater may in turn not flow easily through the well, causing stagnant water conditions and unrepresentative samples. • Rough, jagged edges, metal pipe, not corrosion resistant. • Poor slot control; slots generally cut too large, causing an excessive amount of material to enter the well, unless the screen is cloth wrapped (making development very difficult and is not a good idea). • Slots cannot be closely spaced. • Increased maintenance costs.
Factory slotted perforated pipe	• Good slot control. • Readily available. • Inexpensive.	• Low amount of open area, making development difficult. • Rough, jagged edges. • Lighter stock material (≤ 8 gage) not useful at depths greater than 100 to 150 ft.
Manufactured louvre-type or bridge-slot screen	• Slots accurately sized. • Wire-brushed to remove roughness and irregularities. • Open areas range from 3% to 20%.	• Clogging occurs readily.
Wire-wound perforated pipe (pipe-based screen)	• Useful in clean, rather coarse material with few or no fines. • Superior tensile and collapse strength. • Can be retrieved at great depths.	• Not useful for materials with fines; fines clog space between pipe and wire wrap. • Single-metal alloys needed for both wire and pipe in corrosive waters.
Wire-wound continuous-slot screen (not pipe based)	• Very good slot control. • Wide range of slot sizes available. • High amount of open area, making good development possible and, therefore, best possible samples can be obtained. • Made in both telescoping and pipe sizes. • Most efficient available screen. • Slot sizes can be custom made to aquifer gradations.	• Higher priced than slotted pipe, but still moderately priced.

Source: After WPR, 1981 and Johnson, 1984.

The steel screens have rough ragged edges but their maximum open area can be high, 20 percent. It should be noted that if lighter gage steel is used it does not have the strength for greater depths. The plastic screens of this type have many advantages; they are useful in clay-rich sediments, they are not affected by corrosion, are easy to install, and are relatively inexpensive. However, they provide only low open area (half of continuous-slot plastic pipe), and are only $\frac{1}{6}$ to $\frac{1}{10}$ as strong as stainless steel (Driscoll, 1986).

Wire-Wound Perforated Pipe. Wire-wound perforated pipe have a perforated pipe as a core surrounded by optional longitudinal rods as spacers and then a wire wrap (Figure 6–12). The total area of outer openings is usually greater than the area of inner pipe holes (WPR, 1981). These screens are useful in clean, rather coarse material with few or no fines that can clog the spaces. The only real advantage of a pipe base screen is its superior tensile and collapse strength, which enables retrieval at great depths (WPR, 1981).

Manufactured Louvre-Type or Bridge-Slot Screen. Manufactured louvre-type or bridge-slot screens have openings arranged in rows either parallel or at right angles to the pipe axis (Figure 6–10). The latter presents more problems with screen clogging. The slots of these screens can be accurately sized, can provide open areas from 3 to 20 percent, and are wire-brushed to remove roughness and irregularities.

Wire-Wound Continuous-Slot Screen. Continuous-slot wire-wound screens are considered the best and most efficient screen type in the industry (Figure 6–10). These screens are made by wrapping wire continuously around longitudinal rods and welding the rods to the wire. They have numerous advantages and are available

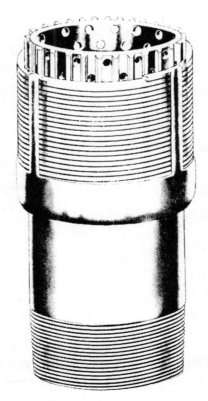

Figure 6–12. Pipe-base screen (after Departments of the Army and the Air Force, 1975).

in both steel and plastic. They are available in almost any slot size, the slot size on a single screen can be varied to meet aquifer gradations, and the open area is the largest available. The major disadvantage is that they are initially expensive; however, in the long run, they usually prove to be more economical. In general, many monitoring wells are constructed with wire-wound screen made of PVC material. At a monitoring site, the openings that are most common range from 10 to 30 slot.

Pipe and Screen Casing Fitting Types

The type of fittings used to connect casing pipe to pipe or to screened pipe vary as their material varies. There are plain square end fittings, threads and couplings, or flush thread fittings. The advantages and disadvantages of each are given in Table 6-6.

Plain Square End Fittings. These are connected by welding if the casing material is metal, or by using solvent cement for plastic casing. If metal pipe is used, it is critical that standardized welding procedures be followed (American Welding Society, 1981). In addition, welding different metals together requires the proper selec-

Table 6-6. Groundwater Monitoring Well Fitting Types.

TYPE	ADVANTAGES	DISADVANTAGES
Plain square ends (no fittings to weld)	• Readily available in pipe and screen. • No need to purchase threads.	• Special equipment and skills needed to field weld metals. • Plastics are welded using solvent cement. • Cementing procedures are very temperature and moisture sensitive. • Cements must be cured after application. • Cements may interfere with groundwater quality; in many cases cementing would be unacceptable because the cement leaches into the groundwater. • Time spent welding may cause this type of fitting to actually cost more than threads.
Threads and couplings	• No solvents needed. • Lengths of pipe and screen joined quickly. • Readily available. • Reasonably priced.	• May be difficult to get filter pack and/or grout past the lip of couplings. • May need to wrap threads with Teflon tape to make connections watertight.
Flush threads or joints[1]	• No solvents needed. • No couplings needed; filter packing and grouting simplified. • Lengths of pipe and screen joined quickly. • Readily available. • Reasonably priced.	• Threads or joints generally not compatible from manufacturer to manufacturer. • Friction of jointed pipe may cause weakness and strain.

[1]Jointed pipe are joined by friction.
Source: After Johnson, 1984 and Driscoll, 1986.

tion of electrodes to prevent structural weakness. Cementing plastic or PVC casing also requires special care. The solvent cement used must meet the specifications for the type of plastic used (WPR, 1981). The procedures for cementing are very temperature and moisture sensitive, and the setting time must be adhered to. Cements also may interfere with groundwater quality, and thus are not recommended for monitoring well programs.

Threads and Couplings.   This method of joining pipe or screen requires a coupling or collar to connect equal or unequal diameter casing. Threaded and coupled pipe has an advantage over plain square ended pipe in that no welding or solvents are needed, although solvents may be used in addition. The disadvantages of this method of joining pipe are: placement of the filter pack and/or grout may get caught up on the coupling lip; the seal is usually not watertight unless threads are wrapped with Teflon tape, or a solvent cement is used on plastic pipe; and threaded and coupled pipe should not be driven when the diameter is larger than two inches (51 mm). (Solvent cement is not recommended for water-quality monitoring programs.)

Flush Threads.   This method of joining pipe has the most advantages over the other two methods discussed. In addition to the advantages described for threads and couplings, there is no lip of a coupling, so filter packing and grouting is simplified. However, if different diameter pipe or pipe from several manufacturers is used, the threads may not be compatible, and thus a coupling may be required.

Filter Packs

There are two types of filter packs used in monitoring wells, a natural filter and an artificial filter pack. In the naturally developed well, fine materials in the formation surrounding the well are removed by well development. In the artificially gravel-packed well, sand or gravel which is coarser than the natural formation is placed immediately surrounding the well screen along the length of the screen, to a few feet above the screen. The purpose of this filter pack method is to create a more permeable zone surrounding the well screen and increase the effective diameter of the well (Driscoll, 1986). The reason that the artificial filter pack is placed a few feet above the screen is because it creates a zone of protection from settling. In most cases, an effective monitoring well will require an artificial filter pack, and it is recommended by the USEPA (1975a).

An artificial filter pack (gravel pack) for a monitoring well consists of clean, uniform, smooth, and well-rounded grains of sand or gravel that are installed in the annular space between the well screen and the wall of the well bore. These characteristics increase the permeability and porosity of the filter pack material, stabilize the aquifer material, and reduce entrance velocity and head losses to the well. The filter material should be purchased from reputable suppliers and should consist of hard, rounded particles having an average specific gravity of not less than 2.5 (Johnson, 1980). The filter should consist of mostly siliceous particles, with a limit of five percent by weight of calcareous material, and should be free of shale, mica, clay, dirt, loam, or organic impurities of any kind. The filter material should also be free of iron or manganese in a form or quantity that will adversely affect the water quality (USEPA, 1975a). It is very important that the material be clean (actually washed before placement in borehole), so as not to interfere with the ground-water quality which is to be monitored.

The design of the artificial filter pack may be elaborate, usually for production or extraction wells, or very simple, as for most monitoring wells. In order to design artificial filter packs for production wells, sieve analysis curves are constructed for the strata comprising the screened formation. The stratum composed of the finest materials to be screened is determined, and then the grading of the filter pack is selected based on the sieve analysis of this material (Driscoll, 1986). Artificial filter packs for monitoring wells, however, are commonly determined as quickly as possible in the field, to facilitate prompt well construction. The filter pack design in this case is based on information on the water-bearing material that is gathered, preferably from previous wells drilled on the site, if available; from boring logs in the area in the records of state or local agencies; or from geologic reports of the area or region or estimates of grain size based on environments of deposition. It should be realized that the gravel packs, well screen, and casing should be at the drill site and already washed before the borehole is drilled. Therefore, the sand or gravel pack must be preselected and its grain size distribution must be estimated. For monitoring wells, the grain size distribution of the filter pack should be based on the slot size of the screen, which should retain 95 percent of the artificial gravel pack.

For monitoring wells, the filter pack material is usually inserted around the screen with a radius thickness of from three to six inches. The thinner the envelope of the filter pack the better. Three inches is a practical dimension on the lower end considering the mechanical difficulties of placing a filter pack which will completely surround the screen in the field (Johnson, 1980). A gravel pack greater than six inches may prevent effective well development. The filter material should extend at least $2\frac{1}{2}$ times the largest diameter of the well above and below the screened interval (USEPA, 1975a). Additional filter pack material may be placed above the top of the well screen before the placement of the seal, but should absolutely not extend beyond the top of the aquifer in which the well is screened. If it does extend beyond the top of the aquifer into another aquifer, the filter material will act as a vertical conduit between the two aquifers. This may result in cross-contamination, and the samples and water levels will not be representative of the aquifer being monitored.

The filter pack is usually placed by gravity using a tremie pipe to avoid bridging (gaps) and segregation of the filter material. The tremie pipe is a long pipe which delivers the filter pack or grout directly to the site of emplacement. The material should be introduced at a uniform and metered rate. Other effective methods of placing the filter pack include pumping or washing the filter material in with water as a slurry. Once the placement of the filter material is started, it should proceed at a uniform rate until the completion of the placement. The placement of the filter pack in relation to the well screen is shown in Figure 6–1.

Grouting Materials for Monitoring Wells

During and after the monitoring well construction, it is necessary to prevent the migration of water from the surface and from overlying or adjacent formations into the monitoring well. This is usually accomplished by placing a bentonite or cement grout in the annular space of the borehole (the space between the well casing and the borehole wall), above the filter pack. The filter pack is extended above the top of the screen (Driscoll, 1986) to prevent downward migration of the grout material into the screen. The filter pack should not extend into overlying formations, because this could lead to downward vertical seepage through the filter pack.

For monitoring wells, usually both bentonite and cement grouts are used. Polymeric fluids are not recommended, due to extremely low solids content and great shrinkage if dried. Table 6–7 shows a comparison of bentonite and cement grouts.

In most cases, a seal of bentonite is placed above the filter pack. This seal creates a buffer zone between the sand or gravel pack and the overlying cement grout, which could otherwise seep through the sand or gravel into the well screen and hence into the well. Such a seal consists of three to five feet of bentonite clay pellets (Figure 6–13), which can swell from 10 to 15 times in volume after wetting with deionized water. Swelling volumes of 25 to 50 percent of the maximum values are common, due to variations in the composition of the contacting solution (Barcelona et al., 1983). Above this seal, the annular space is usually grouted with a cement, bentonite, and water slurry (neat cement). A good mix consists of low-sodium cement, five to seven percent bentonite by weight, and enough water for a pumpable mix. For example, generally 7 gallons of water should be used per 94 pound bag of cement. According to the literature, the amount of water added per 94 pounds of cement should never exceed 10 gallons.

The bentonite pellets can be placed by means of gravity. Care should be taken to insure that there is a clear path for the pellets to flow, or else bridging (gaps) may occur and the pellets may swell before reaching their destination. The cement is

Table 6–7.   Grouting Materials for Monitoring Wells.

TYPE	ADVANTAGES	DISADVANTAGES
Bentonite	• Readily available. • Inexpensive (about $6 per 50 lb.).[1] • Pellets can be used.	• May cause constituent interference due to ionic exchange. • May not give complete seal. • There is a limit to the amount of solids that can be pumped in a slurry without the pump clogging. • Pellets may bridge; they may wet and swell before reaching destination, sticking to formation or casing. • Cannot determine how effectively material has been placed. • Cannot assure complete bond to casing.
Cement	• Readily available. • Inexpensive (about $10 per 94 lb.).[1] • Can use sand and/or gravel filler. • Possible to determine how well the cement has been placed by means of geothermal logs or sonic bond logs.	• May cause constituent interference. • Mixer, pump, and tremie line are required. Generally more cleanup than with bentonite. • May be problems getting the material to set up. • Shrinks when it does set—complete bond to formation and casing not assured.

[1]All prices are based on 1988 prices and may vary somewhat depending upon supplier.
Source: Adapted from Johnson, 1984 and Driscoll, 1986.

Figure 6-13. Bentonite pellets shown in three sizes (photo courtesy of Slope Indicator Co.).

placed via a tremie pipe (Figure 6-14). The tremie pipe is lowered to the bottom of the zone being grouted, and then the grout is pumped through the tremie pipe. The discharge end of the tremie pipe should remain continuously submerged in the grout until the zone to be grouted is filled.

Faulty seals or grouts in a monitoring well can cause water samples from the formation of interest to be nonrepresentative, and can bias the analytical results on those samples. This is particularly true where water-quality conditions vary between formations or surface soils are badly contaminated. Not only can a leaky well bore disrupt sampling programs, but it may also act as a conduit to permit rapid contaminant migration and cross-contamination that may not have otherwise occurred (Barcelona et al., 1983).

Faulty seals or grouts can be caused by many factors, including: pouring a grout mixture that may cause separation or will cause separation if placed below the water table; pumping a bentonite slurry that is too thin into a well, if able to go through the pipe, or if thick enough to form a good seal, will clog the pipe; or if the water contains something to deflocculate the clay pellets, which prevents them from swelling. The placement of the bentonite seal and cement relative to the well casing and screen is shown in Figure 6-15.

Well Completion

Once the borehole has been drilled, the casing and screen have been placed in the borehole, and the filter pack, bentonite seal, and grout have been placed around the casing, the final steps to complete the well at the surface must be carried out.

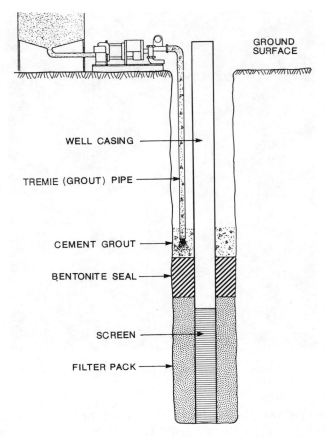

Figure 6–14.   Tremie pipe used for placing cement grout (after Driscoll, 1986).

These steps include: surface protection; installing security devices; slitting and capping the well; marking a measurement point; and labeling the well.

The type of surface protection required depends on whether the well is in a high-traffic locality, which requires below-grade completion, or not. Below-grade completion requires installation of a "Christy-type box" which surrounds the well below grade and is capable of withstanding traffic weight (Figure 6–15). It is installed after the excess casing has been cut off.

Completion of a well above grade is similar from this point on (Figure 6–1), unless wood or metal posts are driven in around the well for surface protection. In this case, the posts should be either angled or placed in such a way as to allow sampling rigs ample room to get close to the well.

Security casing can be installed on both above- and below-grade wells. They are usually made of steel, three to four feet in length, with either a locking lid or a lid with locking capabilities. The entire device is placed around the well prior to grouting the casing and buried two feet into the grout surrounding the casing. During installation, it is necessary to keep the lid closed to make sure that the casing fits within the security box. Note also when selecting the size of the security casing, to allow room for the well cap and the removal of same.

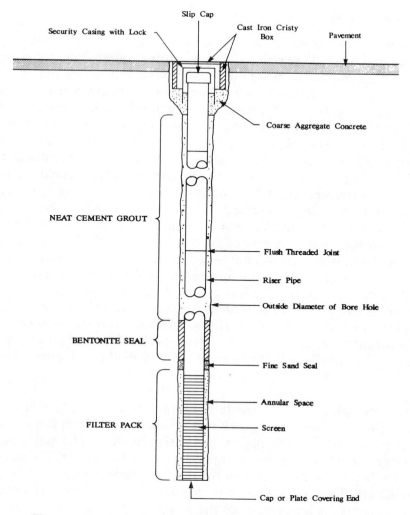

Figure 6–15. Typical monitoring well completed below grade.

Prior to placing the well cap on, it is necessary to ventilate the casing in order to avoid creating a vacuum within the well, yielding incorrect water-level measurements if not given time to equilibrate, and making the removal of the cap very difficult. Plastic casing can be slit to below the cap coverage to avoid this situation.

Wells constructed for groundwater monitoring require a consistent reference point for all measurements taken at the well. Thus a permanent mark should be notched or marked (preferably notched) in the actual well casing, not on the metal security box. The corresponding elevation, in general, should be surveyed to a minimum of 0.001 feet of accuracy and reported to a minimum of 0.01 feet of accuracy. All future measurements should reference this point.

The final step in completing the well involves labeling the well with an identification number. Labels can be stamped into the metal protection box, a name plate welded onto the casing or protection box, or written with a permanent marker on the inside of the cap.

Well Development Methods

Well development is the removal of sand and other fines (including drilling mud) from the aquifer around the well by surging, jetting, intermittent pumping (rawhiding), or other actions which move the fines into the well so that they may be bailed or pumped out. The objective of well development is to create a filter zone around the well screen or gravel pack that is more permeable and more porous than the aquifer itself, and which prevents further sand migration into the filter and ultimately into the well. The purpose of this is to maximize water production efficiency, which in turn allows representative water-quality samples to be obtained. Well development also improves the resistance of well components to corrosion and encrustation, stabilizes the aquifer, and minimizes further sand pumping. Achievement of these objectives, although established for production wells, should also be attempted for monitoring wells. It should be realized that the small diameters of monitoring wells do not allow for many of the development techniques to be carried out, and that the monitoring well will not be as efficient as a production well. The trade off is the much lower cost of constructing the smaller diameter monitoring well.

Care must also be taken regarding placing water and air into the formation. If volatile constituents are to be monitored, developing with air can change the concentrations over a period of time. Similarly, adding water can also affect the concentrations of constituents within the aquifer. Records of volumes of air or water pumped into or out of the formation must be recorded on a Well Development Form (Figure 6–16).

Intermittent Pumping. This procedure, also referred to as rawhiding or pumping and surging, requires pumping a well in a series of steps from low discharge rates to discharge rates considerably higher than design capacity. The well is pumped until the discharge is relatively sand-free, then the power is shut off, allowing the water in the well column to surge back into the well to agitate the fine material surrounding the well. The discharge rate is increased and the procedure is repeated until the final rate is the maximum capacity of the pump or well. For this process, the pump used must have a check valve or rachet to prevent the pump from reversing rotation (WPR, 1981). Another problem may be encountered if the rate of discharge is high initially. This causes "bridging" (clogging of the well screen with sand grains by the sudden pull of the sand towards the well), which can prevent fine material from being removed (Todd, 1980). In areas where groundwater is highly contaminated, plans must be made to deal with the discharge water. This includes storing, removal, and treatment off-site or storing, treating and discharging on-site.

Surge Block. This procedure uses the up and down motion of a surge block attached to the bottom of a drill stem to agitate fine material (Todd, 1980). The surge block is either solid, vented, or spring-loaded, and is fitted to the well screen snugly by either a leather, rubber, or similar flap to prevent damage to the screen. The gentle down strokes, which cause the flap to open and to allow water to move through, cause only sufficient backwash to break up any bridging. The stronger up strokes, where the flap seals the hole, pulls the sand grains, loosened by the destruction of the bridging, into the well (WPR, 1981). Before surging is done, the well should be bailed clean and the surge block cable marked to identify the screened

R. L. STOLLAR & ASSOCIATES, INC.
WELL DEVELOPMENT DATA

SITE TYPE | SITE ID

| WELL | |

DEPTH TO BOTTOM (INITIAL) _____ PROJECT NO. _____

(FINAL) _____ DATE(S) INSTALLED _____

STATIC WATER LEVEL (INITIAL) _____ DATE(S) DEVELOPED _____

(FINAL) _____ PUMP (TYPE) _____

MEASURING POINT _____ (CAPACITY) _____

CASING I.D. _____ BAILER (TYPE) _____

HYDROGEOLOGIST _____ (CAPACITY) _____

DRILLER _____

TIME	VOLUME OF WATER REMOVED	pH	SPECIFIC CONDUCTANCE AT 25°C	TEMP	SAND CONTENT	OTHER PHYSICAL CHARACTERISTICS (CLARITY, ODOR, PARTICULATES, COLOR)

FOPM14 / SEPT 87

Figure 6-16. Well development data form.

interval and bottom of the well. It is the oldest and most effective method of well development and is particularly applicable for use with the cable tool rig.

Surge Block with Air. This procedure, along with backwashing with air, requires a considerable amount of equipment and a skilled operator (WPR, 1981). This method uses compressed air, which is released suddenly in large volumes into the well to produce a strong surge. This loosens the fine material surrounding the perforations (WPR, 1981). Continuous air injection causes the loosened material to come into the well and be pumped out (by an air lift pump) (Todd, 1980). The operation is repeated at intervals along the screened section (by moving the air pipe and discharge pipe vertically) until the sand accretion becomes negligible.

Backwashing with Air.   This method also uses compressed air and requires large amounts of equipment and a skilled operator (WPR, 1981). To develop wells by the backwashing method, compressed air is used to force air and water out of the well through a discharge pipe (air-lift pumping). The air supply is then shut off and the water level recovers. Air is then supplied to the top of the well through a shorter air pipe that backwashes the water from the well through the discharge pipe, and at the same time agitates the sand grains surrounding the well. This supply is shut off when it begins escaping from the discharge pipe (Todd, 1980). This cycle is repeated until the well is fully developed.

Hydraulic Jetting.   This method is most effective in rock holes, those lined with cage-type wire-wound screen or louvre screen, and those wells which are gravel packed (WPR, 1981). Jetting is accomplished through the use of high-velocity water, which sprays out two or more jet nozzles when placed at depth within the well. The nozzles are attached to a rotating head that attaches to a string of pipe and hose, which connects to a high-pressure, high-capacity pump. As the high-velocity water is sprayed out of the rotating jet heads, the device is slowly raised along the length of the well to develop the well. During this process, the water should be pumped from the well, if possible.

Chemicals.   Use of chemicals is not recommended for monitoring wells. Several chemicals are available to develop production wells in specific aquifer materials. For limestone or dolomite aquifers, hydrochloric acid may be added to the water in open-hole wells. It should then be developed by one of the previously described methods (Todd, 1980). Hydrofluoric acid is similarly used for silicate rocks. Polyphosphates are added as deflocculants and dispersant of clays and other fine-grained materials. Blocks of solid carbon dioxide (dry ice) can be added after acidizing and surging to create pressure buildup and the release and expulsion of muddy water to complete well development.

Methods for Small Diameter Wells.   The above methods or combinations are generally used to develop production wells, but can be used to develop monitoring wells. The monitoring well is usually from two to six inches in diameter and, therefore, the equipment for some of the above methods cannot be placed in small diameter wells, although surge blocks have been made for small wells, and certainly the air methods and intermittent pumping have been used. Very common methods of developing monitoring wells also include bailing and/or pumping. If the well is pumpable (if permeability is high enough), a submersible pump may be used. If the permeability of the formation screened is very low and water cannot be pumped, the well development should be carried out with a bailer. The water is pumped with the objective of removing the fines and stabilizing the formation. If well development is successful, the water being pumped will begin to clear up. In addition, the pH, conductivity, and temperature of well water flowing through either a "flow-through box" (discussed below) or from a simple container are usually recorded periodically to determine the changes and whether the values for these parameters are stabilizing. In other words, the objective of developing a monitoring well is to obtain representative samples of water from the aquifer. A secondary objective is to be able to pump aquifer water with very little or no fine-grained material. During development, this is measured by the clarity of the water and whether the parameters of pH, conductivity, and temperature stabilize.

A recommended procedure for monitoring groundwater parameters and collecting groundwater samples includes measurements of pH, Eh, conductivity, and temperature of well water flowing through either an in-line measuring chamber, preferably, if available or from a plastic container using probes. Use of the in-line measuring chamber, or "flow-through box," provides a simple, efficient method of monitoring groundwater quality before exposure of the water to oxygen in the air and the subsequent chemical changes that immediately would occur. Figure 6–17 illustrates the flow-through box. A three-way nylon or Teflon valve inserted in the discharge line will split the flow, providing continuous flow not only to the flow-through box, and subsequently across the probes, but also to the sampling container. With this design, special materials for the flow-through box are not required, since none of the water flowing through this chamber will be sampled. The flow-through box itself consists of a Lucite block approximately 12 inches long, 3 inches wide, and 2 inches high. The block contains several ports for the insertion of the pH, Eh, reference, and conductivity probes, as well as ports for optional measurement using dissolved oxygen and specific ion probes. Temperature is generally measured with the conductivity probe but an optional thermometer can be added. Space

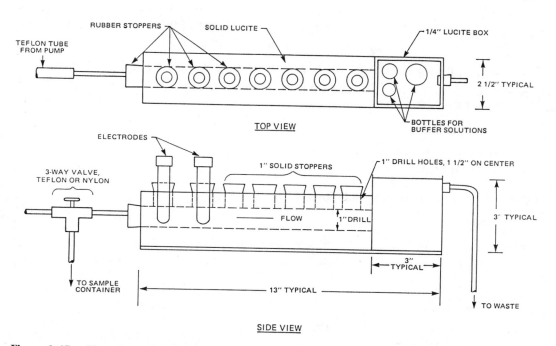

Figure 6–17.   Flow-through box for in-line determination of pH, Eh, conductivity, temperature, and specific ions.

is included at the end of the block for containers of pH buffers and Zobell solution. It is critical to adjust the pH using buffers equilibrated to actual water temperature. Keeping the buffers in the chamber at the end of the flow-through box will assure thermal equilibrium.

When using the flow-through box during water sampling, the indicator parameters are continuously monitored during the establishment of chemical rating curves. Samples are taken after equal volumes of water have been pumped to provide data for these curves. When samples are taken before and after the establishment of rating curves, pH, conductivity, and the temperature are recorded before and after samples are taken. If there is a significant difference in these readings, it is necessary to repeat the above procedure and re-sample the water. Samples for water-quality analysis are not collected through the flow-through box. The water flowing through the flow-through box is discharged to waste. All of the well development activities should be recorded on a Well Development Form for the permanent records (Figure 6–16).

Decontamination of Drilling and Soil Sampling Equipment.   All equipment that comes in contact with potentially contaminated soil or water should be cleaned prior to and after each use. Decontamination should consist of combinations of steam cleaning and/or detergent (trisodium phosphate or Alconox) wash, water rinse, and an acetone or methanol rinse and distilled water rinse, depending on the contaminants present at the site. Some of the decontamination solutions are shown on Table 6–8. It should be noted that the use of acetone, methylene chloride (now a priority pollutant), or hexane may affect the water quality results. The procedures described in this section constitute a cleaning protocol for equipment used during the drilling process.

All drilling equipment, including the drilling rig, all boring tools, and samplers should be washed with potable water and steam cleaned initially, and also after the completion of each boring, according to the following procedure:

- The drilling rig should be washed with potable water and steam cleaned.
- All drill rods, auger flights, dual tubes, bits, stems, and samplers should be cleaned by

  steam cleaning;

  rinsing with potable water; and

  air drying on above-ground racks.

Table 6–8.   Decontamination Solutions.

TYPE OF HAZARD	NAME OF SOLUTION	REMARKS
Amphoteric-acids and bases	Sodium bicarbonate	5–15% aqueous solution
Inorganic acids, metal processing wastes, heavy metals	Sodium carbonate	Good water softener, 10–20% aqueous solution
Solvents and organic compounds, oily, greasy unspecified wastes	Trisodium phosphate	Good rinsing solution or detergent, 10% aqueous solution
Pesticides, fungicides, cyanides, ammonia and other non-acidic inorganic wastes	Calcium hypochlorite	Excellent disinfectant, bleaching and oxidizing agent, 10% aqueous solution

Source: Adapted from Richter and Collentine, 1983 and Geo-Environmental Consultants, Inc., 1983.

- If soil samples are being collected from a soil boring for analysis, the sampler should be cleaned between each sample by

    removing adhering soil particles by scrubbing with potable water;
    scrubbing the inside surfaces with a solution of potable water and detergent;
    rinsing with potable water;
    rinsing with deionized water; and
    drying with clean paper toweling.

- All personnel should use inert plastic gloves, which are to be disposed of after use on a single hole.
- All personnel should handle tools and materials in a coordinated manner, to avoid any possible cross-contamination.

Feed water for the steam cleaner should be obtained from a source of potable water. A water sample from this source should be analyzed for key constituents. The drilling crew should exercise care in adding fuel to the heater tank so that cross-contamination of the water tank is avoided.

For cleaning the inside surfaces of pipes, augers, and dual tubes, a spray wand designed to produce a radial spray pattern should be provided. This permits thorough cleaning of interior surfaces, including annular spaces.

It is recommended that all decontamination procedures be carried out on a specially constructed pad to contain steam cleaning water, rinse water, etc. All of the excess water on the pad can then be pumped and containerized to prevent possible spread of contaminants to the ground surface and eventual infiltration.

Waste and Wastewater Disposal Procedures

Drilling, installation, development, and testing activities will generate borehole materials and fluids that may be contaminated. Contaminated materials or wastes should be properly stored prior to removal from the site and disposal, if necessary, at a hazardous waste facility. These materials should be stored in drums or other suitable containers to prevent contact by any personnel. Temporary storage areas should be away from active areas, isolated, lined, clearly posted, and accessible to the type of equipment needed for material removal. Incompatible materials should be stored at safe distances from each other, appropriately isolated from surface runoff and soft earth surfaces, as well as off-site areas that could be potentially affected by storage accidents. All stored materials should be labeled with the well number, date, job number, contractor's name, contents, hazard, and any other pertinent remarks prior to storage, off-site transport, and disposal.

WATER-QUALITY SAMPLING PROTOCOL

In addition to the precautions taken during the construction of the monitoring system, extreme care must be taken during sample collection, storage, and analysis. Removing water from a well has, in itself, the potential to introduce error. The method of sampling is dependent on the constituent of interest. For example, the sampling method can cause error due to contamination of the sample, degassing, atmospheric invasion, volatilization, adsorption, or desorption. The sampling method also is dependent on the hydraulic conductivity of the formation, volume of water to be flushed from the casing, and accessibility of the site.

Field Preparation

Prior to any sampling, but after each well has been completed and developed, the depth to water and to the bottom of the well should be measured, and the well should be purged of all standing well water. Purging a well ensures that the water sampled represents the true water quality of the aquifer being sampled and not just standing water. The material and equipment to be brought on-site for the sampling and associated field activities generally includes the following:

- Field copy of the sampling and quality assurance and quality control plans, the well construction logs, and the appropriate maps;
- Forms—sampling form, log book;
- Pump of some type or bailer and the associated equipment, if not already permanently installed, other sampling devices, operating manuals;
- pH, conductivity, and dissolved oxygen meters (if necessary), thermometers or temperature probes, as well as a complete set of spare probes, cables, and batteries for each;
- Calibration standard solutions and detailed calibration procedure instructions;
- Wash bottles and extra beakers;
- Water-level measuring device;
- Nylon-clad steel tape;
- Roll of plastic sheeting;
- Health and Safety equipment and plans;
- Spare gloves;
- Plastic bags;
- $\frac{1}{4}$-inch nylon rope;
- Metals filtration kit (peristaltic pump, filter holder, replacement hoses, filters, 50 ml of dilute nitric acid, and pH indicator paper);
- Complete set of spare sample containers; and
- Barrel or drum for purged water storage.

Prior to the onset of field work for groundwater monitoring activities, the on-site supervisor should ensure that all personnel have been fully trained in both safety requirements for the site and in the operation of all field equipment. Each team member should understand the field procedures described in the sampling plan. In the event that procedure modifications are made, or additional equipment or instrumentation is incorporated into an ongoing program, the field supervisor should schedule training sessions introducing these modifications or equipment changes to field personnel, and any changes should be recorded and documented.

All data collected during groundwater monitoring should be recorded on Field Data Sheets and in bound Field Logbooks. Examples of some pertinent forms are shown in Figures 6–16, 6–18, and 6–19. When not in use, all Field Logbooks should be kept in a secured area. Logbooks should be checked out by the field supervisor to the field team leaders on a daily basis.

R. L. STOLLAR & ASSOCIATES. INC.
WATER QUALITY FIELD SAMPLING DATA SHEET

PROJECT _____ SAMPLERS _____

PROJECT NUMBER _____

WELL NO. _____ SUPERVISOR _____

DATE SAMPLED _____ TIME SAMPLED _____

_____
_____
_____
_____
_____
_____
_____
_____
_____
_____
_____
_____
_____
_____
_____
_____

MEASURING POINT _____ WELL DIAMETER (ID) _____

WATER LEVEL _____ WELL DEPTH _____

SCREENED INTERVAL _____ SINGLE WELL VOLUME _____

PURGE PUMPING RATE _____ PURGE VOLUME _____

DEPTH OF SAMPLING _____

SAMPLE PUMPING RATE _____ NO. OF SAMPLES _____

NO. OF SAMPLES RELINQUISHED _____ TO _____ DATE _____

FIELD EQUIPMENT

Eh/pH METER _____ SERIAL NO. _____

EC. METER _____ SERIAL NO. _____

PUMP _____ SERIAL NO. _____

TUBING TYPE _____

WATER LEVEL METER _____ SERIAL NO. _____

FILTER APPARATUS _____ FILTERS _____

BAILER _____ SIZE _____ IN. X _____ IN. _____

FOPM24 / OCT 1987

Figure 6–18. Water-quality field sampling data sheet.

R. L. STOLLAR & ASSOCIATES, INC.
WATER LEVEL MEASUREMENT SHEET

Project Number _____

WELL NUMBER	DATE	TIME	DESCRPT OF MEAS. PT.	MEASURING DEVICE	ORIGINAL WELL DEPTH	TOTAL DEPTH	STICK UP	ELEV. OF MEAS. PT.	DEPTH TO WATER	ELEV. OF WATER TABLE	COMMENTS

FOPM12 / Sept 1987

Figure 6–19. Water level measurement sheet.

218

Water-Level Measurement.   Water-level measurements are taken for two reasons: to ensure data consistency each time that water samples are collected from a well; and to evaluate both water-level fluctuations over time and the groundwater flow system. Water-level measurements for the former purpose should be conducted prior to the initiation of water sampling. The objective of the latter is usually to prepare water table or potentiometric maps and tables. These are usually a synoptic measurement, and all of the wells should be measured within a very short period of time. All water-level measurement and sampling programs should include measuring the bottom of the well to determine if the well is filling with sand. If wells are heavily contaminated, health and safety procedures must be followed. The procedure for collecting water-level measurement data is summarized below:

- Depending on the constituents being monitored, the health and safety program may require respirators and other safety equipment. With respirators on and from an upwind direction, remove the well cap and record background and casing headspace readings with a photoionization detector (PID) or flame ionization detector (FID). Respirators may be removed if readings are below background levels within the breathing zone around the well head;
- Record the well number, date, time, measuring point, and initials of field personnel taking measurements on the Water-Level Measurement Form and in a Field Logbook (Figure 6–19);
- Insert the water-level indicator probe (Figures 6–20 and 6–21) until it reaches water. These indicator probes either buzz, have a light go on, or a needle response on a meter to indicate that the probe has entered the water (the meter should be calibrated). Measure the depth to water and total well depth from the same measuring point (if not marked, mark it)[2] and record the value to the nearest one-hundredth foot;
- There are two other typical methods of measuring water levels. One is the steel tape and chalk method, and the second is attaching a steel tape to the water-level indicator type probe. The steel tape is divided into feet, inches, and hundredths of an inch (or metric divisions). With the first method, chalk is applied to the first few feet of the tape. The tape is lowered to the water table and held at, for example, an exact foot marker at the measuring point. The tape should be lowered into the water where only part of the chalk is washed off. When the tape is removed from the water, the amount "wet" where the chalk is washed off is recorded and subtracted from the measurement held at the measuring point to obtain the depth to water. With the second method, the measuring tape is attached to the probe. The probe allows the water table to be detected, and the depth to water can be more accurately read from the steel tape.
- Compare collected data to previous measurements (where applicable). If discrepancies are observed, perform second measurement verifications and document as such;
- Calculate the volume of water in the well to determine the volume of water to be purged before sampling;

---

[2]The measuring point is usually a notch cut into the top of the casing. Elevation of this point should be surveyed.

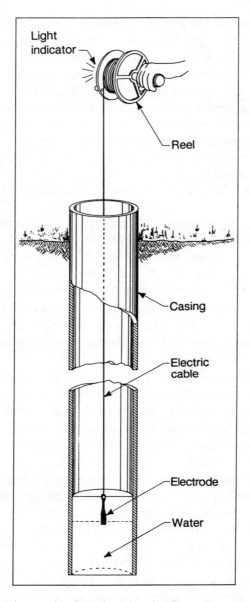

Figure 6-20. Electric sounder for measuring depth to water (after Driscoll, 1986).

- Retrieve the water-level indicator probe and rinse the cable and probe with deionized water as they are withdrawn from the well;
- Record well conditions (cracked casing, missing cap, subsidence features, etc.) and any other pertinent observations;
- Insure that all labels and flagging clearly indicate the well's location and number; and
- Inspect the area to assure that all equipment and materials have been retrieved, no litter is left, and the well cap is secure.

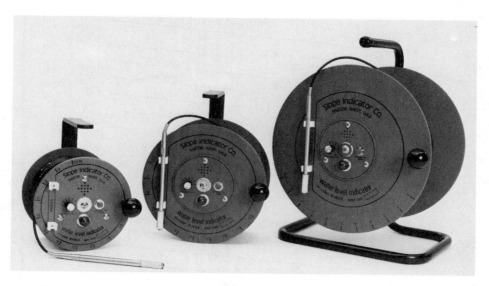

Figure 6-21. Water level indicators (photo courtesy of Slope Indicator Co.).

Purging. After the water level has been measured, the well should be purged prior to sampling. The number of casing volumes of water to be removed from a well varies. However, three to ten casing volumes is generally an acceptable amount. If field parameters such as pH, conductivity, and the temperature of well water flowing through a "flow-through box" are taken, these should reach stable measurements regardless of the number of casing volumes removed. If stabilization occurs rapidly, at least three casing volumes should still be removed. Regulatory agencies who have jurisdiction over the project should be consulted for specific requirements, and the exact field protocol should be predetermined prior to the commencement of field activities. It is essential to establish the sampling protocol before the long-term monitoring program begins. This includes the volume of water to be removed during purging, and how to remove the water from the well.

An example of the importance of protocol is shown in Figure 6-22 A, B, and C. Figure 6-22A illustrates a nonpumping well in a uniform flow field. The flow lines are labeled A through E. As the pumping begins (Figure 6-22B), this figure shows that flow lines A and B are intercepted upstream and slightly downstream of the well. Downstream of the groundwater divide, flow lines A and B continue to flow downstream. Flow lines C and D are also affected by the well (shown by the slight movement toward the well just beneath the well) but are not intercepted by the well.

Figure 6-22C shows the configuration of the flow lines with continued or increased pumpage. Here, flow lines A, B, and C are captured by the well. D and E are affected, but are not captured.

If the contaminants were only in the area of flow line C, then the contaminants would not be captured by the pumping shown in Figure 6-22B. On the other hand, if the contaminants were in the area of flow lines A and B, the concentrations would be greater in the representation of Figure 6-22B than in 6-22C. This is because the concentrations would be diluted in Figure 6-22C.

Attempts must be made to control these types of irregularities. This can be done

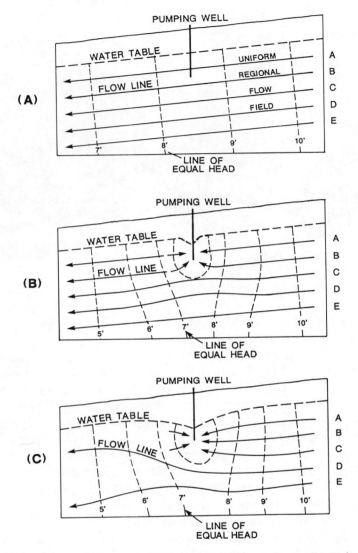

Figure 6–22A, B, & C. Groundwater flow lines during well purging.

by predetermining the volume of water to be purged before sampling. Once a protocol is established for a well, the sampling team should always follow the given protocol.

The placement of the pump during purging is also important. There are generally two positions that are common, as shown in Figure 6–23. Remember that the objective of purging is to remove all standing water in the well, in order to sample water that is representative of the aquifer.

In the first method, the pump is placed within the screen (Figure 6–23). Therefore, the water moving into the screen is certainly representative of aquifer water. However, the exact timing of when all water standing above the pump is completely drained, and the water being pumped is totally aquifer water, is subject to question.

The preferred method is to place the pump slightly below the water table (Figure

WELL EVACUATION METHOD 1

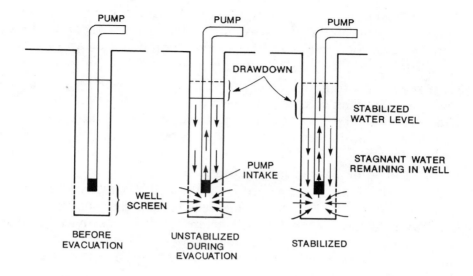

PREFERRED WELL EVACUATION METHOD 2

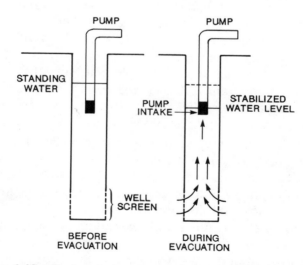

Figure 6–23.   Preferred placement methods of pump for well purging.

6–23). After purging a predetermined casing volume of water from the well, it is assured that all the water in the well is derived from the aquifer and not from standing water in the well, as all water entering the pump is flowing from the aquifer into the well through the screen, up the casing into the pump.

The preferred procedure for purging a well is outlined as follows:

- Record and measure the depth to bottom of well;
- Record and measure water level, as described in the previous section;

- Calculate and record casing volume; compare with previous volumes to ensure relative and comparable results;
- Calibrate field instruments used to monitor pH, temperature, conductivity, and dissolved oxygen (production wells only) against known standards. Record instrument calibration responses, times, and calibration standards used;
- Record field instrument serial or ID numbers;
- Purge and sample from the top of the water column. Bailers should be lowered slowly into the water column to a depth equal to the length of the bailer being used. Pumps should be placed two to three feet below the top of the water column, and repositioned as necessary in response to water level fluctuations during purging;
- Purging should be done as quickly as possible;
- A portion of the water discharged from the well should be collected or measured while in a flow-through box (Figure 6–16), and the following information recorded: field parameter values (pH, temperature, conductivity, and dissolved oxygen) date, time, PID reading, pumping rate, and purged volume removed. This information should be documented at the onset of purging and after each casing volume is removed;
- A minimum of three casing volumes should be removed from each well prior to sampling. As discussed above, if a "flow-through box" is used, samples should not be collected until field parameters have stabilized. Prior to stopping pumping, lift the pump or intake to break suction. This assures that all standing water above the pump or intake is removed from well. Wells that dewater prior to the removal of three casing volumes or stabilization, should be exempt from these requirements. Samples should be collected from these wells once sufficient recovery has been attained. Dewatered wells should be given a maximum of 24 hours to recharge. If sufficient recharge has not been attained within a 24-hour period, as many sample fractions as possible should be collected; and
- All water removed from the well during development or purging should be collected and properly disposed of, depending upon the level of hazard presented by the dissolved constituents. This disposal should be based on the results of laboratory analysis. All purged water, if necessary, should be collected and containerized at the well site, and a sample collected for laboratory analysis. If contained, label containers with well number, date, contents, and any other pertinent remarks, and store at the site until the laboratory analysis is complete and the appropriate disposal method has been determined.

Decontamination of Equipment.   All equipment that comes in contact with potentially contaminated soil or water, such as sampling, water-level measuring, and sample preparation equipment, should be cleaned prior to and after each use. Decontamination should consist of combinations of steam cleaning and/or detergent (trisodium phosphate or Alconox) wash, water rinse, acetone (extremely flammable) or methanol rinse and distilled water rinse, depending on the contaminants present at the site. It should be noted that the use of acetone, methylene chloride (now a priority pollutant), and hexane may affect the water quality results (see the caution in the subsection on the Decontamination of Drilling and Soil Sampling Equipment on page 214.

Decontamination solutions should be designed to react with, and neutralize, specific contaminants found at a hazardous waste site. However, since the contaminants on a particular site will be unknown in the majority of cases, it is necessary to use a decontamination solution that is effective for a variety of contaminants. Several of these general purpose decontamination solutions (some ingredients are available at hardware or swimming pool supply stores) are listed on Table 6–8 with their recommended uses.

The USEPA or other regulatory agencies of hazardous waste sites may require strict regulations for decontaminating equipment. The following paragraphs give examples of specific cleaning requirements for a heavily contaminated site.

Well Purging and Sampling Equipment. The following methods are typical of EPA recommendations for cleaning equipment. Bailers, both PVC and Teflon, used for purging and sampling monitoring wells, should be cleaned prior to use in the field by the following procedure:

- Alconox detergent wash;
- Potable water rinse;
- Deionized water rinse;
- Reagent methanol rinse; or a hexane or acetone rinse if necessary;
- Deionized water rinse; and
- Air dry.

Pumps used for purging wells prior to each sampling event should be decontaminated to prevent cross-contamination between wells. The following protocol is a typical example of an EPA-recommended decontamination procedure. The specified volumes for flushing are considered to be the minimum volumes:

- Potable water rinse   (2 volumes);
- Alconox detergent wash   (2 volumes);
- Potable water rinse   (3 volumes);
- Deionized water rinse   (2 volumes), or 3 times the volume of the pump and hoses;
- Reagent methanol rinse   (1 volume);
- Deionized water rinse   (3 volumes); and
- Air dry.

Additionally, before the pump is placed into the well, the outside of the pump should be wiped down with a methanol rinse solution, followed by a rinse with deionized water. As the pump is being lowered, the lines and cables should be cleaned with a rag soaked in deionized water.

All the resulting waste water should be contained and stored at the well site. Each barrel used to contain well evacuation and equipment decontamination water should be sealed and labeled with an appropriate inventory number, date, job number, and contractor(s) name.

Sampling Containers. Most sampling containers should be obtained from a laboratory which follows strict rules for cleaning and preparation per the specific constituent to be sampled. Some even certify the cleaning process. Thus, in these cases, the cleaning of containers prior to sampling is not necessary, and definitely not applicable where preservatives have been added previously by the laboratory. Some sample containers, such as when large samples are sought, although they have been cleaned, should be rinsed with well water prior to filling with sample. Additional information on sampling containers is discussed later in this section.

Groundwater Quality Sampling Methods and Devices

Successful water-quality monitoring requires collecting samples which are representative of aquifer conditions that exist at the location of the well. Therefore, careful selection of the appropriate sampling device is important, as the process of collecting the sample could change the chemistry of the sample. Examples of the various sampling methods available include grab mechanisms such as bailers and syringes, negative (suction) displacement methods such as jet pumps and suction-lift pumps (centrifugal and peristaltic), and positive displacement methods such as gas-drive devices, bladder pumps, electrical submersible pumps (gear-drive and helical-rotor), and piston pumps. Each method has its own advantages and disadvantages (Table 6-9), which comprise its ability to adapt to certain circumstances and conditions.

Ideally, the sampling system should be able to collect a sample at a specific depth within the well and deliver it to the surface without the alteration of any chemical or physical characteristic. The mechanism must be capable of adapting to any size well, and should be portable, especially in the case of remote field locations. The proper equipment must also be cost-effective.

Chemical alteration of the sample may occur by cross-contamination from other wells if the device is not dedicated to a single site. Equipment that can be easily disassembled and cleaned and that is constructed of inert material can help prevent such contamination. Devices are also designed to obtain a sample from a specific depth within the water column without contamination from shallower depths. Methods such as syringe devices are capable of isolating the sample, while bailers obtain a more integrated sample.

Water-quality samplers are often constructed of inert materials such as Teflon,[3] polypropylene, or polyvinylchloride (PVC) which inhibit absorption or reaction of contaminants on the surface of the mechanism (Pettyjohn et al., 1981).

Volatile-gas stripping, oxidation, degassing, and associated pH shifts due to the degassing of carbon dioxide (CO_2) can result from the introduction of oxygen or inert gas to the sample. This is a potential hazard of air lift sampling methods such as gas-driven pumps. Depressurization and turbulence incurred during transport of the sampling device through the water or transfer of the collected sample to a storage vial can also cause these detrimental changes in sample composition.

---

[3] A registered trademark of DuPont, Inc.

Table 6-9. Groundwater Sampling Devices.

BAILERS[1]

ADVANTAGES	DISADVANTAGES
Constructed of flexible, nonflexible, or chemically inert materials.	Well evacuation difficult and time consuming in deep wells. Possible cross-contamination of samples.
Simple to repair, operate, and clean (Teflon withstands use of strong solvents, can be autoclaved).	Care needed for volatile organics; aeration, degassing, and turbulence during transfer of water to sample bottle.
Inexpensive, allowing dedicated installation for monitoring wells.	User may be exposed to contamination; labor intensive.
No depth limitations.	Nonspecific depth representation.
Adaptable to any size wells, with a wide variety of diameters or lengths.	Check valves may not operate properly in some conditions (i.e., high-suspended solids content and freezing temperatures).
Highly portable.	
No external power source.	Swabbing effect of bailers in tight-fitting casing.
Flexible materials allow passage through non-plumb wells.	Bailing does not supply a continuous flow of water to the surface.
Transparent bailers allow the location of the interface of immiscible contaminants in well water to be estimated.	Line used with bailer must be "noncontaminating" material used only at one designated well, or adequately cleaned after each sampling event.
Readily available.	
Low surface area to volume ratio, resulting in a very small amount of outgassing of volatile organics while sample is contained in bailer.	

SYRINGE DEVICES[2]

ADVANTAGES	DISADVANTAGES
No contact with atmospheric gases.	Inefficient for large volume samples.
Aeration and degassing inhibited by negative pressure exerted on sample.	Can not evacuate a well.
Sample at discrete intervals.	Cross-contamination possible if materials are not properly selected.
Small volume sampling.	Limited to low suspended solids content samples.
Constructed of inert materials.	
Syringe can be used as sample container.	
Inexpensive.	
Portable.	
Easy to operate.	
Use in $1\frac{1}{2}$-inch diameter wells and greater.	

SUCTION LIFT MECHANISMS[3]

ADVANTAGES	DISADVANTAGES
Easily controllable flow-rate.	25 ft maximum sampling depth.
Highly portable.	Degassing may occur due to suction.
Inexpensive.	

(continued)

Table 6-9. (Continued)

SUCTION LIFT MECHANISMS[3]

ADVANTAGES	DISADVANTAGES
Readily available.	
Any diameter well; also used in nonplumb wells; flexible tubing available.	
Simple to construct.	
Easy to flush system.	

CENTRIFUGAL PUMPS

High pumping rates (5–40 gpm possible).	Potential hydrocarbon contamination due to gasoline motor power source.
	Aeration and turbulence.
	Sample contacts non-inert impeller parts.
	Pump must be primed.

PERISTALTIC PUMPS

Pump parts do not contact sample; only tubing must be cleaned.	Sample in contact with silicone tubing may adsorb organic contaminants.
Self-priming and low volume.	Low pumping rates (0.2 to 1 gpm) inhibit well evacuation in reasonable time frame.
	Require electric power source.

GEAR DRIVE ELECTRICAL SUBMERSIBLE PUMPS[4]

ADVANTAGES	DISADVANTAGES
Constructed of inert materials.	No flow-rate control.
Suitable for organics sampling if Teflon discharge line is used.	High solids content will necessitate gear replacement.
Portable.	Potential cavitation at drive mechanism.
Continuous sample.	
2-inch and greater well diameter.	
Well evacuation and sampling.	
150 ft sampling depth (and greater with use of auxiliary power source).	
Easy to operate, clean, and maintain in the field.	
Inexpensive.	

HELICAL ROTOR SUBMERSIBLE PUMP[5]

ADVANTAGES	DISADVANTAGES
Portable.	125 ft pumping lift.
2-inch and greater diameter well.	Turbulence due to high pumping rates.
High pump rates.	Disassembly for cleaning is difficult in the field.
Constructed of inert materials specifically for groundwater contaminant monitoring.	Operational problems if the suspended solids content is high.

Table 6–9. (Continued)

HELICAL ROTOR SUBMERSIBLE PUMP[5]

ADVANTAGES	DISADVANTAGES
	Relatively expensive.
	Pump must be cycled on/off approximately every 20 minutes to avoid overheating of the motor.
	Uncontrollable flow-rate.

GAS-OPERATED DOUBLE-ACTING PISTON PUMP[6]

ADVANTAGES	DISADVANTAGES
Sample isolated from driving gas.	Relatively expensive.
Easy to operate and maintain.	Particulates can damage the valving system if not filtered.
Continuous sample.	
$1\frac{1}{2}$-inch and greater well diameters.	Valving and associated pressure changes may cause degassing of sample.
Uses compressed gas economically.	
Pumping lifts up to 500 feet.	Not portable; must be vehicle mounted.
Easily controllable flow rate.	Tubing bundles are difficult to clean and may cause cross-contamination.
Constructed of inert material.	
High pumping rates at great depths allowing large volume sampling rapidly.	

GAS DRIVE DEVICES[7]

ADVANTAGES	DISADVANTAGES
Capable of being used in a 1-inch diameter well.	Air or oxygen as driving gas can alter sample integrity (oxidation, stripping, pH change).
Portable.	
Inexpensive—may be dedicated.	Reduced portability due to large air compressor or compressed-air tanks.
Descrete depth sampling.	
Controlled, continuous sample delivery rate.	Repair or retrieval of permanently installed devices is difficult.
Use of inert gas (i.e., N_2) minimizes sample alteration (by oxidation, for example).	
Can be permanently installed in uncased wells.	
Constructed of inert materials.	
Depth limited only by burst-strength of materials.	

GAS-LIFT (AIR LIFT) SAMPLERS[8]

ADVANTAGES	DISADVANTAGES
Relatively portable.	Not an appropriate method for the acquisition of water samples for detailed chemical studies, owing to degassing.
Readily available.	
	Changes in CO_2 concentrations make this method unsuitable for sampling for pH-sensitive parameters, such as metals.

(continued)

Table 6-9. (Continued)

GAS-LIFT (AIR LIFT) SAMPLERS[8]

ADVANTAGES	DISADVANTAGES
	Limitations on the amount of gas pressure that can be safely applied to the tubing.
	Oxygenation is impossible to avoid unless elaborate precautions are taken (only a very small amount of oxygen is required to cause a water sample to attain saturation with respect to oxygen).

GAS-OPERATED BLADDER PUMPS[9]

ADVANTAGES	DISADVANTAGES
Constructed of inert materials.	Not cost effective for deep sampling due to large volume of gas required.
Driving gas does not contact sample.	Check valves may fail if suspended solids content is high.
Portable; accessory equipment may be cumbersome.	Expensive.
High pumping rate provides well evacuation and large sample volume collection.	Pumping rate may be too high in some models for the sampling of volatiles.
Pumping lift greater than 200 ft.	
Easy to disassemble for cleaning and repair.	
Most adapt to 2-inch diameter wells, some are for smaller casings.	
$3\frac{1}{2}$-inch outer diameter adaptor for large diameter monitoring wells available.	

JET PUMPS[10]

ADVANTAGES	DISADVANTAGES
Adaptability to small wells (2-inch diameter) in deep-lift installations.	Sample contacts pump unit.
Simplicity.	Changes in pressure during water circulation causes degassing and loss of volatiles.
Accessibility of all moving parts at the ground surface.	Inefficient for large-production purposes.
Relatively low first-cost.	Large volume of water must be pumped before sample is considered representative.
Adaptability to installation of all moving parts offset from the well.	Not suited for areas of seasonal fluctuations.
	Not suited for areas where severe corrosion or incrustation occur; causes enlargement or plugging at the nozzle.
	Pump system must be primed; sample is mixed in.
	Some turbulence.

Sources:
[1]Adapted from Nielsen and Yeates, 1985; Johnson, 1984; and Driscoll, 1986.
[2]Adapted from Nielsen and Yeates, 1985.
[3]Adapted from Nielsen and Yeates, 1985; Scalf et al., 1981; and Driscoll, 1986.
[4]Adapted from Nielsen and Yeates, 1985 and Driscoll, 1986.
[5]Adapted from Nielsen and Yeates, 1985 and Driscoll, 1986.
[6]Adapted from Nielsen and Yeates, 1985.
[7]Adapted from Nielsen and Yeates, 1985.
[8]Adapted from Driscoll, 1986.
[9]Adapted from Nielsen and Yeates, 1985 and Driscoll, 1986.
[10]Adapted from Gillham et al., 1982 and Driscoll 1986.

Depressurization and associated degassing of the volatile components can be reduced by maintaining a low flow rate. Pumps with easily controllable flow rates may provide this service, while additionally being capable of supplying enough power to purge the well.

The pumping lift capacity of the mechanism controls the well depth to which it can adapt. Also, the diameter of the well and its orientation may limit the type of device that is suitable (for example, a well may not be plumb and only a flexible or small device may work). Some devices are designed for ease of transport. For example, the gear-drive electrical submersible pump can be loaded onto a backpack frame.

Excessive repair costs and downtime can be avoided if the system is designed for simple maintenance, is durable, and is constructed of corrosion-resistant materials. The expense of accessory equipment, routine operation costs, and initial installation investment must be considered.

A discussion of how each sampling method operates, and the components involved, are described below. An excellent review of this topic is given by Nielsen and Yeates (1985) and Gillham et al. (1982). Each method's advantages and disadvantages are discussed and are also listed in Table 6–9.

Grab Devices.

Bailer. The conventional bailer is perhaps the oldest and simplest sampling device available. Samples are obtained using this method by lowering the bailer, a cylindrical tube, by an attached line into the well after the well has been developed (Figure 6–24). As the bailer is lowered into the water, the water passes in through the lower check valve and on through the sample chamber. When the desired sample depth is reached, the bailers descent is stopped and the check valve automatically shuts, and the sample of water, which is an integrated sample of the water column, is trapped inside the bailer (Nielsen and Yeates, 1985). The bailer line is then pulled up by lifting on the wire or rope, or by using a reel. The intact sample is then transferred to the appropriate sample container, labeled and shipped for laboratory analysis.

One type of bailer consists of a long cylindrical tube, preferably graduated, that is open at the top and sealed at the bottom. Other bailers have a check-valve at the bottom called a single check-valve bailer (Figure 6–25). The top of the tube is attached to a wire or rope for lowering into the well. The cylinder may be constructed to any size specifications and is generally constructed out of one of four different types of material (i.e., Teflon—a DuPont trademark, polyvinyl chloride (PVC), stainless steel, or a combination of stainless steel and Teflon). The bailer line is generally composed of polypropylene or nylon rope, or stainless steel or Teflon-coated wire.

Other variations available include a double check-valve bailer (Figure 6–26) (or "single source or point source" bailer), which is designed for sampling a discrete interval of the water column in a well (Morrison, 1983 and Nielsen and Yeates, 1985). In this bailer, the second check valve is located at the top opening of the bailer, and shuts automatically at the same time as the lower valve, trapping a specific interval of water. Another device is available for allowing slow and controlled rates of decanting a sample, to avoid aeration. Another modification can be used

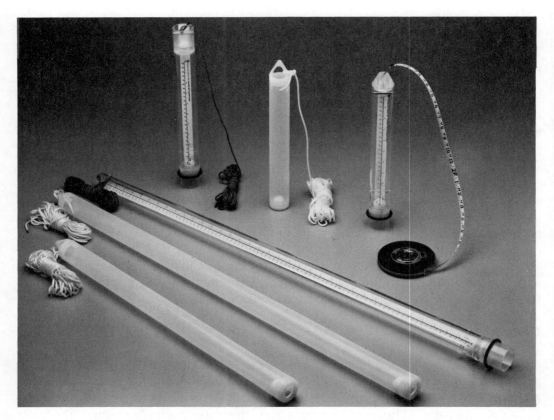

Figure 6-24. Cast acrylic and Teflon (center) bailers (photo courtesy of ORS Environmental Equipment).

to obtain a larger sample by attaching a long, flexible tube onto the bailer, which extends to the surface when the bailer is at the desired sample depth (Gillham et al., 1982). When suction is applied to the tube, the sample can be collected at the surface. However, the probability of degassing and loss of volatile constituents is increased. This latter method, although it uses a bailer, is really a suction lift method, as described in the next section.

Important precautions must be followed if quality samples are to be obtained with the bailer. First, the bailer and bailer line must be thoroughly cleaned prior to each well sampled, the well must be purged prior to sampling by removing the appropriate number of well volumes, and, if purged using a bailer, bailing should be timed to approach a constant pumping rate (Nielsen and Yeates, 1985). The bailer line should be held off of the ground during sampling and be of noncontaminating material, which can be cleaned (i.e., not intertwined rope), and the bailer should be lowered to the same depth in the well every time.

The bailer sampling device is commonly used and can be used for sampling water from any depth, and in small-diameter wells with low yields (Table 6-9). Studies by Gibb et al. (1981) indicate that results using the bailer are as good as any other device for sampling organics. The simplicity of the device generally makes it inexpensive to purchase as well as to operate. Each bailer sample contains a relatively

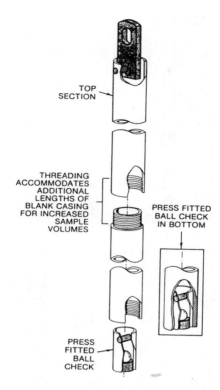

TOP
SECTION

THREADING
ACCOMMODATES
ADDITIONAL
LENGTHS OF
BLANK CASING
FOR INCREASED
SAMPLE
VOLUMES

PRESS FITTED
BALL CHECK
IN BOTTOM

PRESS
FITTED
BALL
CHECK

Figure 6-25. Single check-valve bailer (diagram courtesy of Timco Mfg. Inc.).

small amount of water and thus is inefficient, and in some cases impractical, as with removing water samples from a deep well or from large production wells, which is time consuming and laborious.

Depth-Specific Sampling Device. Another, similar, sampling device available is a mechanically operated depth-specific sample collection device. It consists of a cylindrical tube with spring-activated rubber stoppers at each end, which can be activated to shut when the desired sample depth is reached, entrapping a depth-specific sample. The rubber stoppers and spring assembly could, however, cause a bias in the trace metal and trace organic constituents of the sample. Other devices may be pressurized or put under vacuum, and then lowered into the well for sampling at a specific depth by releasing the pressure or opening the container. This method could cause chemical alteration of the water sample when contacted, with the release of pressurized gas or when put under vacuum.

Syringe. This pneumatic sampling device, which is rapidly being replaced by the bailer, consists of an inexpensive disposable plastic syringe that is modified by cutting off the plunger and the finger grips or a more elaborate stainless steel and teflon syringe, both of which are attached to a gas-line. The device is lowered into the well to the desired depth and, when depressurized, the water sample is collected through the syringe needle (Gillham et al., 1982). The advantages of the plastic syringe are that it may be rinsed down-hole with the water to be sampled, it is portable, inexpen-

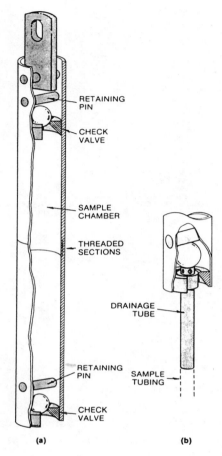

RETAINING
PIN

CHECK
VALVE

SAMPLE
CHAMBER

THREADED
SECTIONS

DRAINAGE
TUBE

RETAINING
PIN

SAMPLE
TUBING

CHECK
VALVE

(a)   (b)

Figure 6–26. Point source bailer, also referred to as double check-valve bailer (courtesy of Timco Mfg. Inc.).

sive, there is decreased possibility of cross-contamination between sample points, and after the tip is sealed the syringe may be used for storage of the sample, reducing the possibility of degassing during sample transfer (Table 6–9). The stainless steel syringe is more expensive, but can be used at depths of hundreds of feet, and is easily cleaned between sampling points.

Negative (Suction) Displacement Methods—Suction Lift Methods. A suction-lift pump operates by the use of negative pressure (suction) applied to the water or to the gas phase above the water (e.g., vacuum flask) to lift the water sample out of a well (Gillham et al., 1982) through a suction line.

The suction-lift components generally consist of a long tube, a sampling device, and a pump. The tube has one end attached to a surface or shallow-depth pump, and the other end is lowered into the well water to collect the sample. The tubing may be an integral part of the well or may have to be installed. The sampling device is either attached to the tube at the surface between the downhole end of the tube and the pump or is used to collect the outflow from the pump. A variety of pumps and sampling devices may be used with this method. The negative pressure can be applied by hand-vacuum pumps connected to a vacuum flask (Figure 6–27), peristaltic pumps, or other shallow-depth pumps, such as piston pumps and centrifugal pumps (Gillham et al., 1982). The sampling devices consist of vacuum flasks (although not useful for collecting samples for volatile organics), direct collection of the sample from pump outflow (can be contaminated from the pump's metal parts), and syringe or similar devices that connect to a tee or three-way valve installed in tubing in advance of the flask or pump. The advantages of the syringe include limited aeration and degassing, the ability to add preservatives to the syringe prior to sampling, and it doubles as a sample container, however inefficient for collecting large sample volumes.

The suction-lift method of sampling is realistically restricted to wells where the surface of the standing water is less than 6 to 8 meters (20 to 26 feet) below the ground surface due to the limit of suction-lift (the fact that water cannot to be lifted by suction a distance greater than that equivalent to one atmospheric pressure). Its major advantages are that it is useful in nonplumb wells, it is simple, readily available, relatively inexpensive, portable, and can provide continuous or variable flow rates (Table 6–9). Caution must be used however, in selecting the appropriate suction-lift method and its components (i.e., pump, sampling device, tubing and set up) when sampling for volatile organics or trace metals.

Positive Displacement Methods.   Positive displacement sample-collection methods use positive pressure (driving force) to drive water from the well to the ground surface (Gillham et al., 1982). These devices include submersible piston and centrifugal pumps, gas-operated bladder pumps, gas-drive devices, and gas-lift devices. The latter two methods allow the gas (the driving positive pressure) to be in contact with the water to be sampled, whereas the former methods do not or do not use gas as the driving force. Positive displacement methods are useful in obtaining water samples at greater depths, thousands of meters, than both suction-lift methods, which are limited to the suction-lift capability (approximately 20 ft to 26 ft), and also bailers, which are laborious and impractical at great depths (Gillham et al., 1982). Another advantage is that positive displacement methods allow for substantial and variable pumping rates (Gillham et al., 1982). Chemical alterations are also less as

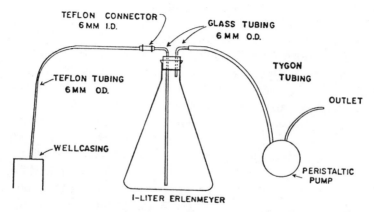

Figure 6–27.   Suction lift device with in-line suction flask (after Pettyjohn et al., 1981).

a result of imposing positive pressure on the water sampled versus applying negative pressure (suction). At great depths, even within tens of meters of the surface, however, samples will undergo a decrease in pressure when raised to the surface, hence degassing could occur.

Submersible Piston Pump. The most common submersible piston pumps operate by use of either a mechanical device (rod pumps) or a gas-drive mechanism. Both of these variations drive a plunger (piston) or set of plungers inside a stationary submerged cylinder. The cylinder is lowered to the desired depth using hose that supports the weight of the pump, conveys the water sample to the surface, and houses the sample-drive mechanism (i.e., rod, electric cable, gas line, etc.).

As the piston moves inside the cylinder, water is drawn into the cylinder through a lower one-way check valve. As the piston moves back, the cylinder's one-way check valve shuts, forcing the water through a discharge line past another one-way check valve and toward the surface. As the piston moves again water is again drawn into the cylinder but the one-way check valve on the discharge line prevents the previously collected water from returning to the cylinder. The process is then repeated.

This type of system is common and is referred to as the stationary barrel (cylinder) piston pump. A variation that operates under the same principle has a moving cylinder with the piston remaining stationary.

Rod Pumps. A submersible piston pump that operates by a mechanical device is called a rod pump (Figure 6–28). In this system, the piston or pistons are attached to wooden or steel rods (sucker rods) which are attached to a mechanical driving mechanism at the surface.

The cylinder is attached to a pipe (drop line) that houses the sucker rods, water discharge line, and carries the weight of the system. The driving mechanism at the surface, which drives the rods (and thus the piston) up and down, can be powered by hand, an electric motor, a gasoline engine, or a windmill (Gillham et al., 1982).

Rod pumps were initially developed for water and petroleum production, and as such are not recommended for water-quality monitoring. The major disadvantages are that it has a great potential for sample contamination, it is not easily portable, and it usually requires an external power source.

Gas-Driven Piston Pumps. Unlike the rod pumps, the gas-driven piston pumps have recently been modified specifically for water-quality sampling. This system consists of a source of gas-pressure at the surface, a gas-line connecting the gas-source to the gas chamber, a gas chamber, a piston, a chamber for the water sample, a sample discharge line, and associated check valves (Figure 6–29). By applying positive and negative pressure, the gas-source drives the piston, rather than rods as in a rod-pump system, but the principles of both systems are the same.

Other, more elaborate, gas-driven single-acting piston pump devices are available and have been described by Bianchi et al. (1962), Smith (1976), and Hillerich (1977). Another, more efficient, piston pump that is recommended for water-quality sampling, is referred to as the "double-acting" piston pump (Signor Pump) or Bennett Pump (Signor, 1978) as shown in Figures 6–30 and 6–31. This device uses the same

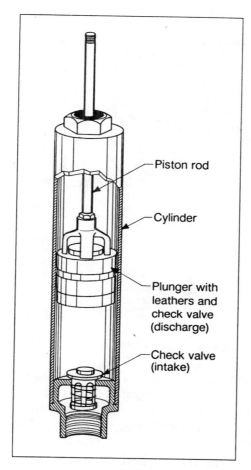

Figure 6-28. Rod pump, a single-acting piston pump (after Driscoll, 1986).

principles, only there are two water chambers—one on either side of a gas chamber, two pistons—one for each gas chamber, and a switching unit to alternate piston movement. With this system, water being sampled is continuously being collected and driven to the surface, and hence is more efficient in driving-gas consumption. Each of the cylinder walls will contact the air used for power and the fluid being pumped during one complete cycle. Similarly, the Bennett Pump has two cylinders isolated by seals. However, one cylinder is exclusively for air used for power (the pump motor), and the other cylinder is exclusively for pumping the fluid. This isolation of air and fluid is significant in sampling applications. The advantages and disadvantages are given in Table 6-9.

The modifications developed thus far have enabled these devices to be portable, to be constructed of inert material reducing the possibility of sample contamination, and have eliminated the gas source from contacting the water sample as in other gas-drive or gas-lift methods, described later in this section.

Submersible Centrifugal Pumps. The centrifugal, or conventional, pump has a rotating element, including an impeller (a set of rotating vanes) and a shaft enclosed

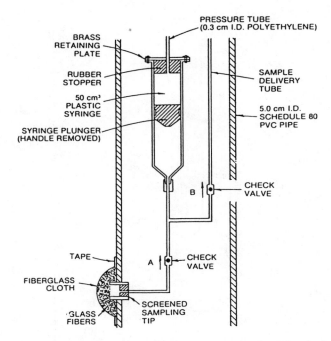

Figure 6–29. Single-acting gas-driven piston pump (after Morrison, 1983).

within a casing, bearings, a water-entry line, a discharge line, and some type of motor or force to drive the impeller (Figure 6–32). In a centrifugal pump, a rotating impeller imparts energy to water through centrifugal force.

Once the pump has been "primed" (the waterways of the pump have been filled with water), the impeller is set in motion, creating a centrifugal force which discharges the water out at the pump's periphery at a higher velocity. This creates a pressure difference in the center of the impeller where the water-entry line is attached. The pressure void draws water up from the well through the water-entry line into the pump, where the impeller again sets the water into centrifugal motion and discharges it at its periphery.

This is a continuous cycle as long as the impeller is in motion and no air enters the system. The discharge is either sent out a volute (volute pump) or through a set of stationary diffusion vanes and then out a discharge line (diffuser or turbine pumps).

The pumping rates of centrifugal pumps are wholly a function of pump size, number of stages, input power, etc. (Nielsen and Yeates, 1985). Estimated pumping rates from 5 to 40 gallons per minute (gpm) are possible with these pumps (Scalf et al., 1981), even in small diameter wells.

The disadvantages of using a centrifugal pump for water-quality sampling are numerous (Table 6–9). Because the pump has to be primed, the water used may be a potential source for sample contamination. The sample is also in contact with the pump housing and impeller, which are usually not made of inert, noncontaminating materials. The sample is also put under significant pressure changes and a high degree of turbulence, allowing for degassing and a loss of volatiles. These pumps are also limited by the suction lift limit, which is approximately 25 feet, and thus are only practical for wells with shallow water. When used as a submersible pump,

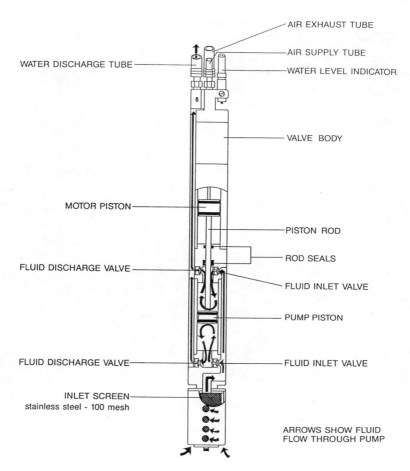

AIR EXHAUST TUBE

AIR SUPPLY TUBE

WATER DISCHARGE TUBE

WATER LEVEL INDICATOR

VALVE BODY

MOTOR PISTON

PISTON ROD

ROD SEALS

FLUID DISCHARGE VALVE

FLUID INLET VALVE

PUMP PISTON

FLUID DISCHARGE VALVE

FLUID INLET VALVE

INLET SCREEN
stainless steel - 100 mesh

ARROWS SHOW FLUID
FLOW THROUGH PUMP

Figure 6–30. Schematic of double-acting piston pump or Bennett sample pump (courtesy of Robert Bennett Co.).

the lift limit may be increased (see submersible pumps in the Positive Displacement subsection, page 236. For pumps using a gasoline motor, the sample may be further contaminated by exposure to gasoline fumes or from gasoline spillage.

These pumps were generally designed for the water-well industry for developing a well, and obtaining large quantities of water where the water is shallow, and not for water-quality sampling.

A modification to the conventional centrifugal pump has been designed specifically for groundwater-quality sampling by Johnson-Keck.[4] The advantages of this pump are that it is smaller in diameter (fits into wells 5 cm (2 inch) or greater), is constructed of stainless steel and Teflon, is reasonably efficient for collecting water from depths below the limit of suction lift, minimizes turbulence, and has an adequate pumping rate, at least for small-diameter installations, to remove standing water from the well (Gillham et al., 1982). The disadvantages of this pump are that it still may produce degassing within the pump or in its course up to the ground surface.

[4]Environmental Products Group, Johnson Division, UOP Inc., P.O. Box 43118, St. Paul, MN 55164.

Figure 6–31.   Double-acting piston pump (Well Wizard™ System) in operation for collection of volatile organic sample (photo courtesy of Q.E.D. Environmental Systems, Inc.).

Gas-Driven Devices.

Gas-drive device (gas phase replaces piston).   The operating principle of a gas-drive sampling device is not unlike that of a single-acting piston pump. However, the gas phase replaces the piston (Gillham et al., 1982). The gas-drive sampling device generally consists of a rigid cylindrical chamber, with a gas-entry tube and a sample

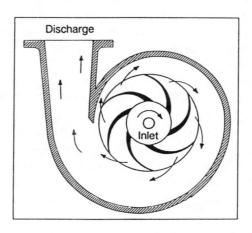

Figure 6–32.   Schematic of volute-type centrifugal pump (after Driscoll, 1986).

discharge tube that attaches to the top of the cylinder and extends to the ground surface, a water-entry check valve at the bottom, and a cylindrical screened intake may be attached below that (Figure 6–33).

Both the gas-entry and sample discharge lines extend into the cylindrical chamber of the sampling device. However, the sample discharge line extends almost to the bottom of the chamber. Many variations to this simple and most common gas-drive sampling device are available.

This sampling method operates by first closing the water-entry check valve with applied positive gas-pressure, then the device is lowered to the desired sample depth, the gas pressure is then released, allowing water to enter through the screened interval (if present) and through the water-entry-check valve into the cylinder's sampling chamber. Applying positive gas-pressure again, above that of the hydrostatic pressure of the sampling depth, closes the check valve and forces water up into the sampling discharge tube to the ground surface.

The gas-drive sampling method is useful for collecting water samples at great depths (hundreds of meters) limited only by the burst strength of the tubing and the fittings (Nielsen and Yeates, 1985). Also, by creating multiple pumping stages, the depth limit may be increased. This method can be utilized in wells of $1\frac{1}{2}$ inch inside diameter, and in existing well installations or permanently in boreholes without casing. When used as permanent installations, they are most applicable in relatively permeable materials and where only a small number of sampling points are required, due to cost. Other variations have been developed to alleviate some of the difficul-

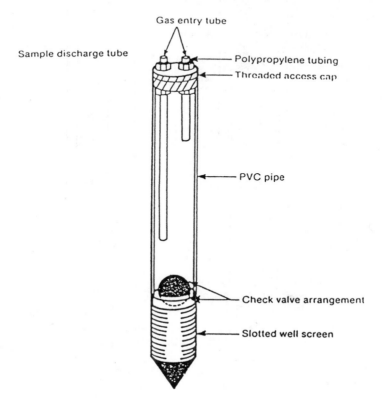

Figure 6–33. Gas-drive sampling device (after Morrison and Brewer, 1981).

ties encountered. One variation is a continuous-flow gas-drive device (Tomson et al., 1980) which consists of two separate, in-line gas-drive devices developed for collecting volatile organics (Figure 6–34). This particular device is fragile, made of glass and Teflon, but other inert materials may be used instead.

Two alternatives described in Robin et al. (1982) of smaller diameter (diameters

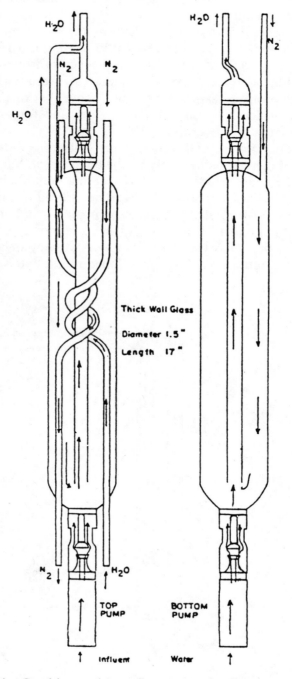

Figure 6–34. Gas-drive continuous flow pump (after Tomson et al., 1980).

as small as 0.64 cm (0.25 in) have been used) can be constructed from a variety of materials and at low cost. Another simplistic variation of the same gas-drive principle is available, using only a sample discharge line dropped into a well, with the annulus between the piezometer and the discharge line used as the gas-line when pressurized (Trescott and Pinder, 1970). This method has disadvantages in that water, and possibly the gas, may be forced back into the formation walls, thus limiting sampling to shallow depths and/or in formations of low hydraulic conductivity.

Gas-lift method (or air-lift sampler). A variation of gas-drive method, the gas-lift method of sampling, also uses gas pressure as the driving force. However, the gas is applied directly to the well water and is not contained in a sampling device that acts as a piston. The gas-lift method (also known as the air-lift method) basically operates by applying gas (usually air and sometimes an inert gas, such as nitrogen) to well water to force water out of a sample discharge line, which may be the well installation itself (Figure 6–35) (Morrison and Brewer, 1981). A variation to this method applies the gas pressure to the well water to form a stream of bubbles entrained in the sample line, thus effectively reducing the specific gravity of the water column in the sample line, causing the water column to rise to the ground surface (Gillham et al., 1982). The equipment required ranges from a simplistic system that requires only some type of hand pump and any reasonably flexible tubing used for either a gas-entry line or sample discharge line or both, to a more elaborate system consisting of a small air compressor and more complex piping arrangements.

This system as described here is not suitable for water-quality sampling, but rather is frequently used in the water-well industry for well development. Because of the presence of the gas phase, which is in contact with the water sampled, the sample may be contaminated by the carrier gas, may be stripped of dissolved gasses, and even if air is not used as the gas medium, atmospheric contamination of the sample will occur in the installation and at the surface (Table 6–9). This method may also cause the contaminated water in the installation or the carrier gas to be driven into the surrounding formation (Gillham et al., 1982).

Compared to samples obtained with a bailer or by the suction-lift method, samples obtained by gas-lift methods generally have more carbon dioxide stripping, as indicated by pH values, and thus are not suitable for sampling pH-sensitive parameters, such as for metals and volatile organics. The primary limitations to this sampling method, in addition to the potential alteration of water-quality parameters, is the amount of air pressure that can be safely applied to the tubing, and finding a suitable source of compressed air.

Gas-operated bladder (squeeze) pump. The gas-operated bladder or squeeze pump operates on a similar principle to the submersible piston pump, except that the gas pressure is applied to a flexible bladder rather than to a piston to drive the water to the surface (Middleburg, 1976, 1978). Most systems of this type consist of a long, rigid cylinder that houses a collapsible membrane and a perforated tube inside the bladder (Figures 6–36 and 6–37). When this sampling device is submerged in the well, water fills the bladder through a screened intake check valve located on the bottom of the housing. After the bladder is filled, gas pressure is applied from a surface compressor through an air line that attaches to the bladder housing. This

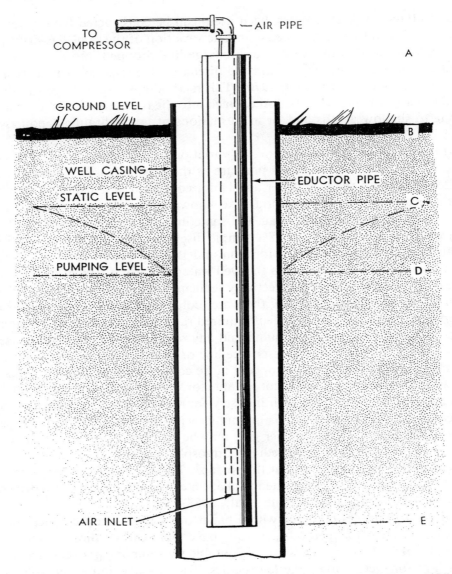

Figure 6-35. Air-lift sampler (after Departments of the Army and Air Force, 1975).

"squeezes" the bladder, forcing water out of the bladder through an upper discharge check valve and through a discharge line to the surface (Middleburg, 1978). With the release of pressure, the bladder refills, and the cycle is repeated.

Gas-operated bladder pumps are useful in relatively deep wells (as deep as 400 feet for some devices), in as small as two-inch inner diameter wells, and due to the capability of controlled variable pumping rates, are useful for efficient well evacuation as well as for sampling volatile organics at low pump rates (i.e., to less than 100 ml/min) (Nielsen and Yeates, 1985).

Many variations to this method have been developed. One device operates by inflating and deflating the bladder with gas, allowing the water to fill and empty the annulus between the bladder and the bladder housing. Another gas-operated

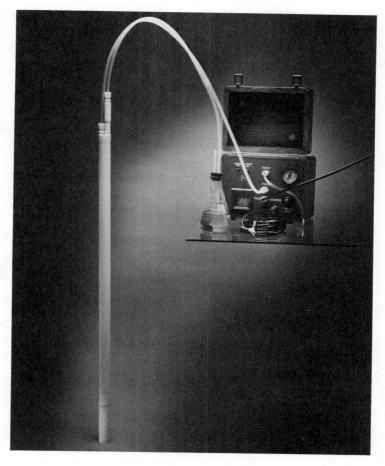

Figure 6–36. Sampling technique of gas-operated bladder pump (courtesy of Timco Mfg. Inc.).

bladder device has been developed at low cost for permanent installation. The system consists of the downhole bladder unit, connecting gas-entry and sample discharge lines (all of which can be made of Teflon or other nearly inert materials) and a locking well cap assembly with quick-connect tube fittings to allow for rapid sampling (Nielsen and Yeates, 1985).

The major disadvantages are: they can be relatively expensive, accessory equipment may be cumbersome, and pumping rates are not as great as with submersible, suction, or jet pumps (Table 6–9).

Jet Pump. A jet pump is actually a combination of an ejector and a centrifugal pump (Figure 6–38). The centrifugal pump is the prime mover of water through the system and initiates the flow. An ejector, located within the system, consists of a nozzle, an intake pipe from the well, and a venturi tube. As water is pumped through the nozzle, which smoothly, but rather abruptly reduces the area through which the flow must pass, the velocity of the flow increases in accordance with a physical law first stated by Bernoulli; water pressure in a pipe decreases in direct relation to an increase in velocity of flow and vice versa (Driscoll, 1986). Thus, the

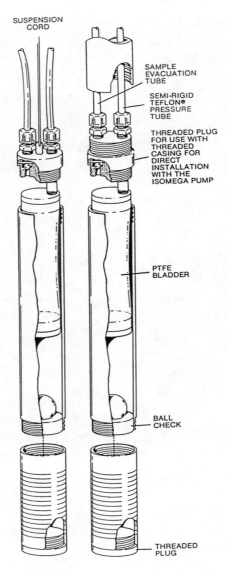

Figure 6–37. Schematic of gas-operated bladder pump, Isomega bladder pump (Courtesy of Timco Mfg. Inc.).

increased velocity out of the nozzle decreases the pressure, and if there is a sufficient pressure decrease, water is drawn into the system at this point from the well intake pipe. The combined water then continues through the venturi within the system, which gradually enlarges the area through which the water must pass to the full diameter of the pipe. According to the physical law stated above, the pressure is now increased and the velocity is reduced (with a minimum of turbulence). From this point upward the centrifugal pump moves the combined water to the control valve and discharge line. Not all of the water is released, as some is kept in the system for recirculating the flow through the system. The major advantages to this pumping system are its adaptability to small diameter wells (down to 2 inches inside

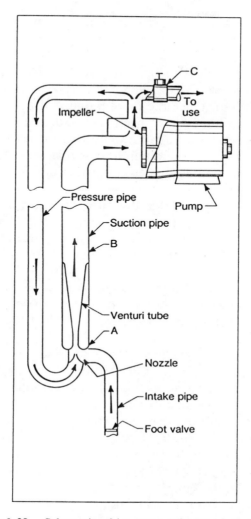

Figure 6–38. Schematic of jet pump (after Driscoll, 1986).

diameter) in deep-lift installations (Driscoll, 1986), its simplicity, and its relatively low first-cost and maintenance. The jet pump's disadvantages are many (Table 6–9), and the most important ones are it is inefficient, a large volume of water must be pumped before the sample is representative of the aquifer, the sample comes in contact with the pump and with the priming water, and the changes in pressure during water circulation causes some degassing (Gillham et al., 1982).

Transporting Devices for Sampling Equipment. The various types of sampling devices useful for groundwater monitoring can be cumbersome to mobilize to field locations. To alleviate this problem, several types of transportation equipment have been designed. One example, designed by R. L. Stollar & Associates, Inc., is a downhole pump rig truck which contains: a 3-inch submersible stainless steel pump for purging the well, a mounted power reel and generator, a flow-through box, steam cleaning equipment, and conventional bailers with tripod reel for collecting water samples. The submersible pump, which is attached to 200 feet of 2-inch hose, can be lowered into the well from the truck-mounted power reel. Once in place, the pump is powered by the truck-mounted generator, which pumps well water through the hose and into a drum.

A small plastic hose attached to the main base outlet diverts a portion of the pumped water to a flow-through box, where the pH, conductance, and temperature of the water are monitored. Once the three monitored parameters have stabilized and at least three well volumes have been removed, the pump is lifted to break suction, assuring total evacuation of the well, then turned off and removed from the well with the power reel. As the pump hose is removed, it is steam cleaned. The pump and any truck-mounted equipment that has come in contact with the water is also steam cleaned.

There is a ballcock in the pump that does not allow any of the water in the hose to drain back into the well as the pump is being removed. Once the pump has been removed from the well, the water in the hose is forced from the hose using filtered compressed air. These precautions are taken to avoid cross-contamination between wells.

The sampling is performed with the portable tripod reel placed over the well and lowering the bailer. This reel is also powered by the generator, which allows ease of sampling. The process of collecting the sample with the bailer is otherwise identical to that described earlier.

Sampling Device Summary.   Successful selection of a sampling device depends on given parameters, such as existing well diameters, sampling depths required, and project budgets, but should ultimately avoid alteration of the chemical and physical characteristics of the groundwater sample. Many sampling devices are available as described above, and in Table 6–10. However, few are as practical and efficient or as widely recommended as the bailer or, for more elaborate systems, the double-acting piston pump (or Bennett pump). Both of these devices minimize aeration and turbulence of the sample, the sample contacts only inert or stainless steel parts, and both are easy to operate, clean, and maintain.

Sampling Procedures

Once the well has been purged, the sample can be collected using the appropriate sampling device for the constituents of interest. The typical sampling devices available and the advantages and disadvantages of each (as discussed in the previous subsection on Groundwater Quality Sampling Methods and Devices, on page 226. should be taken into consideration, as well as the contaminants of interest listed in the sub-subsection on Variations in Sampling Methods for Contaminant Types, on page 251. Each time a well is sampled, water levels and the depth of the well should be measured, and purging procedures should be followed as set forth in the sub-subsection on Purging, on page 221. All previous well data should also be reviewed, in order to have some idea as to data expectations or deviations.

General Groundwater Sampling Procedures.   Groundwater samples should be collected from monitoring wells according to the time frame and intervals stated in the monitoring plan developed specifically for the site. One key method to reduce the

Table 6–10. Summary of Characteristics of Sampling Devices Available for Small-Diameter Monitoring Wells.

DEVICE	MINIMUM WELL DIAMETER (INCHES)	APPROXIMATE MAXIMUM SAMPLING DEPTH (FEET)	TYPICAL SAMPLE DELIVERY AT MAXIMUM DEPTH	FLOW CONTROL-LABILITY	MATERIALS[1] (SAMPLING DEVICE ONLY)	POTENTIAL FOR CHEMICAL ALTERATION	EASE OF OPERATION, CLEANING, AND MAINTENANCE	APPROXIMATE COST FOR COMPLETE SYSTEM[2]
Bailer	$\frac{1}{2}$	Unlimited	Variable	Not applicable	Any	Slight-moderate	Easy	<$100–$250
Syringe samplers	$1\frac{1}{2}$	Unlimited	0.2 gal.	Not applicable	Stainless 316, Teflon, or polyethylene/glass	Minimum-slight	Easy	<$100 (50 mL homemade) $1500 (850 mL commercially available)
Suction-lift (vacuum) pumps	$\frac{1}{2}$	25	Highly variable	Good	Highly variable	High-moderate	Easy	$100–$550
Gear-drive submersible pumps	2	200	0.5 gpm	Poor	Stainless 304, Teflon, Viton	Minimum-slight	Easy	$1200–$2000
Helical rotor submersible pumps	2	125	0.3 gpm	Poor	Stainless 304, EPDM, Teflon	Slight-moderate	Moderately difficult	$3500
Gas-driven piston pumps	$1\frac{1}{2}$	500	0.25 gpm	Good	Stainless 304, Teflon, Delrin	Slight-moderate	Easy to moderately difficult	$3400–$3800
Gas-drive samplers	1	300	0.2 gpm	Fair	Teflon, PVC, polyethylene	Moderate-high	Easy	$300–$700
Bladder pumps	$1\frac{1}{2}$	400	0.5 gpm	Good	Stainless 316, Teflon/Viton, PVC, silicone	Minimum-slight	Easy	$1500–$4000

[1]Materials dependent on manufacturer and specification of optional materials.
[2]Costs highly dependent on materials specified for devices and selection of accessory equipment and are based on 1985 prices.
Sources: Adapted from Nielsen and Yeates, 1985.

potential of cross-contamination is to sample the cleanest wells first, progressively sampling more contaminated wells. The most contaminated well will then be sampled last. Typical procedures for the collection of samples are outlined below (each sampling program may develop slightly different procedures and should be outlined in detail in the sampling plan):

- Applicable field parameter values should usually be collected using a "flow-through box," and recorded for each well immediately prior to sample collection. Sample labels should include time, date, sampler's initials.
- When pumps are used, samples should be collected directly from pump discharge lines at low flow rates to avoid agitating samples and the possible degassing of volatiles. The pumps should be placed near the top of the water column. If bailed, samples should be collected from bottom decanting bailers from the top of the water column (See discussion in the sub-subsection on Purging, page 221).
- Sample bottles should be rinsed with well water prior to filling unless laboratory-provided clean bottles are used or preservatives have been previously added. Sample fractions should be filled in the following sequence: 1) volatile organics (VOA's); 2) nonvolatile organics; and usually last 3) metals, since they usually have to be filtered. If the water is cloudy, however, samples for metals can be taken prior to the VOA's, to allow time for settling prior to the filtering process. The VOA sample fractions should be filled completely and capped tightly to avoid the presence of air bubbles. Except for metals, all remaining sample fractions are filled to a minimum of 90 percent capacity. Metals fractions should be filtered in the field (unless unfiltered samples are requested by the regulatory agency) using 0.45 micron microcellulose or cellulose acetate filters, filled to a minimum of 700 milliliters, and preserved with dilute nitric acid to a pH of 2. Unfiltered nitrate fractions should be similarly fixed with sulfuric acid to a pH of 2. Specific storage and preservation methods are discussed in the sub-subsection on Sample Preservation/Storage/Shipping, on page 259. The laboratories that will be performing the analysis may also have specific requirements or information and should be consulted prior to sampling. All sample fractions should be placed on ice immediately upon filling. Sampling technique, sample depth, and fractions collected should be recorded on the Sample Data Sheet, Chain-of-Custody Record Form, and in the Field Logbook.
- Fresh gloves should be worn at each sample site.
- The field personnel, or field supervisor when required to be present, should sign and date the Sample Data Sheet (Figure 6–18) after ensuring that the sheet has been fully completed and that the information has also been documented in the Field Logbook. The field supervisor completes the Chain-of-Custody Record Form when relinquishing the samples for storage and shipment.
- All sampling equipment should be thoroughly decontaminated at the well site prior to storage (See the sub-subsection on Decontamination of Equipment on page 224). All of the resulting waste water should be contained and stored in barrels or drums at the well site, if necessary. All cleaned equipment should be wrapped and stored in clean plastic sheeting.

- Each barrel or drum used to contain well evacuation and equipment decontamination water should be sealed and labeled appropriately (See the subsection on Waste and Waste Water Disposal Procedures, on page 215.
- The final activity at the well site will be to inspect the area for cleanliness and for well security.

Variations in Sampling Methods for Contaminant Types. Special considerations are necessary for designing a sampling program when certain contaminants are being tested for or when certain water properties are present. The list of contaminants being tested for today is practically limitless. However, they can be grouped into two categories for sampling techniques, inorganics and organics. The organics can be subdivided into nonvolatile, semivolatile, and volatile organics. The inorganics include metals.

Metals. At present, the regulatory agencies are not in total agreement about filtering samples for metals in the field. However, it is recommended at this time that, when sampling for metals, the samples be filtered on-site (to avoid desorption or adsorption), and then prepared by adding nitric acid (HNO_3), to a pH of less than two. The addition of the acid prevents any dissolved metals from precipitating out. If filtration is not possible in the field, the sample should not be acidified and filtration should be requested of the laboratory.

Some regulatory agencies require total metals to be determined, and therefore require that samples are not filtered. If there is sediment in the water, the metals can be desorbed from the sediment. The concentration of the metals in sediments can lead to results where metals are above regulatory limits. However, these values may be representative of natural metal concentrations in the soil. It is very difficult to determine background soil concentrations. Therefore, it is recommended that the water be filtered or the process be negotiated with the regulatory agency.

Iron is another metal that, if present in the groundwater, needs to be evaluated prior to material selection for well installation and pumping and sampling devices. Iron contributes to well maintenance and well efficiency problems. It also may affect water sampling data if the sample is not buffered with HNO_3.

Organics.

Volatile organics. Sampling for volatile organics probably requires the most care of all of the contaminants, since they have a tendency to alter their chemical and physical properties rapidly. They usually require specific pump and sampling device designs, and special handling considerations during sampling, storage, and shipment.

Depressurization, turbulence, and oxidation must be minimized or avoided during sampling and during transfer of the sample to a storage vial. The subsection on Groundwater Quality Sampling Methods and Devices, on page 226, describes the different sampling techniques and devices in more detail and recommends those appropriate for collecting volatile organic samples. Evaporation of volatile organic samples can be minimized by rapidly transferring the sample to 40 ml Teflon-lined amber septum vials. The vials should be completely filled and capped to prevent air entrapment. (Once the vial is filled and capped, turn the bottle over, if an air bubble is obvious, discard vial and fill a new one.) Placement of volatile organic samples under refrigeration (4°C) inhibits bacterial growth, which could otherwise further degrade sample integrity. Refer to the sub-subsection on Sample Preservation/Storage/Shipping, on page 259, and Tables 6–11 and 6–12 for specific details on preservation methods.

Nonvolatile organics. Most of these contaminants do not require special devices for pumping or sampling. However, well installation materials, sampling device materials, and sample container materials may need to be selected accordingly to avoid adverse chemical reactions such as adsorption or desorption. The subsection on Material Selection for Groundwater Monitoring Casings and Screens, on page 194, discusses the pros and cons of material types, some in relation to specific chemical constituents being tested for.

Sample Documentation. Because all data and procedures for data collection may be used for evidence, a stringent system is needed for data and activity documentation and recordkeeping. The EPA Contract Laboratory Program (CLP) has established standard operating procedures for sampling documentation (USEPA, 1985). All paperwork (sample sheets, labels, shipping forms, logbooks, etc.) should be identified in the sampling plan, and forms obtained well before field work commences. Sampling team members should be familiar with the required documentation before they go into the field and should allow adequate time (20 to 25 percent

Table 6–11. Preservatives Used to Retard Chemical and Biological Changes In Water Samples.

PRESERVATIVE	ACTION	APPLICABLE TO
Mercuric chloride (HgCl₂)	Bacterial inhibitor	Nitrogen forms, phosphorus forms
Nitric acid (HNO₃)	Metals solvent, prevents precipitation	Metals
Sulfuric acid (H₂SO₄)	Bacterial inhibitor	Organic samples, oil and grease, organic C, TOC, Ammonia, COD, organic N, phosphorus
	Salt formation with organic bases	Ammonia, amines
Sodium hydroxide (NaOH)	Salt formation with volatile compounds	Cyanides, organic acids
Refrigeration	Bacterial inhibitor	Acidity-alkalinity, organic materials, BOD, color, odor, organic P, organic N, organic C, etc., biological organisms (coliform, etc.)
Hydrochloric acid (HCl)	Oxidation inhibitor	Organic C, purgeable aromatic hydrocarbons, TOH or TOX
Sodium thiosulfate (Na₂S₂O₃)	Retards microbial activity	Bacterial tests, purgeable halocarbons, purgeable aromatic hydrocarbons, acrolein and acrylonitrile, phenols, benzidines, nitrosamines, nitroaromatics, polynuclear aromatic hydrocarbons, haloethers, TOH

Source: USEPA, 1974.

Table 6–12. Recommendations for Sampling and Preservation of Samples According to Measurement[1]

MEASUREMENT	CONTAINER VOLUME (ml)	CONTAINER[2]	PRESERVATIVE[3]	MAXIMUM HOLDING TIME[4]
Acidity	200	P,AG	Cool, 4°C	14 days
Alkalinity	200	P,AG	Cool, 4°C	14 days
Alpha/	100	P,G	HNO_3 to pH<2	6 mos.
Beta	125	P,G	HNO_3 to pH<2	6 mos.
Arsenic	100	P,G	HNO_3 to pH<2	6 mos.
BOD	1000	P,G	Cool, 4°C	48 hrs.
Bromide	100	P,G	None required	28 days
COD	50	P,G	H_2SO_4 to pH<2 Cool, 4°C	28 days
Chloride	50	P,G	None required	28 days
Chlorine, total residual	50	P,G	None required	Analyze immediately
Coliform, fecal and total	25	P,G	0.008% $Na_2S_2O_3$ Cool, 4°C	6 hrs.[5]
Color	50	P,G	Cool, 4°C	48 hrs.
Cyanide, total and amenable to chlorination	500	P,G	Cool, 4°C NaOH to pH>12[6] 0.6g ascorbic acid	14 days
Dissolved oxygen/ probe	300	G only	None required	Analyze immediately
Dissolved oxygen/ winkler	300	G only	Fix on site and store in dark	8 hrs.
Fluoride	60	P	None required	28 days
Hardness	100	P,G	HNO_3 to pH<2 H_2SO_4 to pH<2	6 mos.
Herbicides	80 oz[7]	G	Cool, 4°C	40 days after extraction
Metals, dissolved	500	P,G	Filtration on site, HNO_3 to pH<2. If filtration is not possible in the field, *DO NOT* acidify sample, request filtration from lab.	6 mos.
Metals, total	600	P,G	HNO_3 to pH<2	6 mos.
Mercury, dissolved and total	400	P,G	Filter; HNO_3 to pH<2 HNO_3 to pH<2	28 days 28 days
Nitrogen ammonia	400	P,G	Cool, 4°C H_2SO_4 to pH<2	28 days[8]
Kjeldahl	500	P,G	Cool, 4°C H_2SO_4 to pH<2	28 days[8]

(continued)

Table 6–12. (Continued)

MEASUREMENT	CONTAINER VOLUME (ml)	CONTAINER[2]	PRESERVATIVE[3]	MAXIMUM HOLDING TIME[4]
Nitrate	125	P,G	Cool, 4°C	48 hrs.[8]
Nitrate-nitrite		P,G	Cool, 4°C; H_2SO_4 to pH < 2	28 days[8]
Nitrite	50	P,G	Cool, 4°C	48 hrs.[8]
Oil and Grease	100	G only	Cool, 4°C H_2SO_4 to pH < 2	28 days
Organic carbon	25	P,G	Cool, 4°C HCl or H_2SO_4 to pH < 2	28 days
Pesticides and PCBs	200 or 80 oz	AG G	Cool, 4°C	40 days after extraction
pH	25	P,G	Determine on site	Analyze immediately
Phenolics	500	G only	Cool, 4°C H_2SO_4 to pH < 2 1.0 g $CuSO_4$/l	24 hrs.
Phosphorus (elemental)	50	G	Cool, 4°C	48 hrs.[8]
Orthophosphate	50	P,G	Filter on site Cool, 4°C	48 hrs.[8]
Phosphorus, total	50	P,G	Cool, 4°C H_2SO_4 to pH < 2	28 days[8]
Radium	100	P,G	HNO_3 to pH < 2	6 mos.
Residue				
Filterable and nonfilterable (TSS)	100	P,G	Cool, 4°C	7 days
Total	100	P,G	Cool, 4°C	7 days
Volatile	100	P,G	Cool, 4°C	7 days
Settleable matter	1000	P,G	Cool, 4°C	48 hrs.
Selenium	50	P,G	HNO_3 to pH < 2	6 mos.
Silica	50	P only	Cool, 4°C	28 days
Specific conductance	100	P,G	Cool, 4°C	28 days
Sulfate	50	P,G	Cool, 4°C	28 days
Sulfide	50	P,G	Cool, 4°C and NaOH to pH > 9	7 days
Sulfite	50	P,G	None required	Analyze immediately
Surfactants	250	P,G	Cool, 4°C	48 hrs.
Temperature	1000	P,G	None required	Determine immediately
Turbidity	100	P,G	Cool, 4°C	48 hrs.

Table 6–12. (Continued)

MEASUREMENT	CONTAINER VOLUME (ml)	CONTAINER[2]	PRESERVATIVE[3]	MAXIMUM HOLDING TIME[4]
Organic Tests				
Purgeable halocarbons	two 40-ml vials	G, Teflon-lined septum	Cool 4°C, 0.008% $Na_2S_2O_3$	14 days
Purgeable aromatic hydrocarbons	two 40-ml vials	G, Teflon-lined septum	Cool, 4°C, 0.008% $Na_2S_2O_3$, HCl to pH<2	14 days
Acrolein and acrylonitrile	two 40-ml vials	G, Teflon-lined septum	Cool, 4°C, 0.008% $Na_2S_2O_3$, Adjust pH to 4–5	14 days
Phenols	1000	G, Teflon-lined cap	Cool, 4°C, 0.008% $Na_2S_2O_3$	7 days until extraction, 40 days after extraction
Benzidines	1000	G, Teflon-lined cap	Cool, 4°C, 0.008% $Na_2S_2O_3$	7 days until extraction
Phthalate esters	1000	G, Teflon-lined cap	Cool, 4°C	7 days until extraction, 40 days after extraction
Nitrosamines	1000	G, Teflon-lined cap	Cool, 4°C, store in dark, 0.008% $Na_2S_2O_3$	40 days after extraction
PCBs	2000	G, Teflon-lined cap	Cool, 4°C	40 days after extraction
Nitroaromatics and isophorone	1000	G, Teflon-lined cap	Cool, 4°C, 0.008% $Na_2S_2O_3$, store in dark	40 days after extraction
Polynuclear aromatic hydrocarbons	1000	G, Teflon-lined cap	Cool, 4°C, 0.008% $Na_2S_2O_3$, store in dark	40 days after extraction
Haloethers	1000	G, Teflon-lined cap	Cool, 4°C, 0.008% $Na_2S_2O_3$	40 days after extraction
Chlorinated hydrocarbons	1000	G, Teflon-lined cap	Cool, 4°C	40 days after extraction
TCDD	1000	G, Teflon-lined cap	Cool, 4°C, 0.008% $Na_2S_2O_3$	40 days after extraction
Total organic halogens	250	G, Teflon-lined cap	Cool, 4°C, H_2SO_4 to pH<2	7 days

[1]More specific instructions for preservation and sampling are found with each procedure as detailed in the USEPA manual, (1986).
[2]Plastic, glass, or amber glass.
[3]If the sample is stabilized by cooling, it should be warmed to 25°C for reading, or temperature correction made and results reported at 25°C.
[4]It has been shown that samples properly preserved may be held for extended periods beyond the recommended holding time.
[5]If samples cannot be returned to the laboratory in less than six hours, and holding time exceeds this limit, the final reported data should indicate the actual holding time.
[6]Treatment for chlorine or other known oxidizing agents may be necessary. Test a drop of the sample with potassium iodide starch test paper (K-I starch test paper). Blue color indicates treatment needed; add ascorbic acid, a few crystals at a time, until no color on indicator paper, then add an additional 0.06g of ascorbic acid for each liter of sample volume.
[7]USEPA, 1975b.
[8]Mercuric chloride may be used as an alternate preservative at a concentration of 40 mg/l, especially if a longer holding time is required. However, the use of mercuric chloride is discouraged whenever possible.

increase in sampling time) and labor for handling the paperwork associated with field exercises. The following documents, forms, labels, and other records have been found useful by CLP, and should be specified in the sampling plan:

- Sample Labels (Figure 6–39)/Custody Seals;
- Chain-of-Custody Forms (Figure 6–40)/Shipping Forms;
- Field Log Books; and
- Other special logs and/or forms.

The following may be added to the sampling plan if the data are being used in litigation and/or the project is using the CLP program.

- Organic traffic reports (Field Sample Record and Transmittal/Submission Forms for samples for organic analyses);
- Inorganic traffic reports (Field Sample Record and Transmittal/Submission Forms for samples for inorganic analyses);
- High-hazard traffic reports (Field Sample Record and Transmittal/Submission Forms for any sample suspected of containing at least 15 percent contamination);

Because legal actions may not occur for years after the data have been gathered, it is crucial that records be sufficiently detailed to provide a complete and accurate history of data gathering and results. Each field program will be slightly different, depending on the objectives. The following discussion is for the more stringent programs where litigation is expected. The precautions and steps to be taken in data processing and storage are discussed below.

Documenting Field Measurements and Observations. All field measurements and observations should be recorded in project logbooks, field data records, or similar

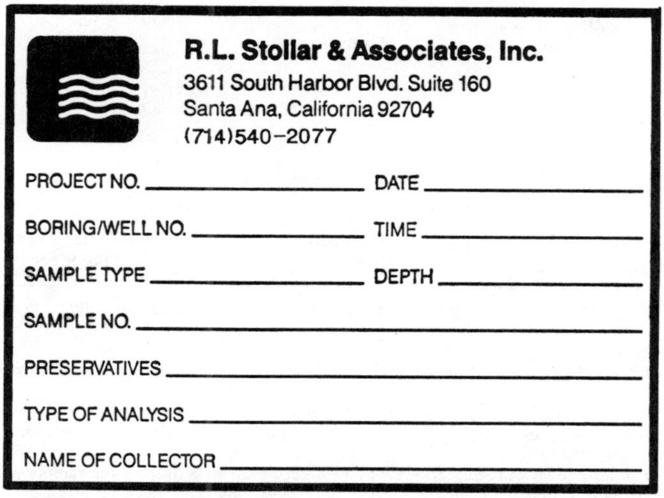

R.L. Stollar & Associates, Inc.
3611 South Harbor Blvd. Suite 160
Santa Ana, California 92704
(714)540–2077

PROJECT NO. ______ DATE ______

BORING/WELL NO. ______ TIME ______

SAMPLE TYPE ______ DEPTH ______

SAMPLE NO. ______

PRESERVATIVES ______

TYPE OF ANALYSIS ______

NAME OF COLLECTOR ______

Figure 6–39. Sample label.

Figure 6-40. Chain-of-Custody form.

CHAIN-OF-CUSTODY RECORD

R. L. STOLLAR & ASSOCIATES, INC.
3611 South Harbor Boulevard, Suite 160
Santa Ana, CA 92704
(714) 540-2077

PROJECT NAME

PROJECT NO.

COLLECTED BY (Name)

(Signature)

ANALYSES TO BE PERFORMED

SITE ID	LAB SAMPLE NUMBER	COLLECTORS SAMPLE NUMBER	DATE	TIME	TYPE OF SAMPLE	601/8010 Halogenated Volatiles	602/8020 Aromatic Volatiles	8015 Fuel Hydrocarbons	624/8240 GC/MS Volatile Cmpds	625/8270 GC/MS Base/Neutral/Acid Cmpds	418.1 Petroleum Hydrocarbons	413.2 Oil & Grease	604/8040 Phenols, sub phenols	Metals ( )	Number of Containers	REMARKS

RELINQUISHED BY (Signature)	ORGANIZATION	DATE/TIME	RECEIVED BY (Signature)	ORGANIZATION	DATE/TIME	REMARKS
RELINQUISHED BY (Signature)	ORGANIZATION	DATE/TIME	RECEIVED BY (Signature)	ORGANIZATION	DATE/TIME	
RELINQUISHED BY (Signature)	ORGANIZATION	DATE/TIME	RECEIVED FOR LABORATORY BY (Signature)		DATE/TIME	

FORM 16 JAN 88

types of recordkeeping books. Field measurements include well depth, depth to water, pH, temperature, conductivity, water flow, and certain air quality parameters measured with sampling equipment. All data must be recorded directly and legibly in Field Log Books and sometimes require all entries to be signed and dated. If entries must be changed, the change should not obscure the original entry. The reason for the change should be stated, and the change and explanation should be signed and dated or identified at the time the change is made. Field data records should be organized into standard formats whenever possible, and retained in permanent files that are well-defined and have efficient document control, inventory tracking, and filing systems (USEPA, 1985).

Sample Identification and Chain-of-Custody.   Field samples should be identified by a Sample Tag or other appropriate labeling technique. The information on the Sample Tag should include: the date and time the sample was collected, the sampling location or station and cross-reference to the sampling plan, the name of the individual collecting the sample, the analysis to be performed, preservation, and any pertinent remarks. Copies of the sample tags for the more stringent programs should be stored in a permanent file maintained for the site.

Samples and data from samples are often used as legal evidence. Therefore, sample possession must be traceable from the time that the sample is collected or developed until the derived data are introduced as evidence in legal proceedings. Chain-of-custody procedures should be followed to document sample possession and the chain-of-custody document (Figure 6–40) should remain with the samples at all times. A sample is considered under your custody if:

- It is in your possession, or
- It is in your view, after being in your possession, or
- It is in your possession and you locked it up, or
- It is in a designated secure area.

Sample identification and chain-of-custody procedures are established in the National Enforcement Investigations Center Policies and Procedures Manual (USEPA, 1981); this document should be consulted in establishing such procedures. Any documentation associated with these procedures (e.g., Chain-of-Custody Records or Receipts for Sample Forms) should also be placed in a permanent project file (USEPA, 1985).

Sample Integrity Documentation (Blanks, Splits, and Duplicates).   If organics are part of one sampling program and the program is a CLP-type program, the following is necessary. Each shipment of water samples should contain a blank water sample of certified organic-free water. The blank should be collected in a 40ml vial, sealed, labeled, packed, and stored in a manner identical to the other water samples collected. The identity of the blank water sample should not be known to the laboratory performing the analysis. Observations such as if the bottle appears to be leaking, has bubbles, etc. should be noted in the Field Log Book and on the Chain-of-Custody Form.

Duplicate water samples for analysis should be included in each sampling round.

The duplicates should be collected, sealed, labeled, packed, and stored in a manner identical to the other water samples collected. Duplicate samples should comprise at least ten percent of the samples sent to the laboratory, or at least one duplicate per sample day. Again, the identity of the duplicate water samples should not be known to the laboratory performing the analysis.

Water samples are generally collected and sent to one laboratory. However, to verify the laboratory results, it is often necessary or required to collect a water sample, split it into two containers and ship them to two separate laboratories. The tests requested should be identical. These samples should, again, not be identified to the laboratories performing the analysis. The number of splits needed varies with the extent and regulation of the project, generally they are taken only upon request.

Sample Preservation/Storage/Shipping.

Preservation. It is impossible to completely preserve samples of industrial waste water or natural water. Regardless of the nature of the sample, complete stability for every constituent can never be achieved. Preservation techniques can only retard the chemical and biological changes that inevitably continue after the sample is removed from the sampling source. According to the USEPA (1971), the changes that take place in a sample are either chemical or biological. In the former case, certain changes occur in the chemical structure of the constituents that are a function of physical conditions. Metal cations may precipitate as hydroxides or form complexes with other constituents; cations or anions may change valence states under certain reducing or oxidizing conditions; other constituents may dissolve or volatilize with the passage of time. (Metal cations, such as iron and lead, may also adsorb onto surfaces (glass, plastic, quartz, etc.). Biological changes taking place in a sample may change the valence of an element or a radical to a different valence. Soluble constituents may be converted to organically bound materials in cell structures, or cell lysis may result in the release of cellular material into solution. The well-known nitrogen and phosphorus cycles are examples of biological influence on sample composition.

Methods of preservation are intended generally to 1) retard biological action, 2) retard hydrolysis of chemical compounds and complexes, and 3) reduce the volatility of constituents. Preservation methods are generally limited to pH control, chemical addition, refrigeration, and freezing. Table 6–11 (USEPA, 1974) shows the various preservatives that may be used to retard changes in samples.

This table is a basic reference for monitoring water and wastes in compliance with the requirements of the Federal Water Pollution Control Act Amendments of 1972, although other test procedures may be used. A note of caution is needed, however, in that since these methods were selected to be applicable to the widest range of sample types, significant interferences may be encountered in certain isolated samples. For specialized projects, the means of preservation should be determined with the laboratory and the regulatory agency reviewing the data.

In summary, refrigeration at temperatures near freezing (and below, when recommended by laboratory or regulatory agency) is the best preservation technique available, but it is not applicable to all types of samples.

Many wastewater samples are unstable. In situations where the interval between sample collection and analysis is long enough to produce changes in either the concentration or the physical state of the constituent to be measured, the preservation practices in Table 6–12 for various constituents are recommended (USEPA, 1986). These choices are based on the accompanying references to Table 6–12, and on information supplied by various Regional Analytical Quality Control Coordinators.

The laboratory selected for sample analysis should provide containers that have been cleaned according to USEPA procedures, and analysis should be according to federal or state laboratory protocols. As noted in Table 6–12, the container size and the types of preservatives required depend on the analyte and sample-matrix types. Preservatives and the correct containers should be available in ample quantities for the required number and type of samples. Therefore, it is wise to take a bottle inventory at the beginning of each field session and when a new shipment arrives (include extras for bottle breakage). Some containers also come prepared from the laboratory with preservatives already added, so do not overpreserve (add more preservatives), do not allow overflow, and do not rinse with well water.

Storage and Shipment.   Each water sample should be labeled in the field with the well or boring number, date and time of sampling, collector's name and company, and the analyses to be performed. All pertinent data concerning each sample should be recorded in a Field Log Book. The sample should be immediately placed in ice chests, packed in ice, and remain in the custody of the sample collector until transport to the laboratory. Blue ice, a commercially available ice substitute (a gel-like material which can be refrozen), is sometimes used; however, do not allow the sample jars to contact blue ice or they may break. Letters of transmittal, chain-of-custody documentation, and laboratory schedules for analyses to be performed should be prepared at the end of each sampling event, and sealed inside each shipment to the laboratory. Water samples should be transported to the laboratory within 24 hours of collection. The time of transport will depend on holding times (See Table 6–12). Obviously, holding times cannot be exceeded. Holding times and the tracking of holding times should be established and discussed with the laboratory.

Quality Assurance/Quality Control (QA/QC).   Decisions concerning the control and management of hazardous substances documented in a study or the need for legal actions are based on analytical data generated during the site investigation. Because such decisions can be no better than the data on which they are based, the quality of the data must be ensured. A comprehensive and well-documented quality assurance (QA) program is essential to obtaining precise and accurate data that are scientifically and legally defensible. The concepts outlined in the QA program must be considered in decisions about the following:

- selection of sites for sampling;
- frequency of sampling;
- number of samples to be collected;
- collection and preservation procedures; and
- transport of samples.

It is also essential for the laboratory or QA officer to develop reporting mechanisms for the calibration and maintenance of instruments: and for the processing, verification, and reporting of the data. The quality control plan should provide records of traceability, adherence to prescribed protocols, nonconformity events, corrective actions, and inherent data deficiencies. Specific QA/QC requirements apply to several sampling and site characterization investigation activities. EPA references include USEPA 1974, 1975b, 1980, 1981, 1982, 1983a, 1983b, and 1985.

Sampling QA. The objectives of sampling quality assurance are: 1) to ensure that the procedures used will not detract from the quality of results, and 2) to ensure that all activities, findings, and results follow an approved plan and are documented. These objectives dictate that much of the sampling quality assurance effort be made *before* the field work. Activities that should precede sampling include:

- Preparing written protocols for all field activities and for sampling each well.
- Training all field team members.
- Ensuring that all containers and equipment have been properly cleaned and are appropriate for matrices and analytes of interest.
- Preparing a list of all field equipment required, ensuring the operability, serviceability, and safety of it, and calibrating it before and after shipping.
- Ensuring coordination with the laboratory.

A distinction should be made between field quality control and laboratory quality control. Any laboratory analyzing samples from hazardous waste sites will have an associated quality control program, and it is tempting to rely on the laboratory for all quality control. However, the laboratory's program provides adequate quality control *for the analytical function* only and cannot be used to ensure the quality of the entire sampling and analysis process. Consequently, the sampling plan should provide for adequate "field quality control" to permit evaluation of the validity of results.

In addition to provisions for quality control, sampling quality assurance should specify a system of quality assurance procedures, checks, audits, and corrective actions that is specific to the site activities.

Site Characterization QA. The purpose of site characterization quality assurance and control is to ensure that the data collected are of known and sufficient quality to assess contamination at the site qualitatively and quantitatively. QA/QC control for site characterization encompasses two important aspects:

- Records of traceability and adherence to prescribed protocols, complete descriptions of relaxed or lax quality control, and corrective actions.
- Data on the quality of the data collection and analyses, deficiencies that may affect quality, and the uncertainty limits for results.

Thus, the quality assurance/quality control plan should address at least the following elements:

- Objectives of QA/QC;
- QA/QC aspects of measurements, sampling, and analytical procedures;
- Instrument calibration, preventive maintenance, and corrective maintenance procedures;
- Data reduction and interpretation procedures;
- Quality assurance/quality control performance audits, corrective actions, and verifications;
- Documentation and document control for QA/QC; and
- Personnel responsible for QA/QC tasks.

Document Control, Inventory, and Filing Systems.   Precautions should be taken in the analysis and storage of the data collected during a site investigation to prevent the introduction of errors or the loss or misinterpretation of data. The data storage and information system should be capable of:

- Receiving all data;
- Screening and validating data to identify and reject outliers or errors;
- Preparing, sorting, and entering data into the data storage files (either computerized (preferably) or manual);
- Providing stored data points with associated quality assurance/quality control (QA/QC) "labels," which can indicate the level of confidence or quality of the data. These labels should:

    Indicate what QA/QC activities were included in the major steps of the monitoring process;
    Quantitatively describe the precision/accuracy of the analysis; and
    Assure efficiency in data security and disclosure.

Specific requirements and procedures for these aspects of data processing should be described in the QA plan prepared for the project. A member of the project team should be designated to establish and maintain the document control system for the duration of the investigation.

Analytical Process

It is apparent that there are a large number of factors to be considered, not only when designing a monitoring program, but also in evaluating groundwater quality data. Some of the factors are interactive, constituent specific, and site specific. For example, sampling methods can vary from being acceptable to unacceptable depending on the depth to water; and some sampling methods can give erroneous results for heavy metals if iron and manganese are present in the groundwater environment, and accurate results if iron and manganese are not present.

The data collected must, therefore, be scrutinized carefully before accurate analysis can be performed. It is just as important to promptly review all data when it becomes available, in order to avoid or correct problems before they become serious or while they still can be corrected.

Evaluation of Data. The design of sampling plans ensure that data will be acceptable and usable. Analysis and evaluation of data collected in accordance with an approved sampling plan are essential site characterization activities. However, it may also be necessary to evaluate existing data before sampling begins in order to develop an effective sampling plan. Analysis should be performed to show the need for additional data by examining the validity, sufficiency, and relevancy of existing data and ultimately should be applied to the results of the remedial investigation site characterization effort (USEPA, 1985).

Determining Data Validity. Because the site characterization and remedial investigation process depends on data collected at the site, quantitative evaluation of the validity of the data is essential. Validation analysis should be performed on all existing data before the sampling plan is developed to ensure that errors are identified and any necessary resampling is scheduled, as well as on all currently collected data.

Before existing data are used, the data and supporting documentation should be evaluated. This evaluation should be similar to a quality assurance audit. Data may be considered invalid if the following information is not available:

- Sampling date;
- Identity of sampling teams or person in charge;
- Sampling location and description;
- Sampling depth increment;
- Collection technique;
- Field preparation technique;
- Laboratory preparation technique;
- Laboratory analytical methods; and
- Laboratory detection limits.

Determining Data Sufficiency. Determining data sufficiency means answering the question, "Does the data adequately characterize the site?" The evaluation of existing data will identify what remains to be clarified about the types and extent of contamination, pathways of contaminant migration, and receptors. The limitations identified in the data should include:

- Hazardous waste sources; including location, quantities, concentrations, and characteristics;
- Migration pathways, including information on geology, pedology, hydrogeology, physiography, hydrology, water quality, meteorology, and air quality;
- Receptors, including demography, land use, and ecology;
- Engineering aspects, including soils, etc; and
- Aquifer restoration, if necessary.

The most important criterion in determining whether the information within a particular category is sufficient is that the data must be complete enough to allow the investigator to fully evaluate the need for source control or the management of mitigation measures and the alternatives for meeting these needs.

SUMMARY

As discussed, groundwater monitoring is an imprecise science. A groundwater monitoring program is designed based upon a conceptualization of the groundwater flow system, the physical and chemical characteristics of potential contaminants and their interaction. The results of the groundwater monitoring program can be affected by the well-drilling and well-installation processes, the selected well material, the water-sampling techniques, and the laboratory analyses. This chapter discussed the appropriate procedures, methods, equipment, and materials necessary for drilling and constructing a monitoring well, collecting soil and groundwater samples for chemical and physical analyses, and ultimately implementing a sound groundwater quality monitoring system. More detailed descriptions for each technique, method, and the materials are discussed in the cited references.

REFERENCES

Acker III, W. L. 1974. *Basic Procedures for Soil Sampling and Core Drilling,* p. 246, Acker Drill Co. Inc., Scranton, Pennsylvania.

American Society for Testing and Materials (ASTM). 1974. ASTM Designation: D 1586-67 (Reapproved 1974). Standard method for penetration test and split-barrel sampling of soils. In *Annual Book of ASTM Standards.* pp. 287–289. Philadelphia, PA: American Society for Testing and Materials.

American Society for Testing and Materials (ASTM). 1974. ASTM Designation: D 1587-74. Standard method for thin walled tube sampling of soils. In *Annual Book of ASTM Standards.* pp. 290–292. Philadelphia, PA: American Society for Testing and Materials.

American Society for Testing and Materials. 1977. ASTM Designation: D 2113-70 (Reapproved 1976). Standard method for diamond core drilling for site investigation. In *Annual Book of ASTM Standards.* pp. 324–326. Philadelphia, PA: American Society for Testing and Materials.

American Society for Testing and Materials. 1977. ASTM Designation: D 2488-69 (Reapproved 1975). Standard recommended practice for description of soils (visual-manual procedure). In *Annual Book of ASTM Standards.* pp. 379–387. Philadelphia, PA: American Society for Testing and Materials.

American Welding Society. 1981. Structural Welding Code AWS D1.1. Miami, FL: American Welding Society.

Anderson, E. G. 1977. New groundwater contaminant sampler for contaminant plume mapping. *Canadian Water Well,* p. 36.

Barcelona, M. J., Gibb, J. P., and Miller, R. A. August, 1983. *A Guide to the Selection of Materials for Monitoring Well Construction and Groundwater Sampling.* Urbana, IL: Illinois State Water Survey, ISWS Contract Report 327. 78 pp.

Bianchi, W. C., Johnson, C. E., and Haskell, E. E. 1962. A positive action pump for sampling small bore wells. *Soil Science Society of America Proceedings,* Vol. 26, No. 1, pp. 86–87.

Clements Associates, Inc. 1983. *J.M.C. Soil Investigation Equipment,* Catalog No. 6.

Driscoll, F. G. 1986. *Groundwater and Wells,* second edition. St. Paul, MN: Johnson Division, UOP Inc. 1089 pp.

Everett, L. G. and Wilson, L. G. 1977. Unsaturated Zone Monitoring for Hazardous Waste Land Treatment Units. Las Vegas, NV: Office of Research and Development U.S. Environmental Protection Agency. Contract No. 68-03-3090.

Geo-Environmental Consultants, Inc. October, 1983. Ground water investigations at hazardous waste sites: Safety and liability considerations. In *Proceedings National Water Well Association Symposium,* San Diego, CA (Geo-Environmental Consultants Inc., Port Chester, New York).

Gibb, J. P., Schuller, R. M., and Griffin, R. A. 1981. Procedures for the Collection of Representative Water Quality Data from Monitoring Wells. Cooperative Ground Water Report 7: Illinois State Water Survey, Illinois State Geologic Survey.

Gillham, R. W., Robin, M. L., Barker, J. F., and Cherry, J. A. 1982. *Ground Water Monitoring and*

Sample Bias. Washington, DC: Prepared for the Ground Water Monitoring Task Force, American Petroleum Institute. 194 pp.

Hillerich, M. S. 1977. Air-driven piston pump for sampling small diameter wells. *Bulletin, Water Resources Division, U.S. Geol. Survey.* October–December, pp. 46–47.

Johnson Division, UOP Inc. December, 1984. *Materials Selection for Ground Water Monitoring Wells.* pp. 1–19. St. Paul, MN: Johnson Division, UOP Inc.

Knirsch, K. F. and Blose, K. T. 1983. A comparison of sampling devices for trace organic constituents in the ground water. (not published.)

Middleburg, R. F. 1976. Air operated pump for sampling small diameter wells. *Bulletin, Water Research Division, U.S. Geol. Survey.* April–June.

Middleburg, R. F. 1978. Methods for sampling small diameter wells for chemical quality analysis. In *Proceedings National Conference on Quality Assurance of Environmental Measurements,* Denver, CO, November 27–29. Silver Springs, MD: Information Transfer Inc.

Miller, G. D. 1982. Uptake and release of lead, chromium and trace level volatile organics exposed to synthetic well casings. In *Proceedings Second National Symposium on Aquifer Restoration and Ground Water Monitoring,* Columbus, OH.

Morrison, R. D. 1983. *Ground Water Monitoring Technology.* Prairie Du Sac, WI: Timco Manufacturing Inc. 111 pp.

Morrison, R. D. and Brewer, P. E. 1982. Air-Lift Samplers for Zone-of-Saturation Monitoring. *Ground Water Monitoring Review,* 1(1):52–55.

National Water Well Association (NWWA) and Plastic Pipe Institute. 1980. *Manual on Selection and Installation of Thermoplastic Water Well Casing.* Worthington, OH: The National Water Well Association, and New York, NY: The Plastic Pipe Institute.

Nielsen, D. M. and Yeates, G. L. 1985. A comparison of sampling mechanisms available for small-diameter ground water monitoring wells. *Ground Water Monitoring Review,* 5(2):83–99.

Pettyjohn, W. A., Dunlap, W. J., Cosby, R., and Keeley, J. W. 1981. Sampling ground water for organic contaminants. *Ground Water,* 19(2):180–189.

Richter, H. R. and Collentine, M. G. 1983. Will my monitoring wells survive down there?: Design and installation techniques for hazardous waste studies. In *Proceedings of the Third National Symposium on Aquifer Restoration and Groundwater Monitoring.* Nielson, D. M. (Ed.). Worthington, OH: National Water Well Association.

Robin, M. J. L., Dytynyshyn, D. J. and Sweeney, S. J. 1982. Two gas-drive sampling devices. *Ground Water Monitoring Review,* 2(1):63–66.

Scalf, M. R., McNabb, J. F., Dunlap, W. J., Cosby, R. C., and Fryberger, J. 1981. *Manual of Groundwater Sampling Procedures.* NWWA/EPA Series. Worthington, OH: National Water Well Association. pp. 93.

Schmidt, R. G. 1981. Considerations in choosing drilling, casing methods for ground water monitoring wells. In: *Practical Considerations in the Design and Installation of Ground Water Monitoring Wells. A National Short Course.* Columbus, OH: The National Water Well Association.

Signor, D. C. July, 1978. Gas-driven pump for ground water samples. Water Resources Investigations, U.S. Geol. Survey, 78-72 Open-file report.

Smith, A. J. July–August, 1976. Water sampling made easier with new device. *The Johnson Drillers Journal.*

Speedstar Division, Koehring Co. *Well Drilling Manual.* Columbus, OH: The National Water Well Association.

Stollar, R. L. and Hume, M. 1983. Ground water monitoring industrial facilities, or what do we have to do now? In *Proceedings of the 1983 Annual Conference, "Environmental Management: An Industrial Perspective."* Zitney, G. R. (Ed.). Pacific Grove, CA: Association of Environmental Professionals, pp. 87–126.

Stollar, R. L., Pellissier, R., and Studebaker, P. November, 1983. Management of a ground water problem. *Journal Water Pollution Control Federation,* 55(11):1393–1403.

Todd, D. K. 1980. *Ground Water Hydrology,* second edition. New York, NY: John Wiley & Sons, Inc. 336 pp.

Tomson, M. B., Hutchins, S., King, J. M. and Ward, C. H. 1980. A nitrogen powered continuous delivery, All-Glass-Teflon pumping system for ground water sampling from below 10 meters. *Ground Water,* 18(5):444–446.

Trescott, P. C. and Pinder, G. F. 1970. Air pump for small diameter piezometers. *Ground Water,* **8**(1):10–14.

Unwin, J. P. 1982. A guide to ground water sampling, Technical Bulletin 362. New York, NY: National Council of the Paper Industry for Air and Stream Improvement Inc.

U.S. Department of the Army and the Air Force. June, 1975. *Well Drilling Operations Manual,* FM5-166, AFR 85-23. Baltimore, MD: U.S. Army AG Publications Center.

(USDI) Water and Power Resources Service (WPR). 1981. *Ground Water Manual, A Water Resources Technical Publication,* revised edition. Denver, CO: U.S. Department of Interior, U.S. Government Printing Office. 480 pp.

U.S. Environmental Protection Agency. 1971. *Methods for Organic Pesticides in Water and Wastewater,* 479-301/2113, Cincinnati, OH: National Environmental Research Center. U.S. Government Printing Office. 47 pp.

U.S. Environmental Protection Agency. 1974. *Manual of Methods for Chemical Analysis of Water and Wastes,* EPA-625-1-6-74-003. Cincinnati, OH: Methods Development and Quality Assurance Research Laboratory, National Environmental Research Center. 298 pp.

U.S. Environmental Protection Agency. 1975a. *Manual of Water Well Construction Practices,* Report EPA 570/9-75-001. Washington, DC: Office of Water Supply. 156 pp.

U.S. Environmental Protection Agency. 1975b. National interim primary drinking water regulations, Park IV Water Program. *Federal Register,* 40FR59566, Dec. 24. Washington, DC: Government Printing Office.

U.S. Environmental Protection Agency (EPA). August, 1977. *Procedures Manual for Ground Water Monitoring at Solid Waste Disposal Facilities,* Report EPA 530/SW-611. Washington, DC: U.S. Government Printing Office. 269 pp.

U.S. Environmental Protection Agency. 1979. Amendments to the national interim primary drinking water regulations: Control of trihalomethanes in drinking water. *Federal Register,* 44FR68624, Nov. 29. Washington, DC: Government Printing Office.

U.S. Environmental Protection Agency. 1980. Interim guidelines and specifications for preparing quality assurance project plans, QAMS-005/80. Washington, DC: Office of Monitoring Systems and Quality Assurance, Office of Research and Development.

U.S. Environmental Protection Agency. 1981. *National Enforcement Investigations Center Policies and Procedures Manual,* EPA Report No. 330/9-78-001-R. Washington, DC.

U.S. Environmental Protection Agency. 1982. *Handbook for Sampling and Sample Preservation of Water and Wastewater,* EPA 6004-82-029. Washington, DC.

U.S. Environmental Protection Agency. 1983a. Addendum to *Handbook for Sampling and Sample Preservation,* EPA 600/4-83-039. Washington, DC.

U.S. Environmental Protection Agency. 1983b. *Methods for Chemical Analysis of Water and Wastes,* EPA 600/4-79-020, revised March 1983. Washington, DC.

U.S. Environmental Protection Agency. 1984. Guidelines establishing test procedures for the analysis of pollutants under the Clean Water Act. *Federal Register,* 49FR43250, Oct. 26. Washington, DC: Government Printing Office.

U.S. Environmental Protection Agency. June, 1985. *Guidance on Remedial Investigations Under CERCLA,* EPA 540/G-85-002, Washington, DC.

U.S. Environmental Protection Agency. 1986. *Test Methods for Evaluating Solid Waste,* Volume 1A: *Laboratory Manual, Physical/Chemical Methods.* SW-846, third edition. Washington DC: Office of Solid Waste and Emergency Response.

U.S. Environmental Protection Agency. 1987. *A Compendium of Superfund Field Operations Methods,* EPA 540/P-87/001. Washington, DC.

WPR. See USDI.

Zapico, M. M., Vales, S. and Cherry, J. A. 1987. A wireline piston core barrel for sampling cohesionless sand and gravel below the water table. *Ground Water Monitoring Review,* **VII**(3)74–82.

7

Soil Core Monitoring

Lorne G. Everett, Ph.D.

Chief Scientist
Metcalf & Eddy
Santa Barbara, CA

VADOSE ZONE DESCRIPTION

The purposes of this chapter are twofold: (1) to describe representative devices for obtaining soil cores during unsaturated zone monitoring, and (2) to describe procedures for obtaining soil samples using these devices.

The geological profile extending from the ground surface to the upper surface of the principal water-bearing formation is called the *vadose zone*. As pointed out by Bouwer (1978), USEPA (1986), and Everett et al. (1988), the term "vadose zone" is preferable to the often-used term "unsaturated zone" because saturated regions are frequently present in the vadose zone. The term *zone of aeration* is also often used synonymously. Davis and DeWiest (1966) subdivided the vadose (unsaturated) zone into three regions designated as: the soil zone, the intermediate unsaturated zone, and the capillary fringe (Figure 7–1).

Soil Zone

The surface soil zone is generally recognized as the region that manifests the effects of weathering of native geological material. The movement of water in the soil zone occurs mainly as unsaturated flow caused by infiltration, percolation, redistribution, and evaporation (Klute, 1965). In some soils, primarily those containing horizons of low permeability, such as heavy clays, saturated regions may develop during waste spreading, creating shallow perched water tables (Everett, 1980 and 1988).

The physics of unsaturated soil-water movement has been intensively studied by soil physicists, agricultural engineers, and microclimatologists. In fact, copious literature is available on the subject in periodicals (*Journal of the Soil Science Society of America, Soil Science*) and books (Childs, 1969; Kirkham and Powers, 1972; Hillel, 1971; Hillel, 1980; Hanks and Ashcroft, 1980; and Klute, 1986). Similarly, a number of published references on the theory of flow in shallow perched water tables are available (Luthin, 1957; van Schilfgaarde, 1970; and Donigian and Rao,

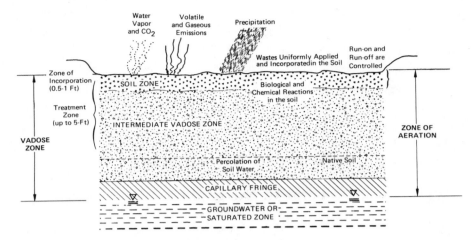

Figure 7-1.   Diagrammatic land treatment cross section in the vadose zone.

1986). Soil chemists and soil microbiologists have also tempted to quantify chemical-microbiological transformations during soil-water movement (Bohn et al., 1979; Rhoades and Bernstein, 1971; Dunlap and McNabb, 1973; and USEPA, 1987).

Intermediate Unsaturated Zone

Weathered materials of the soil zone may gradually merge with underlying deposits, which are generally unweathered, comprising the intermediate unsaturated zone. In some regions, this zone may be practically nonexistent, the soil zone merging directly with bedrock. In alluvial deposits of western valleys, however, this zone may be hundreds of feet thick. Figure 7-2 shows a geologic cross section through an unsaturated zone in an alluvial basin in California. By the nature of the processes by which such alluvium is laid down, this zone is unlikely to be uniform throughout, but may contain micro- or macrolenses of silts and clays interbedding with gravels. Water in the intermediate unsaturated zone may exist primarily in the unsaturated state, and in regions receiving little inflow from above, flow velocities may be negligible. Perched groundwater, however, may develop in the interfacial deposits of regions containing varying textures. Such perching layers may be hydraulically connected to ephemeral or perennial stream channels so that, respectively, temporary or permanent perched water tables may develop. Alternatively, saturated conditions may develop as a result of deep percolation of water from the soil zone during prolonged surface application. Studies by McWhorter and Brookman (1972) and Wilson (1971) have shown that perching layers intercepting downward-moving water may transmit the water laterally at substantial rates. Thus, these layers serve as underground spreading regions transmitting water laterally away from the overlying source area. Eventually, water leaks downward from these layers and may intercept a substantial area of the water table. Because of dilution and mixing below the water table, the effects of waste spreading may not be noticeable until a large volume of the aquifer has been affected (Everett et al., 1984).

The number of studies on water movement in the soil zone greatly exceeds the

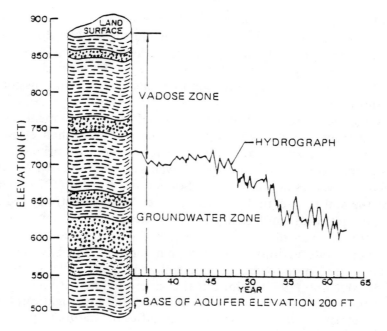

Figure 7–2. Cross section through the unsaturated zone (vadose zone) and groundwater zone.

studies in the intermediate zone. Reasoning from Darcy's equation, Hall (1955) developed a number of equations to characterize mound (perched groundwater) development in the intermediate zone. Hall also discusses the hydraulic energy relationships during lateral flow in perched groundwater. Freeze and Cherry (1979) attempted to describe the continuum of flow between the soil surface and underlying saturated water bodies. Bear et al. (1968) described the requisite conditions for perched groundwater formation when a region of higher permeability overlies a region of lower permeability in the unsaturated zone.

Capillary Fringe

The base of the unsaturated zone, the capillary fringe, merges with underlying saturated deposits of the principal water-bearing formation. This zone is not characterized as much by the nature of geological materials as by the presence of water under conditions of saturation or near-saturation. Studies by Luthin and Day (1955) and Kraijenhoff van deLeur (1962) have shown that both the hydraulic conductivity and flux may remain high for some vertical distance in the capillary fringe, depending on the nature of the materials. In general, the thickness of the capillary fringe is greater in fine materials than in coarse deposits. Apparently, few studies have been conducted on flow and chemical transformations in this zone. Taylor and Luthin (1969) reported on a computer model to characterize transient flow in this zone and compared results with data from a sand tank model. Freeze and Cherry (1979) indicated that oil reaching the water table following leakage from a surface source flows in a lateral direction within the capillary fringe in close proximity to the water table. Because oil and water are immiscible, oil does not penetrate below the water table, although some dissolution may occur.

The overall thickness of the unsaturated zone is not necessarily constant. For example, as a result of recharge at a water table during a waste disposal operation, a mound may develop throughout the capillary fringe extending into the intermediate zone. Such mounds have been observed during recharge studies (e.g., Wilson, 1971) and efforts have been made to quantify their growth and dissipation (Hantush, 1967; Bouwer, 1978).

As already indicated, the state of knowledge of water movement and chemical-microbiological transformations is greater in the soil zone than elsewhere in the unsaturated zone. Renovation of applied wastewater occurs primarily in the soil zone. This observation is borne out by the well-known studies of McMichael and McKee (1966), Parizek et al. (1967), and Sopper and Kardos (1973). These studies indicate that the soil is essentially a "living filter" that effectively reduces certain microbiological, physical, and chemical constituents to safe levels after passage through a relatively short distance (e.g., Miller, 1973; Thomas, 1973). As a result of such favorable observations, a certain complacency may have developed with respect to the need to monitor only in the soil zone.

Dunlap and McNabb (1973) point out that microbial activity may be significant in the regions underlying the soil. They recommend that investigations be conducted to quantify the extent to which such activity modifies the nature of pollutants travelling through the intermediate zone.

For the soil zone, numerous analytical techniques were compiled by Black (1965) into a two-volume series entitled *Methods of Soil Analysis*. Monitoring in the intermediate zone and capillary fringe will require the extension of the technology developed for use in both the soil zone and in the groundwater zone. Examples are already available where this approach has been used. For example, Apgar and Langmuir (1971) successfully used suction cups developed for *in situ* sampling of the soil solution at depths of up to 50 feet below a sanitary landfill. J. R. Meyer (1979) reported that suction cups were used to sample at depths greater than 100 feet below the land surface at cannery and rock phosphate disposal sites in California.

Flow Regimes

Soil-core and soil pore-liquid monitoring are used in the unsaturated zone. These two monitoring procedures are intended to complement one another. At a hazardous waste landfarm, for example, soil-core monitoring will provide information on the movement of "slower-moving" hazardous constituents (such as heavy metals), whereas soil pore-liquid monitoring will provide additional data on the movement of fast-moving, highly soluble hazardous constituents.

Current literature on soil water movement in the unsaturated zone describes two flow regimes, the classical wetting front infiltration of Bodman and Colman (1943) and a transport phenomena labeled as flow-down macropore, noncapillary flow, subsurface storm flow, channel flow, and other descriptive names, but hereafter referred to as macropore flow. The classical concept of infiltration depicts a distinct, somewhat uniform, wetting front slowly advancing in a Darcian flow regime after a precipitation event. The maximum soil moisture content approaches field capacity. Contemporary models combine this classical concept with the macropore flow phenomena (Mercer et al., 1983).

Darcian Flow.   The fundamental principle of unsaturated and saturated flow is Darcy's Law. In 1856, Henry Darcy, in a treatise on water supply, reported on experiments of the flow of water through sands. He found that flows were proportional to the head loss and inversely proportional to the thickness of sand traversed by the water. Considering a generalized sand column with a flow rate Q through a cylinder of cross-sectional area A, Darcy's law can be expressed as:

$$Q = KA \frac{hL}{L} \qquad (7\text{-}1)$$

More generally, the velocity

$$v = \frac{Q}{A} = K \frac{dh}{dL} \qquad (7\text{-}2)$$

where
dh/dL is the hydraulic gradient. The quantity K is a proportionality constant known as the *coefficient of permeability,* or *hydraulic conductivity.* The velocity in Equation (7-2) is an apparent one, defined in terms of the discharge and the gross cross-sectional area of the porous medium. The actual velocity varies from point to point throughout the column.

Darcy's law assumes one-dimensional, steady-state conditions, and is applicable only within the laminar range of flow, where resistive forces govern flow. As velocities increase, inertial forces, and ultimately turbulent flows, cause deviations from the linear relation of Equation (7-2). Fortunately, for most natural groundwater motion, Darcy's law can be applied in the equation of continuity.

Macropore Flow.   The macropore flow phenomena involves the rapid transmission of free water through large, *continuous* pores or channels to depths greater than one would expect if the flow was evenly distributed. It is important to note that this secondary porosity is made up of continuous fractures or fissures, and should not be confused with flow through large porous media. The observation that a significant amount of water movement can occur in soil macropores was first reported by Lawes et al. (1982). Reviews of subsequent work are provided by Whipkey (1967) and Thomas and Phillips (1979). Macropore flow can occur in soils at moisture contents of less than field capacity (Thomas et al., 1978). The concept of field capacity, however, is not relevant to this type of flow regime. The depth of macropore flow penetration is a function of initial water content, the intensity and duration of the precipitation event, and the nature of the macropores (Aubertin, 1971; Quisenberry and Phillips, 1976). Macropores need not extend to the soil surface for flow-down to occur, nor need they be very large or cylindrical (Thomas and Phillips, 1979). Exemplifying the role of macropores, Bouma et al. (1979) reported that planar pores with a effective width of 90μm, occupying a volume of 2.4 percent were primarily responsible for a relatively high hydraulic conductivity of 60cm day^{-1} in a clay soil. Aubertin (1971) found that water can move through macropores very quickly to depths of 10m or more in sloping forested soils. Liquid moving in the macropore flow regime is likely to bypass the soil solution in entrapped or matrix pores surrounding the macropores, and result in only partial displacement or dispersion of dissolved constituents (Quisenberry and Phillips, 1978; Wild, 1972; Shuford et al., 1977; Kissel et al., 1973; Bouma and Wosten, 1979; Anderson and Bouma, 1977).

The current concept of infiltration in well-structured soils combines both classical wetting front movement and macropore flow. Aubertin (1971) found that the bulk of the soil surrounding the macropores was wetted by radial movement from the macropores sometime after macropore flow occurred. A number of researchers have presented mathematical models in an attempt to explain the macropore flow phenomena (Beven and Germann, 1981; Edwards et al., 1979; Hoogmoed and Bouma, 1980; Skopp et al., 1981, Hern and Melancon, 1986).

GENERAL EQUIPMENT CLASSIFICATION

Soil sampling devices and systems for unsaturated zone sampling are divided into two general groups, namely: (1) those samplers used in conjunction with multipurpose or auger drill rigs, and (2) those samplers used in conjunction with hand-operated drilling devices. In most cases, the hand-operated drilling device is also the sampler.

Sampling with Multipurpose Drill Rigs

For most circumstances, the use of hollow-stem augers with some type of cylindrical sampler will provide a greater level of assurance that the soil being sampled within the unsaturated soil zone was not carried downward by the hole excavating or sampling process. For some situations, such as sampling dense to very dense, or stiff to very hard ground, the use of multipurpose auger-core-rotary drills will be necessary. For some geologic circumstances, the use of continuous flight augers will provide an adequate drilling method.

Multipurpose Auger-Core-Rotary Drill Rigs.   Multipurpose auger-core-rotary drill rigs are generally manufactured with rotary power and vertical feed control to advance both hollow-stem augers and continuous-flight (solid-stem) augers to depths greater than 100 ft (30m). These same drills have secondary capability for rotary and core drilling. The larger of these drills have 90 to 130 HP power sources, and are typically mounted on 20,000 to 30,000 lb GVW trucks. The same multipurpose drill rigs are readily available in North America on both rubber-tired and track-driven all-terrain carriers. The smaller of the multipurpose drills have 40 to 60 HP power sources and are typically mounted on trailers or one-ton, 4 × 4 trucks.

Auger Drills.   Augure drill rigs are similar to multipurpose auger-core-rotary drill rigs. They are manufactured specifically for efficient auger drilling but do not have the pumps and hoists that are required for efficient core or rotary drilling. There are relatively few auger drills available in comparison to multipurpose auger-core-rotary drills.

Hollow-Stem Auger Drilling and Sampling.   The tools used for hollow-stem auger drilling (Figure 7–3) consist of outer components: hollow auger sections, hollow auger head, and drive cap; and inner components: pilot assembly, center rod col-

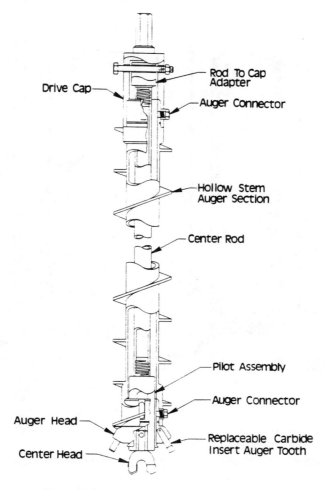

Figure 7-3. Hollow-stem auger drilling tools.

umn, and rod-to-cap adaptor. Auger sections are typically 5 ft in length, and are interchangeable for assembly in an articulated but continuously flighted column. Drilling progresses in 5 ft or shorter increments. Sampling can be accomplished at any depth within a 5 ft drilling increment. On completion of a 5ft increment of drilling, another 5 ft section of hollow auger and center rod is added. Hollow-stem augers are readily available with inside diameters of 2.25 in., 2.75 in., 3.25 in., 3.75 in., 4.25 in., 6.25 in., and 8.25 in. In general, sampling is accomplished by removing the pilot assembly and center rod and inserting the sampler through the hollow axis of the auger (Figure 7-4).

Continuous-Flight Auger Drilling and Sampling.   When continuous-flight (solid-stem) augers are used for sampling, the complete articulated column of 5 ft sections must be removed from the borehole (Figure 7-5). This method can provide an adequately clean borehole in some fine-grained soils. When the continuous-flight auger method is used in caving or squeezing ground (Figure 7-6), the quality of sample and the origin of the recovered sample is questionable.

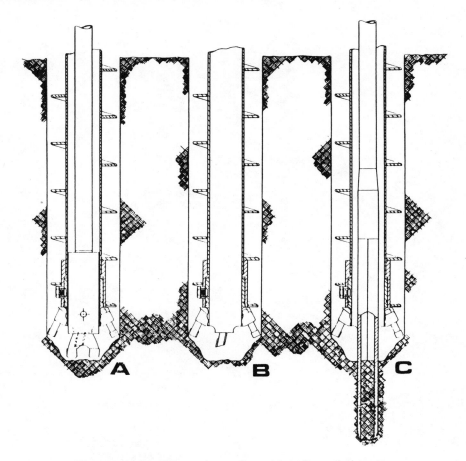

Figure 7–4. Drilling and sampling with hollow-stem augers.

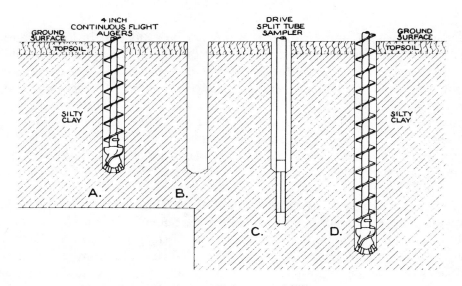

Figure 7–5. Continuous flight auger drilling.

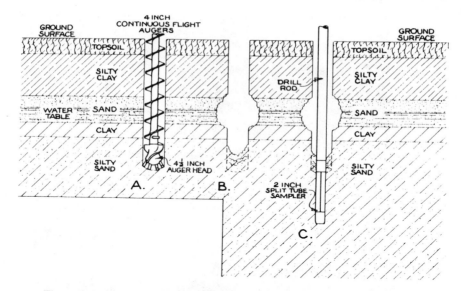

Figure 7–6. Continuous flight auger drilling through coring material.

Cylindrical Soil Samplers. Cylindrical samplers are either pushed or driven in sequence with an increment of drilling or advanced simultaneously with the advance of a hollow auger column.

Thin-Walled Volumetric Samplers. Thin-walled volumetric (Shelby tube) samplers (Figure 7–7) are readily available in 2-in., 3-in., and 5-in. OD, and are commonly 30 in. in length. The 3-in. OD × 30-in. length sampler is the most commonly used one. During the manufacturing process, the advancing end of the sampler is rolled inwardly and machined to a cutting edge that is usually smaller in diameter than the tube ID. The cutting edge ID reduction, defined as a "clearance ratio," is usually in the range of 0.0050 to 0.0150 in. or 0.50 percent to 1.50 percent (Refer to ASTM D1587).

When Shelby tubes are pushed into soil, the sample recovered is often less than the distance pushed, i.e., the recovery ratio is less than 1.0. The recovery ratio is usually less than one because the friction between the soil and the tube ID becomes greater than the shear strength of the soil in front of the tube; consequently, soil in front of the advancing end of the tube is displaced laterally rather than entering the tube (see Hvorshev, 1949). The sampler tube is usually connected with set screws to a sampler head, which, in turn, is threaded to connect with standard drill rods. The sampler head usually has a ball check valve for sampling below the water level. Plastic sealing caps (Figures 7–8A and 7–8B) and other soil sealing devices are read-

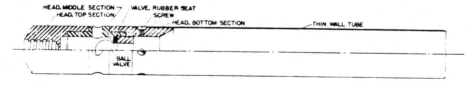

Figure 7–7. Thin-walled (Shelby tube) sampler.

Figure 7–8A. Shelby tube with acetal plastic soil seal inserted.

Figure 7–8B. Trimming tool, applicator rod, and seals with cut-away view of soil seal in place.

ly available for the 2-,3-, and 4-in. diameter tubes. Shelby tubes are commonly available in carbon steel, but can be manufactured from other metal tubing.

Split-Barrel Drive Samplers. A split-barrel drive sampler consists of two split-barrel halves, a drive shoe, and a sampler head containing a ball check valve, all of which are threaded together for the sampler assembly. The most common size has a 2-in. OD and a 1.5 in. ID split barrel, with a 1.375-in. ID drive shoe. This sampler is used extensively in geotechnical exploration (Refer to ASTM D1586). When fitted with a 16-gage liner, the sampler has a 1.375 in. ID throughout. A 3-in. OD × 2.5-in. ID split-barrel sampler, with a 2.375-in. ID drive shoe, is commonly available. Other split-barrel samplers in the size range of 2.5-in. OD to 4.5-in. OD are manufactured, but are less common.

Continuous Sample Tube System. Continuous sample tube systems that fit within a hollow-stem auger column (Figure 7–9) are manufactured and readily available in

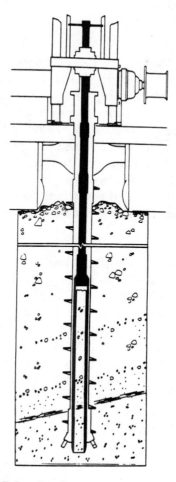

Figure 7-9. Continuous sample tube system.

North America. These sample barrels are typically 5 ft in length, fit within the lead auger of the hollow auger column and for many ground conditions provide a continuous, 5 ft sample. The soil sample enters the sampling barrel as the hollow auger column is advanced. The barrel can be "split" or "solid," and can be used with or without liners of various metallic and nonmetallic materials. Clear plastic liners are often used. Usually, two 30-in. liner sections are used.

Peat Sampler.   At some sites, the soils may contain sufficient organics such that a peat sampler may provide an adequate sample. This sampler consists of a sampling tube and an internal plunger containing a cone-shaped point, which extends beyond the sampling tube, and spring catch at the upper end. Prior to sampling, the unit is forced to the required depth, then the internal plunger is withdrawn by releasing the spring catch via an actuating rod assembly. The next step is to force the cylinder down into the undisturbed soil to the required depth, and then withdrawing the assembly with the collected sample. According to Acker (1974), the sample removed is $\frac{3}{4}$ in. in diameter and $5\frac{1}{2}$ in. in length.

Hand-Operated Drilling and Sampling Devices

Hand-operated drilling and sampling devices include all devices for obtaining soil cores using manual power. Historically, these devices were developed for obtaining soil samples during agricultural investigations (e.g., determining soil salinity and soil fertility, characterizing soil texture, determining soil-water content, etc.) and during engineering studies (e.g., determining bearing capacity). For convenience of discussion, these samplers are categorized as follows: (a) screw-type augers, (b) barrel augers, and (c) tube-type samplers. Soil samples obtained using either the screw-type sampler or barrel augers are disturbed and not truly core samples, as obtained by the tube-type samplers. Nevertheless, the samplers are still suitable for use in detecting the presence of pollutants. It is difficult to use these drilling and sampling devices in contaminated ground without transporting shallow contaminants downward.

Screw-Type Augers. The screw or flight auger essentially consists of a small-diameter (e.g., $1\frac{1}{2}$ in.) wood auger from which the cutting side flanges and tip have been removed (Soil Survey Staff, 1951). The auger is welded onto a length of tubing or rod. The upper end of this extension contains a threaded coupling for attachment to extension rods (Figure 7–10). As many extension rods are used as are required to reach the total drilling and sampling depth. A wooden or metal handle fits into a tee-type coupling, screwed into the upper-most extension rod. During sampling, the handling is twisted manually and the auger literally screws itself into the soil. Upon removal of the tool, the soil is retained on the auger flights.

According to the Soil Survey Staff (1951), the spiral part of the auger should be about 7 in. long, with the distances between flights about the same as the diameter (e.g., $1\frac{1}{2}$ in.) of the auger, to facilitate measuring the depth of penetration of the tool. The rod portion of the auger and the extensions are circumscribed by etched marks in even increments (e.g., in 6-in. increments) above the base of the auger.

Screw-type augers operate more favorably in wet rather than dry soils. Sampling in very dry (e.g., powdery) soils may not be possible with these augers.

Barrel Augers. Basically, barrel augers consist of a short tube or cylinder within which the soil sample is retained. Components of this sampler consists of (1) a penetrating bit with cutting edges, (2) the barrel, and (3) two shanks welded to the barrel at one end and a threaded section at the other end (see Figure 7–11). Extension rods are attached as required to reach the total sampling depth. The uppermost extension rod contains a tee-type coupling for attachment of a handle. The extensions are

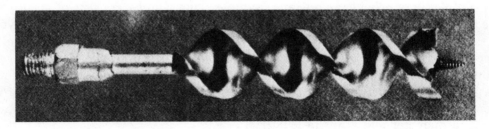

Figure 7–10.   Screw-type auger/spiral auger.

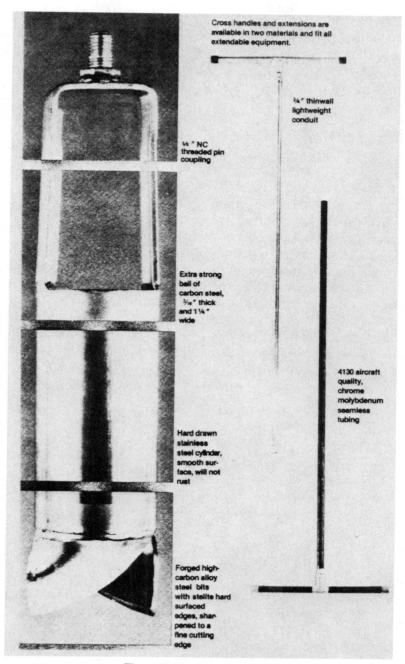

Cross handles and extensions are available in two materials and fit all extendable equipment.

¾" thinwall lightweight conduit

¼" NC threaded pin coupling

Extra strong bail of carbon steel, ³⁄₁₆" thick and 1¼" wide

Hard drawn stainless steel cylinder, smooth surface, will not rust

4130 aircraft quality, chrome molybdenum seamless tubing

Forged high-carbon alloy steel bits with stelite hard surfaced edges, sharpened to a fine cutting edge

Figure 7-11. Regular auger.

marked in even depth-wise increments above the base of the tool (Soilmoisture Equipment Corporation, 1988).

In operation, the sampler is placed vertically into the soil surface and turned to advance the tool into the ground. When the barrel is filled, the unit is withdrawn from the soil cavity and the soil is removed from the barrel. Barrel augers generally provide a greater sample size than the spiral-type augers.

Post-Hole Augers. The simplest and most readily available barrel auger is the common post-hole auger (also called the Iwan-type auger, see Acker, 1974). As shown in Figure 7–12, the barrel part of this auger is not completely solid, and the barrel is slightly tapered toward the cutting bit. The tapered barrel, together with the taper on the penetrating segment, help to retain soils within the barrel.

Dutch-Type Auger. The so-called Dutch-type auger is really a smaller variation of the post-hole auger design. As shown in Figure 7–13, the pointed bit is attached to two narrow, curved body segments, welded onto the shanks. The outside diameter of the barrel is generally only about 3 in. These tools are best suited for sampling in heavy (e.g., clay), wet soils.

Regular or General-Purpose Barrel Auger. A version of the barrel auger commonly used by soil scientists and county agents is depicted in Figure 7–11. As shown, the barrel portion of this auger is completely enclosed. As with the post-hole auger, the cutting blades are arranged so that the soil is loosened and forced into the barrel as the unit is rotated and pushed into the soil. Each filling of the barrel corresponds to a depth of penetration of about 3 to 5 in. (Soil Survey Staff, 1951). The most popular barrel diameter is $3\frac{1}{2}$ in., but sizes ranging from $1\frac{1}{2}$ in. to 5 in. are available (Art's Machine Shop, 1983).

The cutting blades are arranged to promote the retention of the sample within the barrel. Extension rods can be made either from standard black pipe or from light-

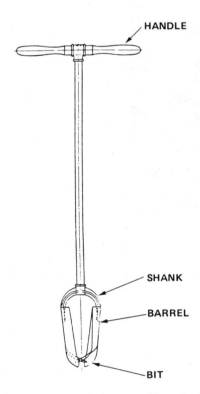

Figure 7-12. Post-hole type of barrel auger.

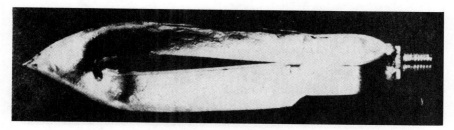

Figure 7-13. Dutch auger.

weight conduit or seamless steel tubing. The extensions are circumscribed by evenly-spaced marks to facilitate determining sampling depth.

Sand Augers.   The regular type of barrel auger described in the last paragraphs is suitable for core sampling in loam-type soils. For extremely dry sandy soils, it may be necessary to use a variation of the regular sampler, which includes a specially-formed penetrating bit to retain the sample in the barrel (Figure 7-14).

Mud Augers.   Another variation on the standard barrel auger design is available for sampling heavy, wet soils or clay soils. As shown in Figure 7-15, the barrel is designed with open sides to facilitate extraction of the samples. The penetrating bits are the same as those used on the regular barrel auger (Art's Machine Shop, 1983).

Tube-Type Samplers.   Tube-type samplers differ from barrel augers in that the tube-type units are generally of smaller diameter, and their overall length is generally greater, than the barrel augers. These units are not as suitable for sampling in dense, stoney soils as are the barrel augers. Commonly used varieties of tube-type samplers include soil-sampling tubes, Veihmeyer tubes (also called King tubes), thin-walled drive samplers, and peat samplers. The tube-type samplers are preferred if an undisturbed sample is required.

Soil Sampling Tubes.   As depicted in Figure 7-16, soil sampling tubes consist of a hardened cutting tip, a cut-away barrel, and an uppermost threaded segment. The sampling tube is attached to sections of extension rods (tubing) to attain the requisite sampling depth. A cross-handle is attached to the uppermost segment.

The cut-away barrel is designed to facilitate examining soil layering and to allow for the easy removal of soil samples. Generally, the tubes are constructed from high strength alloy steel (Clements Associates Inc., 1983). The sampler is available in

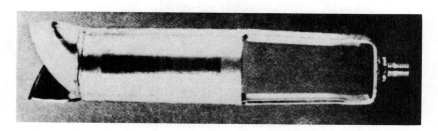

Figure 7-14.   Sand auger.

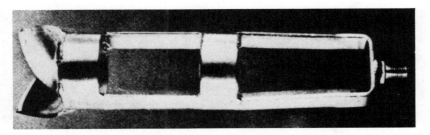

Figure 7-15. Mud auger.

Figure 7-16. Soil sampling tube.

three common lengths, namely, 12 in., 15 in., and 18 in. Two modified versions of the tip are available for sampling in either wet or dry soils. Depending on the type of cutting edge, the tube samplers obtain samples varying in diameter from $11\frac{11}{16}$ in. to $3\frac{3}{4}$ in.

Extension rods are manufactured from lightweight, durable metal. Extensions are available in a variety of lengths, depending on the manufacturer. Markings on the extensions facilitate determining sample depths.

Sampling with these units requires forcing the sampling tube in vertical increments into the soil. When the tube is filled at each depth the handle is twisted and the assembly is then pulled to the surface. Commercial units are available with attachments which allow foot pressure to be applied to force the sampler into the ground. A vibratory head has also been developed to advance tube-type samplers (VI-COR Technologies, 1988).

Veihmeyer Tube. In contrast to the soil probe, the Veihmeyer tube consists of a long, solid tube which is driven to the required sampling depth. Components of the Veihmeyer tube are depicted in Figure 7–17. As shown, these units consist of a bevelled tip which is threaded into the body tube. The upper end of the cylinder is threaded into a drive head. A weighted drive hammer fits into the tube to facilitate driving the sampler into the soil. Slots in the hammer head fit into ears on the drive head. Pulling or jerking up on the hammer forces the sampler out of the cavity. The components of this sampler are constructed from hardened metal. The tube is generally marked in even, depth-wise increments.

Hand-Held Power Augers

A very simple, commercially available auger consists of a flight auger attached to, and driven by, a small air-cooled engine. A set of two handles are attached to the head assembly to allow two operators to guide the auger into the soil. Throttle and clutch controls are integrated into grips on the handles. It is important that, if the auger "hangs up" and the operator loses control of the machine, the operator should not attempt to stop rotation of the machine by grabbing the handles.

CRITERIA FOR SELECTING SOIL SAMPLERS

Important criteria to consider when selecting soil-sampling tools for soil monitoring include: (1) the capability to obtain an encased core sample, an uncased core sample, a depth-specific representative sample, or just a sample according to the requirements of the chemical analyses; (2) suitability for sampling various soil types; (3) suitability for sampling soils under various moisture conditions; (4) accessibility to the sampling site and general site trafficability; (5) sample size requirements; and (6) personnel requirements and availability. The sampling techniques described in the previous sections were evaluated for these criteria, and the results are summarized in Table 7–1. This section briefly reviews the selection criteria. The important capability of being able to obtain a sample at depth that is not contaminated by shallow sources is greatly enhanced by using the hollow-stem auger drilling method.

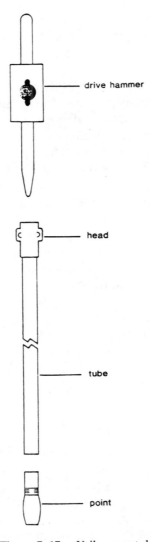

Figure 7–17.   Veihmeyer tube.

Capability for Obtaining Various Sample Types

An encased core sample can be obtained by using the continuous sample tube system, Shelby tube or piston samplers, and split-barrel drive samplers of the type that can be fitted with sealable liners. The continuous sample tube system must be used with the hollow auger drilling method. Shelby tube, piston, and split-barrel drive samplers are best used with hollow auger drilling systems, to minimize contamination of otherwise uncontaminated samples.

An uncased core sample can be obtained with the same sampling equipment and procedures that provide an encased core sample. The continuous sample tube system and split-barrel drive samplers can be used without liners to provide an uncased core sample.

Table 7-1. Criteria for Selecting Soil Sampling Equipment.

TYPE OF SAMPLER	OBTAINS CORE SAMPLE		MOST SUITABLE CORE TYPES		OPERATION IN STONY SOILS		MOST SUITABLE SOIL MOISTURE CONDITIONS			ACCESS. TO SAMPL. SITES DURING POOR SOIL CONDITIONS		RELATIVE SAMPLE SIZE		LABOR REQ'MTS	
	YES	NO	COH	COH'LESS	FAV	UNFAV	WET	DRY	INTER	YES	NO	SM	LG	SNGL	2/MORE
A. Power drilling															
1. Multipurpose drill rig	X		X	X	X		X	X	X	X		X	X		X
2. Drive sampler	X		X	X	X			X			X		X	X	
3. Thin-walled tube sampler	X		X			X			X	X					X
4. Peat sampler	X		X			X	X			X			X		X
5. Continuous sample tube system	X		X		X		X	X		X		X			X
6. Hand-held screw-type power auger		X			X				X	X			X		X
B. Hand auger															
1. Screw-type auger		X	X			X	X			X		X		X	
2. Barrel auger															
a. Post-hole auger		X	X		X		X			X				X	
b. Dutch auger		X	X		X		X						X	X	
c. Regular barrel auger		X	X		X							X		X	
d. Sand auger		X	X		X					X			X	X	
e. Mud auger		X	X		X		X			X			X	X	
3. Tube-type sampler															
a. Soil probe															
(1) Wet tip	X					X	X			X		X		X	
(2) Dry tip	X					X	X			X		X		X	
b. Veihmeyer tube	X								X			X		X	

285

A representative sample can be obtained with almost any sampling device, if contaminated, or even uncontaminated, soil has not fallen to the bottom of the borehole, or has not been transported downwardly by the drilling process. Use of the hollow auger drilling method provides greater assurance that contamination has not occurred from the drilling or sampling processes. When representative samples are desired, and continuous flight (solid-stem) auger drilling or one of the hand-operated drilling methods is used, the borehole must be made large enough to insert the sampler and extend it to the bottom of the borehole without touching the sides of the borehole. It is suggested that, if a hand-operated auger sampling method is used, a larger auger be used to advance and clean the borehole than the auger-sampler that is used to obtain the retained sample.

Sampling Various Soil Types

A split-barrel drive sampler can be used in all types of soils if the larger grain sizes can enter the opening of the drive shoe.

Shelby tubes and the continuous sample tube system are best used in fine-grained (silts and clays) and in fine granular soils. Shelby tubes can be pushed with the hydraulic system of most drill rigs in fine granular soils that are loose to medium dense, or in fine-grained soils that are soft to medium stiff. If denser or stiffer soils are encountered, driving of the tube sampler may be required. The continuous sample tube system can be used to sample soils that are much denser or harder than can be sampled with Shelby tubes, pushed, or driven.

Hand-operated samplers can be used in almost any soil type if there is enough time available—eventually the hole will be completed. Within the above sections, there is guidance provided on which hand-operated drilling device works best according to the soil types and moisture condition.

Site Accessibility and Trafficability

Site accessibility depends upon what the owner will permit. Trafficability relates to the capability of various vehicles to reach a drilling location. The availability of multipurpose drill rigs on 4 × 4 or 6 × 6 trucks or on all-terrain carriers, or the use of helicopters, negates the problems of trafficability except in exceptionally steep or wooded terrain. The relative advantages of using hand-operated drilling and sampling devices involve a comparison of the difference in the costs of decontaminating a drill rig and tools with the difference in quality of samples that can be obtained with two general methods.

Relative Sample Size

When multipurpose drill rigs are used, the sample size will depend only on the size of the drilling tools used. Hollow-stem augers with 6.25-in. ID allow the use of 5-in. OD Shelby tubes, 6-in. OD continuous sample tubes, and 4.5-in. OD split barrel drive samplers. If hand-operated tools are used, the use of larger diameter models will facilitate obtaining large samples.

Personnel Requirements

Generally, it is good practice to have at least two people in the field on all types of drilling and sampling operations. When multipurpose auger-core-rotary drills are used, the speed of drilling and sampling is much greater than the speed of drilling and sampling with hand-operated equipment and may require a larger crew to efficiently handle, log, identify, and preserve the samples.

Compositing Samples

For some of the sampling tools, such as soil probes and Veihmeyer tubes, the sample size is generally small enough that the overall size of the composite is not cumbersome. Other techniques, such as barrel augers, will provide so much sample that a composite will be of a much larger mass than required for analysis. In this case, the sample size should be reduced to a manageable volume. A simple method is to mix the samples thoroughly by shovel, divide the mixed soil into quarters, and place a sample from each quarter into a sample container. Mechanical sample splitters are also available. The USEPA (1982) recommends using the riffle technique. A riffle is a sample-splitting device consisting of a hopper and series of chutes. Materials poured into the hopper are divided into equal positions by the chutes, which discharge alternately in opposite directions into separate pans (Soiltest Inc., 1976). A modification of the basic riffle design allows for quartering of the samples.

Compositing with a Mixing Cloth.   A large plastic or canvas sheet is often used for compositing samples in the field (Mason, 1982). This method works reasonably well for dry soils, but has the potential for cross-contamination problems. Organic chemicals can create further problems by reacting with the plastic sheet. Plastic sheeting, however is inexpensive and can, therefore, be discarded after each sampling site.

This method is difficult to describe. It can be visualized if the reader will think of this page as a plastic sheet. Powder placed in the center of the sheet can be made to roll over on itself if one corner is carefully pulled up and toward the diagonally opposite corner. This process is done from each corner. The plastic sheet acts the same way on the soil as the paper would on the powder. The soil can be mixed quite well if it is loose. The method does not work on wet or heavy plastic soils. Clods must be broken up before attempting to mix the soil.

After the soil is mixed, it is again spread out on the cloth to a relatively flat pile. The pile is quartered. A small scoop, spoon, or spatula is used to collect small samples from each quarter until the desired amount of soil is acquired (this usually is about 250 to 500 grams of soil, but can be less if the laboratory desires a smaller sample). This is mixed and placed in the sample container for shipment to the laboratory.

Compositing with a Mixing Bowl.   An effective field compositing method has been to use large stainless steel mixing bowls. These can be obtained from scientific, restaurant, or hotel supply houses. They can be decontaminated and are able to stand rough handling in the field. Subsamples are placed in the bowls, broken up, then mixed using a large stainless steel scoop. The rounded bottom of the mixing bowl was designed to create a mixing action when the material in it is turned with the scoop. Careful observance of the soil will indicate the completeness of the mixing.

The soil is spread evenly in the bottom of the bowl after the mixing is complete. The soil is quartered and a small sample taken from each quarter. The subsamples are mixed together to become the sample sent to the laboratory. The excess soil is disposed of as waste.

An alternative method of compositing is to collect measured quantities of sub-samples from individual core segments. This eliminates the possibility of disproportionately sampling individual cores, and gives each core roughly equivalent weight in the composite sample. A plastic or stainless steel measuring cup is recommended to collect equal volumes from each core.

Preliminary Activities

In preparation for sample collection, it is strongly suggested that a checklist (see Table 7-2 for a typical checklist) be prepared, itemizing all of the equipment neces-

Table 7-2. Example Checklist of Materials and Supplies.

- Borebrush for cleaning.
- 10 to 12 ten-quart stainless steel mixing bowls.
- Safety equipment as specified by safety officer.
- One-quart Mason-type canning jars with Teflon liners (order 1.5 times the number of samples. Excess is for breakage and contamination losses.).
- A large supply of heavy-duty plastic trash bags.
- Sample tags.
- Chain-of-Custody Forms.
- Site Description Forms.
- Logbook.
- Camera with black-and-white film.
- Stainless steel spatulas.
- Stainless steel scoops.
- Stainless steel tablespoons.
- Caps for density sampling tubes.
- Case of duct tape.
- 100-foot steel tape.
- 2 chain surveyor's tape.
- Tape measure
- Noncontaminating sealant for volatile sample tubes.
- Supply of survey stakes.
- Compass.
- Maps.
- Plot plan.
- Trowels.
- Shovel.
- Sledge hammer.
- Ice chests with locks.
- Dry ice.
- Communication equipment.
- Large supply of small plastic bags for samples.
- Large supply of paper towels or lint-free rags.
- Large supply of distilled water.
- Work gloves.

sary, both for sampling and for maintaining quality assurance. Thus, all of the tools needed for sampling should be itemized and located in the transporting vehicle. Similarly, all of the documentation accessories, such as the field book, maps, labels, etc., should be checked off. A few minutes of preliminary preparation will ensure that all equipment is on hand that time will not be wasted in returning to the operations base for forgotten items.

Careful site preparation will also take a few minutes, but is absolutely necessary to ensure that the samples are representative of in-situ conditions. Specifically, a severe problem with some of the sampling methods described elsewhere in this chapter is that "contamination" of the sample may occur by soil falling in the cavity either from the land surface or from the walls of the borehole. Thus, to minimize contamination from surface soils, loose soils and clods should be thoroughly scraped away from each site prior to sampling. A shovel or rake will facilitate this operation. Under some geologic circumstances with some hand-operated drilling methods, perfect site preparation will not eliminate downward transport of contaminants.

It is recommended that a soil profile description be taken with each soil core sampling event. The profile description will provide information on the spatial variable properties and will assist in the interpretation of monitoring results. For instance, it is quite possible that sandy conduits (e.g., stump-holes or root channels) may contain different levels of a hazardous constituent than surrounding soil.

Sample Collection with Multipurpose Drill Rigs

There are three principal advantages in using multipurpose auger-core-rotary drill rigs for unsaturated zone sampling: (1) the work can be performed rapidly in the most adverse environments such as extremely hot or extremely cold and wet weather, (2) borings can be readily made in the densest or hardest soil conditions, and (3) there is a greater capability for preventing downward movement of contaminants during drilling and sampling. Also, with some samplers, the sample is encased as it is taken in a protective enclosure with minimal atmospheric contamination or loss of volatile constituents. The only disadvantage is the cost of decontaminating the drill and the tools.

It is suggested that the *Drilling Safety Guide* (no date) published by the National Drilling Federation (NDF) be read and studied in-depth by all drilling and sampling personnel before using auger-core-rotary drills.

Hollow-Stem Auger Drilling and Sampling. The general process of using hollow-stem augers to simultaneously advance and case a borehole was previously presented (refer to Figures 7–3 and 7–4). The following is a detailed, yet generalized, procedure.

1. The outer and inner hollow-auger components (Figure 7–4A) are assembled and connected by the shank on top of the drive cap to the rotary drive of the drill rig.
2. This assembly is advanced to the desired sampling depth using the rotary action and ram forces of the drill rig. The auger head cuts into the soil at the

bottom of the hole and directs the cuttings to the spiral flights, which convey the cuttings to the surface (Figure 7–4A).

3. The drive cap is disconnected from the auger column assembly. The pilot assembly with the center rod column is then removed, usually with a hoist line (Figure 7–4B).

4. A sampling device attached to a sampling rod column is inserted and lowered within the hollow axis of the auger column to rest on the soil at the bottom of the hole. The sampling device is then pushed with the hydraulic feed system of the drill or driven with a hammer assembly into the relatively undisturbed soil below the auger head (Figure 7–4C).

5. The sampler is then retracted from the hollow axis of the auger column. The sampler is either retracted with a hoist line or by connecting the sampling rod column to the hydraulic feed (retract) system of the drill rig. "Back-driving" may be required to remove some samplers that are driven to obtain a sample. In some soils, back-driving will cause some or even all of the sample to be released from the sampler and remain in the bottom of the borehole. Back-driving should not be used when a hoist or the hydraulic feed of the drill can be used to retract the sampler.

6. The pilot assembly and center rod column is reinserted, the drive cap is reconnected to the auger column and the rotary drive of the drill rig. The hollow auger column is then advanced to the next sampling depth.

7. For sampling required at depths greater than about 4.5 ft, plus the length of the sampler below the auger head, additional 5-ft hollow auger sections and center rod sections are added. The flights are timed and mated at the coupling to provide a continuous conveyance of cuttings.

8. Fill in the cavity with soil, tamping to increase the bulk density of the added soil. Fill the hole to ground surface.

For some types of samplers, it is difficult to retain the sample in the sampler because of the "vacuum" within (or apparent tensile strength of) the soil at the bottom of the sample. After the sampler is pushed or driven, the hollow augers can be advanced downward to the bottom of the sampler to "break" the vacuum.

Continuous Flight Auger Drilling and Sampling. Continuous flight augers have hexagonal shank and socket connections, which prevent sampling through the usually small-diameter axial tubing; consequently, the complete auger column must be retracted and reinserted for each sampling increment.

1. The continuous flight auger assembly, for example, the auger head and 5-ft flight auger section, is connected by the top shank of the auger to the rotary drive of the drill.

2. The auger assembly is advanced to the desired sampling depth using the rotary action and ram forces of the drill rig (Figure 7–5A).

3. After rotation is stopped and the rotary power train of the drill is placed in neutral, all cuttings are carefully removed from the zone adjacent to the bore-

hole. This will minimize the amount of material that will fall to the bottom of the borehole when the augers are removed.

4. The auger column is then removed from the borehole without further rotation (Figure 7–5B). The augers should be immediately removed from the area of drilling to prevent cuttings from the auger flights falling into the borehole, and it may be necessary to remove cuttings from the area adjacent to the borehole as the auger column is retracted.

5. The sampling device on a sampling rod column is inserted and lowered into the open borehole to rest on the soil at the bottom. Care should be taken to minimize the contact of the sampler and sampling rod column with the side of the open borehole. The sampler is then pushed with the hydraulic feed system of the drill or driven with a hammer assembly, through whatever cuttings may have accumulated at the bottom of the borehole, into the undisturbed soil (Figure 7–5C).

6. The sampler is then retracted from the borehole, using the same procedures and care described above for hollow auger drilling.

7. If additional samples are required, the auger column assembly is reinserted and the drilling and sampling sequence is continued (Figure 7–5D).

8. For sampling required at depths greater than about 4.5 ft, plus the length of the sampler, additional 5-ft auger sections are added.

9. Usually, the top of the sample should be "discarded" to assure that cuttings that fall into the borehole do not provide false data or contaminate the remainder of the sample.

10. Fill in the cavity with soil, tamping to increase the bulk density of the added soil. Fill the hole to ground surface.

Samplers. Various types of samplers and complete sampling systems are available for use with hollow auger, continuous-flight auger, and other appropriate drilling methods. The sampler used will depend upon economic availability, the type of drill rig being used, the general nature of the project, and specific sampling requirements. The following are some of the samplers and related procedures commonly used in North America.

Thin-Walled Volumetric Tube Samplers. Thin-walled volumetric tube samplers are commonly called Shelby tubes (from the original manufacturer's nomenclature). Shelby tube samplers are described earlier. Shelby tube samplers can be used in most soft to stiff fine-grained soils, and in some granular soils. The Shelby tube is a rather ideal sampler in that the soil can remain in the sample tube for transportation to a testing facility. Also, Shelby tubes can be predrilled with smaller circular "sampling ports" that are "taped over" during sampling and transportation to a test facility. At the testing facility, the sealing tape can be removed as required to obtain a small cylindrical "plug sample" from the side of the larger sample. The procedure for the general use of Shelby tubes follows:

1. The borehole is advanced to the sampling depth by the selected method. When hollow-stem augers are used, the auger ID should be at least 0.20 in.

greater than the Shelby tube OD When an open hole drilling method is used, the diameter of the drilled hole should be at least 1.00 in. greater than the Shelby tube OD

2. The Shelby tube sampler is attached to the sampler head, which is, in turn, connected to a sampling rod column.

3. The Shelby tube sampler assembly is lowered within the hollow auger axis or open borehole, to rest on the bottom.

4. The sampling rod column is extended upward to contact the retracted base of the drill rig rotary box.

5. The sampler is then pushed into the soil at the bottom of the borehole by using the hydraulic feed of the drill. The Shelby tube should be pushed at a rate of about 3 to 6 in. per second. Care should be taken to assure that the top of the sampling rod column is squarely against a flat surface of the rotary box, and that there are no loose tool joints in the sampling rod column. All members of the drilling and sampling crew should stand away from the sampling rod as the sampler is being pushed.

6. The sampler should be allowed to "rest" within the soil for at least one minute to allow the soil to expand laterally against the inside of the Shelby tube. This surface contact will improve sample recovery.

7. The sampler is then pulled upward with a hoist line and hoisting swivel, or by connecting to and using the hydraulic feed system of the drill rig. In some cases, sample recovery may be improved by rotating the sampler after it has been pushed and allowed to expand against the inside of the Shelby tube.

8. The Shelby tube with sample enclosed is detached from the sampler head.

9. Any loose material on the "top" of the sample should be removed with a large spoon, a putty knife, or a similar tool.

10. If the sample is to be shipped to a testing facility within the tube, the tube ends should be sealed immediately. Sealing is best accomplished by using expanding soil seals (Figures 7–8A, 7–8B) and then capping the ends of the tubes with "plastic" caps and sealing tape.

11. It may be appropriate to extrude the samples in the field, in which case a hydraulic extruder (Figure 7–18) is used. Following extrusions, the samples are then placed in large, wide-mouthed jars or other sealable containers.

12. Fill in the cavity with soil, tamping to increase the bulk density of the added soil. Fill the hole to ground surface.

Piston Samplers.   Piston samplers usually consist of a Shelby tube sampler with a sampling head that contains a piston follower. The piston follower rests on the soil surface within the Shelby tube prior to and during pushing of the tube into the soil. The piston is then "locked" in position to provide a vacuum on top of the sample to react against the vacuum at the bottom of the sample which develops as the tube and soil sample is pulled out of the soil. Sampling procedures for piston samplers are identical to those for common Shelby tube samplers except for the activities involving the locking of the piston and the breaking of the piston vacuum to remove the sample tube and sample from the sampler head. There are different types of

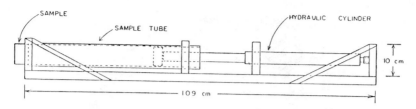

Figure 7-18. Core sample extruding device.

piston sampler heads according to the piston locking mechanism. Generally, it is only advantageous to use a piston sampler over a common Shelby tube sampler in soft, wet soils. Piston samplers will often provide optimum sample recovery in soft, wet organic soils.

Split-Barrel Drive Samplers. The split-barrel drive sampler assembly consists of a drive shoe, two split-barrel halves, and a sampler head (as described earlier). Split-barrel samplers are used with the same procedures as thin-walled volumetric samplers, except that, in almost all cases, the sampler is driven into the soil using a hammer assembly. The common 2-in. O.D. sampler is typically driven with 140 lb drive weight. Larger samplers are often driven with 300 lb, 340 lb, or 350 lb drive weights. Granular samples are often retained with the aid of various spring and flap-valve retainers (Figure 7-19).

Continuous Sample Tube Systems. The "continuous sample tube system" is a patented sampling system that consists of a 5-ft long sample barrel, as described earlier (Figure 7-9). The continuous sample tube system works best in fine-grained soils, but has been used with success in granular soils. The sample barrel is used in conjunction with hollow-stem augers, as follows:

1. The sample barrel assembly is inserted within the first hollow auger to be advanced and connected to a hexagonal extension, which passes through the drill spindle with a bearing assembly to a stabilizer plate above the rotary box.
2. The hollow auger is coupled to a flightless auger section, which is connected to the drill spindle. The cutting shoe of the auger barrel will extend a short distance in front of the auger head when the assembly is completed.
3. The cutting shoe advances into the soil as the augers are rotated and advanced into the soil.
4. The hollow augers and sampler assembly is usually advanced until the drill spindle "bottoms out."
5. The auger is then disconnected at the top from the flightless auger section.
6. The sample barrel is then hoisted upward, leaving the hollow auger in place.
7. The sample is then removed from the sample barrel. Treatment of the sample will generally be like the treatment of Shelby tube samples, but will depend specifically on whether or not a "split" or "solid" outer barrel is used, or whether or not liners are used. Typically, clear "plastic" liners are used within a split outer barrel for efficient processing of samples. These liners with soil

(a)

(b)

Figure 7-19. Soil core retainers for sampling in very wet soils and cohensionless soils.

can be processed for transportation using the same procedures that are used for Shelby tube samples.

8. When greater sampling depths are required, additional 5-ft auger sections and hexagonal drill stem extensions are used. Obtaining optimum recovery with the continuous sample tube system requires some trial-and-error adjustments by the driller. Generally, recovery approaching 100 percent is readily obtainable in fine-grained soils. In some angular granular soils, it is advisable to only advance the system in 2.5-ft increments, to obtain optimum recovery.

9. Fill in the cavity with soil, tamping to increase the bulk density of the added soil. Fill the hole to ground surface.

Peat Sampler. The peat sampler is seldom used. However, under some circumstances, it may provide the optimum sampling method.

1. Place the sampler tip on the soil surface at the exact sampling location.

2. With the tube in an exactly vertical position, force the sampler into the soil to the desired depth of sampling. (*Note:* during this step, the internal plunger is

held in place within the sampling cylinder by a piston attached to the end of the push rods).

3. Jerk up on the actuating rod to allow the plunger to move upward in the cylinder. (The snap catch will prevent the plunger from moving back downward in cylinder.)

4. Push the assembly downward to force the cylinder into undisturbed soil.

5. Extrude the sample into a clean sample container. Label the container.

7. Fill in the cavity with soil, tamping to increase the bulk density of the added soil. Fill the hole to ground surface.

Sample Collection with Hand-Operated Equipment

In the following section, step-by-step sample collection procedures are described for each of the major soil-sampling devices.

Screw-Type Augers.

1. Locate the tip of the auger on the soil surface at the exact sampling location.

2. With the auger and drill stem in an exactly vertical position, turn and pull down on the handle.

3. When the auger has reached a depth equivalent to the length of the auger head, pull the tool out of the cavity.

4. Gently tap the end of the auger on the ground or on a wooden board to remove soil from the auger flights. For very wet, sticky soils, it may be necessary to remove the soil using a spatula or by hand. In the latter instance, the operator is advised to wear disposable rubber gloves for protection from organic contaminants.

5. Clean loose soil away from the auger flights and soil opening.

6. Insert the auger in the cavity and repeat Steps 2 through 5. Keep track of the sampling depth using the marks on the drill rod or by inserting a steel tape in the hole.

7. When the auger has reached a depth just above the sampling depth, run the auger in and out of the hole several times to remove loose material from the sides and bottom of the hole.

8. Advance the auger into the soil depth to be sampled.

9. Remove the auger from the cavity and gently place the head on a clean board or other support. Remove soil from the upper flight (to minimize contamination). Using a clean spatula or other tool, scrape off soil from the other flights into the sample container. Label the sample container pursuant to the information presented in Appendix A.

10. Pour soil back into the cavity. Periodically use a rod to tamp the soil to increase the bulk density. Fill the hole to land surface.

Barrel Augers. The sampling procedures for each of the barrel augers are basically the same with minor variations. Only the procedure for the post-hole auger is presented in detail.

1. Locate the auger bit on the soil surface at the exact sampling location.
2. With the auger and extension rod in an exactly vertical position, turn and pull down on the handle (see Figure 7–20).
3. When the auger has reached a depth equivalent to the length of the auger head, pull the assembly out of the cavity.
4. Gently tap the auger head on the ground or on a wooden board to remove the soil from the auger. For very wet and sticky soils, it may be necessary to remove the soil using a spatula or rod, or by hand. In the latter instance, the operator is advised to wear disposable rubber gloves for protection from organic contaminants.
5. Remove all loose soil from the interior of the auger and from the soil opening.
6. Insert the auger back into the cavity and repeat Steps 2 through 5. Keep track

Figure 7–20. Barrel auger sampling method.

of the sampling depth using the marks on the extension rod or by extending a steel tape into the hole.

7. When the auger has reached a depth just above the sampling depth, run the auger in and out of the hole several times to remove loose materials.

8. Advance the auger into the soil depth to be sampled.

9. Carefully remove the auger from the cavity and gently place the barrel head on a clean board or other support. Using a clean spatula or other tool, scrape then soil from the control part of the head into the sample container. Discard remaining soil. Label the sample container pursuant to the information presented in Appendix A.

10. Pour soil back into the cavity. Periodically use a rod to tamp the soil to increase the bulk density. Fill hole to land surface.

Tube-Type Samplers: Soil Probe. The general procedure for soil sampling using soil probes is presented, together with the modified approach when a "back saver" attachment is used. The basic technique is described first.

1. Place the sampler tip on the soil surface at the exact sampling location.

2. With the sampling point and extension rod in an exactly vertical position, push or pull down on the handle to force the sampler into the soil.

3. When the auger has reached a depth equivalent to the length of the sampling tube, twist the handle to shear off the soil. Pull the tube out of the soil.

4. Gently remove the soil from the tube using a spatula or rod, or by hand. If the tool is cleaned by hand, the operator should wear rubber gloves for protection from organic contaminants.

5. Remove loose soil and soil stuck to the walls of the tool. Similarly, gently remove loose soil around the soil opening.

6. Insert the probe back into the cavity and repeat Steps 2 through 5. Keep track of the sampling depth using the marks on the rod or by extending a steel tape into the hole. If necessary, screw on an additional extension rod.

7. When the auger has reached a depth just above the sampling depth, run the probe in and out of the hole several times to remove loose material from the cavity walls.

8. Advance the auger into the soil depth to be sampled.

9. Carefully remove the unit from the hole and gently place the tube on a clean board. Scrape the soil out of the tube or force the sample out of the tube by pushing down on the top of the sample. Again, rubber gloves should be used. Using a clean spatula, gently place soil samples into sample containers. Label the sample container pursuant to the information presented in Appendix A.

10. Pour soil back into the cavity, periodically tamping to increase the bulk density. Fill the hole back to land surface.

A modified version of the basic sampling procedure for tube samplers provided with a so-called "back saver" handle is described in Figure 7-21.

HOW DOES THE BACKSAVER HANDLE WORK

Procedure used to pull a soil core with a sampling tube equipped with the "Backsaver Handle" or the "Backsaver N-3 Handle."

(1) Steady the soil probe in a nearly vertical position by grasping the handgrip with both hands. Force the sampling tube into the soil by stepping firmly on the footstep.

(2) Remove the first section of the core by pulling upward on the handgrip. Empty the sampling tube and clean. (see "cleaning of the soil sampling tube")

(3) Place the sampling tube in the original hole and push into the soil until the footstep is within an inch or two of the surface of the ground.

(4) While maintaining a slight pressure on the footstep pull upward on the handgrip, until the footstep has been elevated 6 to 8 inches above the surface of the ground.

(5) Maintain a slight upward pressure on the handgrip and step downward on the footstep. The footstep now grips the rod and the sampling tube can be pushed into the soil until the footstep is within 1 or 2 inches above the ground.

(6) Steps 4 and 5 are repeated until the sampling tube is full. The depth of penetration can be determined by the position of the rod end which can be seen through the viewing holes in the side of the square portion of the Backsaver Handle. It is important not to push the sampling tube into the soil to a depth that exceeds the holding capacity of the tube as this jams the sample and can make removal from the ground extremely difficult.

(7) Remove the full sampling tube by lifting upward on the handgrip. After the sampling tube has been elevated 6 to 8 inches, push downward on the handgrip returning the footstep to within 1 or 2 inches of the surface of the ground.

(8) Empty the sampling tube and clean.

(9) Steps 3 through 8 are repeated until the desired depth is reached.

Procedure used to pull a soil core with a sampling tube equipped with the "Backsaver N-2 Handle."

Same as steps 1 and 2 above.

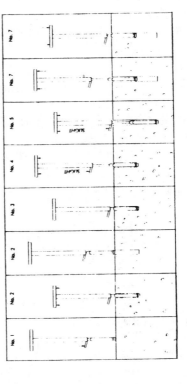

Figure 7-21.   Operation of "backsaver" handle with soil sampling tube.

Tube-Type Samplers: Veihmeyer Tubes.

1. Place the sampler tip on the soil surface at the exact sampling location. Position the tube in an exactly vertical position.

2. Place the tapered end of the drive hammer into the tube. Place one hand around the tube and the other around the hand grip on the drive hammer. While steadying the tube with one hand, raise and lower the hammer with the other. Eventually, a depth will be reached where both hands can be used to control the handle.

3. Drive the sampler to the desired depth of penetration. For some soils, the tube may be extremely difficult to remove because of wall friction. In such a case, the operator may choose to reduce the depth of penetration during advance of the hole.

4. Remove the drive hammer from the tool and place the opening in the hammer above the tube head. Rotate the hammer as required to allow the slots in the opening to pass through the ears on the head. Drop the hammer past the ears and rotate the hammer so that the unslotted opening rests against the ears. Pull the hammer upward to force the tube out of the ground. (In some cases, it may be necessary to jar the hammer head against the ears, or have another person pull up on the hammer.)

5. Gently place the side of the tube against a hard surface to remove soil from the tube. If this procedure does not work, it may be necessary to insert a long rod inside the tube to force out the soil.

6. Scrape off the side of the tube to remove loose soil. Similarly, remove loose soil from the soil cavity.

7. Insert the tube back into the soil cavity and repeat Steps 1 through 6. Keep track of the sampling depth by the marks on the tube or by extending a steel tape into the hole.

8. When the tip has reached a depth just above the sampling depth, gently run the tube in and out of the hole several times to remove loose material from the cavity walls.

9. Drive the tube to the depth required for sampling.

10. Carefully remove the unit from the hole and gently place the tip on a clean board. Force the sample out of the tube using a clean rod or extraction tool. Using a clean spatula, spoon the soil sample into a sample container. As a matter of precaution, the uppermost one or two inches of soil should be discarded on the chance that this segment has been contaminated by soil originating from above the sampling depth. Label the sample container pursuant to the information presented in Appendix A.

11. Pour soil back into the cavity, periodically tamping to increase the bulk density. Fill the hole back to ground surface.

Since the augers, probes, and tubes may pass through contaminated surface soils before reaching the sampling depth, cross-contamination is a real possibility. Soil is compacted into the threads of the auger and must be extracted with a stainless steel spatula. Probes and tubes are difficult to decontaminate without long bore brushes and some kind of washing facility. One possible way to minimize cross-contamination is to use the auger, probe, or tube to open up a bore hole to the desired depth, clean the bore hole out by repeatedly inserting the auger, probe, or tube, and finally using a separate, decontaminated auger, probe, or tube to take a soil sample through the existing open bore hole.

Miscellaneous Tools

Hand tools such as shovels, trowels, spatulas, scoops, and pry bars are helpful for handling a number of sampling situations. Many of these can be obtained in stainless steel for use in sampling hazardous contaminants. A set of tools should be available for each sampling site where cross-contamination is a potential problem. These tool sets can be decontaminated on some type of schedule in order to avoid having to purchase an excessive number of these items.

A hammer, screwdriver, and wire brushes are helpful when working with the split-spoon (split-barrel) samplers. The threads on the connectors often get jammed because of soil in them. This soil can be removed with the wire brush. Pipe wrenches are also a necessity, as is a pipe vise or a plumber's vise.

DECONTAMINATION

One of the major difficulties with soil sampling arises in the area of cross-contamination of samples. The most reliable methods are those that completely isolate one sample from the next. Freshly cleaned or disposable sampling tools, mixing bowls, sample containers, etc. are the only way to insure the integrity of the data.

Field decontamination is quite difficult to carry out, but it can be done. Hazardous chemical sampling adds another layer of aggravation to the decontamination procedures.

Laboratory Cleanup of Sample Containers

One of the best containers for soil is the glass canning jar fitted with Teflon or aluminum foil liners placed between the lid and the top of the jar. These items are cleaned in the laboratory prior to taking them into the field. All containers, liners, and small tools should be washed with an appropriate laboratory detergent, rinsed in tap water, rinsed in distilled water, and dried in an oven. They are then rinsed in spectrographic-grade solvents if the containers are to be used for organic chemical analysis. Those containers used for volatile organics analysis must be baked in a convection oven at 105°C, in order to drive off the rinse solvents.

The Teflon or aluminum foil used for the lid liners is treated in the same fashion as the jars. These liners must not be backed with paper or adhesive.

Field Decontamination

Sample collection tools are cleaned according to the following procedure (Mason, 1982).

1. Wash and scrub with tap water using a pressure hose or pressurized stainless steel fruit tree sprayer.
2. Check for adhered organics with a clean laboratory tissue.
3. If organics are present, rinse with the waste solvents from below. Discard contaminated solvent by pouring into a waste container for later disposal.
4. Air dry the equipment.
5. Double rinse with deionized distilled water.
6. Where organic pollutants are of concern, rinse with spectrographic-grade acetone, saving the solvent for use in Step 3 above.
7. Rinse twice in spectrographic-grade methylene chloride or hexane, saving the solvent for use in Step 3.
8. Air dry the equipment.
9. Package in plastic bags and/or precleaned aluminum foil.

The distilled water and solvents are flowed over the surfaces of all the tools, bowls, etc. The solvent should be collected in some container for disposal. One technique that has proven to be quite effective is to use a large glass or stainless steel funnel as the collector below the tools during flushing. The waste then flows into liter bottles for later disposal (use the empty solvent bottles for this). A mixing bowl can be used as a collection vessel. It is then the last item cleaned in the sequence of operations.

The solvents used are not readily available. Planning is necessary to insure an adequate supply. The waste rinse solvent can be used to remove organics stuck to the tools. The acetone is used as a drying agent prior to use of the methylene chloride or hexane.

Steam cleaning might prove to be useful in some cases, but extreme care must be taken to insure public and worker safety by collecting the wastes. Steam alone will not provide assurance of decontamination. The solvents will still have to be used.

SAFETY PRECAUTIONS

Safety problems may arise when operating power equipment and when obtaining soil cores at sites used to dispose of particularly toxic or combustible wastes.

Explosive gases may be given off from areas used to dispose of combustible wastes (USEPA, 1983). For such wastes, extreme caution must be taken when sampling to avoid creating sparks or the presence of open flames. Sparks will be of particular concern when sampling with power-driven equipment. Workers should not be permitted to smoke.

Protective clothing that should be worn during sample collection must be decided on a case-by-case basis. As a guide, the alternative levels of protective equipment recommended by Zirshky and Harris (1982) for use during remedial actions at hazardous waste sites could be employed. Specific items for each level are itemized in Table 7–3. Level 1 equipment is recommended for workers coming into contact with extremely toxic wastes. Such equipment items offer the maximum in protection. Level 2 equipment can be used by supervising personnel who do not directly contact the waste. Level 3 equipment applies primarily to sampling on background areas or

Table 7-3. Personnel Protective Equipment.

LEVEL	EQUIPMENT
1	3-M White Cap with air-line respiration PVC chemical suit Chemical gloves taped to suit, leather gloves as needed Work boots with neoprene overshoes taped to chemical suit Cotton coveralls, underclothing/socks (washed daily) Cotton glove liners Walkie-talkies for communications Safety glasses or face shield
2	Hard hat Air purifying respirator with chemical cartridges PVC chemical suit and chemical gloves Work boots with neoprene overshoes taped to chemical suit Cotton coveralls/underclothing/socks (washed daily) Cotton glove liners Walkie-talkies for communications Safety glasses or face shield
3	Hard hat Disposable overalls and boot covers Lightweight gloves Safety shoes Cotton coveralls/underclothing/socks (washed daily) Safety glasses or face shield
4	Positive pressure self-contained breathing apparatus PVC chemical suit Chemical gloves, leather gloves, as needed Neoprene safety boots Cotton coveralls/underclothing/socks (washed daily) Walkie-talkie for communications Safely glasses or face shield

Source: Zirshky and Harris, 1982.

sites used to dispose of fairly innocuous wastes. Level 4 equipment could be used during an emergency situation, such as a fire.

OSHA is the principal Federal agency responsible for worker safety. This agency should be contacted for information on safety training procedures and operational safety standards (USEPA, 1983a).

REFERENCES

Acker, W. L. 1974. *Basic Procedures for Soil Sampling and Drilling.* Scranton, PA: Acker Drill Co. Inc.

Anderson, J. L. and Bouma, J. 1977. Water movement through pedal soils: I. Saturated flow. *Soil Sci. Soc. Am. J.* **41**:413–418.

Apgar, M. A. and Langmuir, D. 1971. Ground-water pollution potential of a landfill above the water table, *Ground Water* **9**(6):76–93.

Art's Machine Shop. 1983. Personal communication. American Falls, ID.

Art's Machine Shop. 1982. Soil sampling augers. *Ground Water Monitoring Review* **2**(1):20.

Aubertin, G. M. 1971. Nature and extent of macropores in forest soils and their influence on subsurface water movement. U.S.D.A. Forest Serv. Res. Paper NE-912. Upper Darby, PA: Northeast Forest Exp. Stn.

Ayers, R. S. and Branson, R. L. (Eds.) 1973. Nitrates in the Upper Santa Ana River Basin in relation to groundwater pollution. California Agric. Exp. Station, Bulletin 861.

Bear, J., Zaslavsky, D., and Irmay, S. 1968. *Physical Principles of Water Percolation and Seepage.* United Nations Educational, Scientific and Cultural Organization.

Beven, K. and Germann, P. 1981. Water flow in soil macropores: 2. A combined flow model. *J. Soil Sci.* **32**:15–29.

Black, C. A. (Ed.). 1965. *Methods of Soil Analysis* (in two parts). Agronomy No. 9. Madison, WI: American Society of Agronomy.

Bodman G. B. and Colman, E. A. 1943. Moisture and energy conditions during entry of water into soils. *Soil Sci. Soc. Am. Proc.* **8**:116–122.

Bohn, H. L., McNeal, B. L., and O'Connor, G. A. 1979. *Soil Chemistry.* New York: Wiley Interscience. 326 pp.

Bouma, J., Jongerius, A., and Schoondebeek, D. 1979. Calculation of hydraulic conductivity of some saturated clay soils using micromorphometric data. *Soil Sci. Soc. Am. J.* **43**:261–265.

Bouma, J. and Wosten, J. H. M. 1979. Flow patterns during extended saturated flow in two undisturbed swelling clay soils with different macrostructures. *Soil Sci. Soc. Am. J.* **43**:16–22.

Bouwer, H. 1978. *Groundwater Hydrology.* New York: McGraw-Hill. 480 pp.

Briggs, L. J. and McCall, A. G. 1904. An artificial root for inducing capillary movement of soil moisture. *Science* **20**: 566–569.

Childs, E. C. 1969. *An Introduction to the Physical Basis of Soil Water Phenomena.* New York: Wiley Interscience. 280 pp.

Clements Associates, Inc. 1983. J. M. C. Soil Investigation Equipment, Catalog No. 6.

Davis, S. N. and DeWiest, R. J. M. 1966. *Hydrogeology.* New York: John Wiley and Sons. 287 pp.

Donigian, A. S. and Rao, P. S. C. 1986. Overview of terrestrial processes and modeling. In *Vadose Zone Modeling of Organic Pollutants,* Hern, S. C. and Melancon, S.M. (Eds.). Chelsea, MI: Lewis Publishers, Inc. 314 pp.

Dunlap, W. J. and McNabb, J. F. 1973. *Subsurface Biological Activity in Relation to Ground Water Pollution,* EPA-660/2-73-014. Corvallis, OR: USEPA.

Edwards, W. M., Van Der Ploeg, R. R., and Ehlers, W. 1979. A numerical study of the effects of non-capillary sized pores upon infiltration. *Soil Sci. Soc. Am. J.* **43**:851–856.

Everett, L. G. 1988. Moderator session IV, vadose zone investigations. In *Proceedings Conference on Southwestern Groundwater Issues.* Association of Groundwater Scientists and Engineers, March 23–25, Albuquerque, NM. 421 pp.

Everett, L. G., McMillion, L. G., and Eccles, L. A. 1988. Suction lysimeter operation at hazardous waste sites. In *Groundwater Contamination-Field Methods,* ASTM, STP-963. Collins/Johnson (Eds.). Philadelphia, PA: ASTM.

Everett, L. G., Wilson, L. G., and Hoylman, E. W. 1984. *Vadose Zone Monitoring at Hazardous Waste Sites.* Park Ridge, NJ: Noyes Data Corporation.

Everett, L. G. 1980, *Groundwater Monitoring.* Schenectady, NY: General Electric Co. 440 pp.

Freeze, R. A., and Cherry, J. A. 1979. *Groundwater.* Englewood Cliffs, NJ: Prentice-Hall, Inc. 376 pp.

Hall, W. A. 1955. Theoretical aspects of water spreading, *Am. Soc. Ag. Eng.* **36**(6): 394–397.

Hanks, R. J. and Ashcroft, G. L. 1980. *Applied Soil Physics.* Berlin, FRG: Springer-Verlag. 295 pp.

Hantush, M. A. 1967. Growth and decay of ground-water mounds in response to uniform percolation, *Water Resour. Res.* **3**(1): 277–324.

Hern, S. C. and Melancon, S. M. 1986. *Vadose Zone Modeling of Organic Pollutants.* Chelsea, MI: Lewis Publishers Inc. 268 pp.

Hillel, D. 1980. Applications of Soil Physics, New York: Academic Press. 346 pp.

Hillel, D. 1971. *Soil and Water Physical Principles and Processes.* New York: Academic Press. 295 pp.

Hoogmoed, W. B. and Bouma, J. 1980. A simulation of model for predicting infiltration into cracked clay soil. *Soil Sci. Soc. Am. J.* **44**:485–462.

Hvorshev, M. J. 1949. *Subsurface Exploration and Sampling of Soils for Civil Engineering Purposes.*

Edited and printed by U.S. Army Corps of Eng. Vicksburg, MS: Waterways Experimental Station. 242 pp.

Kirkham, D. and Powers, W. L. 1972. *Advanced Soil Physics.* New York: Wiley Interscience. 298 pp.

Kissel, D. E., Ritchie, J. T., and Burnett, Earl. 1973. Chloride movement in undisturbed swelling clay soil. *Soil Sci. Soc. Am. Proc.* **37**:21-24.

Klute, A. 1986. *Methods of Soil Analysis,* Part 1, Physical and Mineralogical Methods, Agronomy, No. 9, Second ed. Madison, WI: American Society of Agronomy, Soil Science of America. 460 pp.

Klute, A. 1965. Laboratory measurement of hydraulic conductivity of unsaturated soil. In *Methods of Soil Analyses,* Black, C. A. (Ed.), Agronomy No. 9, 253-261. Madison, WI: American Society of Agronomy.

Kraijenhoff van deLeur, D. A. 1962. Some effects of the unsaturated zone on nonsteady free-surface groundwater flow as studies in a scaled granular model. *J. Geophys. Res.* **67**(11), 4347-4362.

Lawes, J. B., Gilbert, J. H., and Warrington, R. 1982. *On the Amount and Composition of the Rain and Drainage Waters Collected at Rothamsted.* London, U.K.: William Clowes and Sons.

Luthin, J. M. (Ed.). 1957. *Drainage of Agricultural Lands.* Madison, WI: American Society of Agronomy.

Luthin, J. M. and Day, P. R. 1955. Lateral flow above a sloping water table, *Soil Sci. Soc., Am. Proc.* **19**:406-410.

Mason, B. J. 1982. *Preparation of Soil Sampling Protocols,* EPA 600/4-83/020. Las Vegas, NV: U.S. Envir. Prot. Agency.

McMichael, F. C. and McKee, J. E. 1966. *Wastewater Reclamation at Whittier Narrows,* Water Quality Publication No. 33. Sacramento, California: Water Resources Control Board.

McWhorter, D. B. and Brookman, J. A. 1972. Pit recharge influenced by subsurface spreading. *Ground Water* **10**(5), 6-11.

Mercer, J. W., Rao, P. S. C., and Marine, I. W. 1983. *Role of the Unsaturated Zone in Radioactive and Hazardous Waste Disposal.* Ann Arbor, MI: Ann Arbor Science. 294 pp.

Meyer, J. R. 1979. Personal communication. USC, Riverside, CA.

Miller, R. H. 1973. The soil as a biological filter. In *Recycling Treated Municipal Waste Water and Sludge Through Forest and Cropland,* Sopper, W. E. and Kardos, L. T. (Eds.). Penn. State U. Press. 264 pp.

National Drilling Federation. (n.d.). *Drilling Safety Guide.* 300 Milkwood Ave. Columbia, SC. 29205.

Parizek, R. R., Kardos, L. J., Sopper, W. E., Myers, E. A., Dairs, D. E., Farrell, M. A., and Nesbitt, J. B. 1967. *Waste Water Renovation and Conservation.* Pennsylvania State Studies, No. 23. State College: PA: Penn. State U. Press. 98 pp.

Quisenberry, V. L. and Phillips, R. E. 1978. Displacement of soil water by simulated rainfall. *Soil Sci. Soc. Am. J.* **2**:675-679.

Quisenberry, V. L., and Phillips, R. E. 1976. Percolation of surface applied water in the field. *Soil Sci. Soc. Am. J.* **40**:484-489.

Rhoades, J. D. and Bernstein, L. 1971. Chemical, physical, and biological characteristics of irrigation and soil water. In *Water and Water Pollution Handbook,* Vol. 1, Ciaccio, L. L. (Ed.), pages 141-222. New York: Marcel Dekker, Inc.

Shuford, J. W., Fritton, D. D., and Baker, D. E. 1977. Nitrate-nitrogen and chloride movement through undisturbed field soil. *J. Environ. Qual.* **6**:736-739.

Skopp, J., Gardner, W. R., and Tyler, E. J. 1981. Solute movement in structured soils: Two-region model with small interaction. *Soil Sci. Soc. Am. J.* **45**:837-842.

Soil Survey Staff. 1951. *Soil Survey Manual,* U.S. Dept. of Agric., Washington, D.C.: Superintendent of Documents. 198 pp.

Soilmoisture Equipment Corp. 1988. Sales division, *Catalog of Products,* Santa Barbara, CA.

Soiltest Inc. 1976. *Soil Testing Equipment.* Evanston, IL: Soiltest Inc. Sopper, W. E. and Kardos, L. T. (Eds.). 1973. *Recycling Treated Municipal Waste Water and Sludges Through Forest and Cropland.* State College, PA: Penn. State U. Press. 68 pp.

Taylor, G. S. and Luthin, J. N. 1969. The use of electronic computers to solve subsurface drainage problems. *Water Resour. Res.* **5**(1), 144-152.

Thomas, G. W. and Phillips, R. E. 1979. Consequences of water movement in macropores. *J. Environ. Qual.* **8**:149-152.

Thomas, G. W., Phillips, R. E., and Quisenberry, V. L. 1978. Characterization of water displacement in soils using simple chromatographic theory. *J. Soil Sci.* **29**:32–37.

Thomas, R. E. 1973. The soil as a physical filter. In *Recycling Treated Municipal Wastewater and Sludge Through Forest and Cropland.* Sopper, W. E. and Kardos, L. T. (Eds.). State College, PA: Penn. State U. Press. 62 pp.

U.S. Environmental Protection Agency. 1987. *Processes Effecting Subsurface Transport of Leaking Underground Tank Fluids.* Las Vegas, NV: Environmental Monitoring Systems Laboratory. 154 pp.

U.S. Environmental Protection Agency. 1986. *Permit Guidance Manual on Unsaturated Zone Monitoring for Hazardous Waste Land Treatment Units,* EPA/530-SW-86-040. Washington, D.C.: Office of Solid Waste and Emergency Response. 129 pp.

U.S. Environmental Protection Agency, 1983. *Hazardous Waste Land Treatment,* SW-846. Washington, D.C.: U.S. Envir. Prot. Agency. 111 pp.

U.S. Environmental Protection Agency, 1982. *Test Methods for Evaluating Solid Waste Physical/Chemical Methods,* SW-874. Washington, D.C.: U.S. Envir. Prot. Agency. 430 pp.

van Schilfgaarde, J., 1970. Theory of flow to drains. *Advances in Hydroscience* **6**:43–106. Ven te Chow (Ed.). Vol. 6, pp. 43–106. New York, NY: Academic Press.

VI-COR Technologies, Inc. 1988. Sales division, *Catalog of Products,* Bellvue, WA.

Whipkey, R. Z. 1967. Theory and mechanics of subsurface storm flow. In *Proc. Int. Symp. on For. Hydrol.*, Sopper, W. E. and Lull, H. W. (Eds.). pp. 155–260. Natl. Sci. Found., 29 Aug–10 Sept. 1965, Penn. State U., University Park, PA. New York: Pergamon Press.

Wild, A. 1972. Nitrate leaching under bare fallow at a site in northern Nigeria. *J. Soil Sci.* **23**:315–324.

Wilson, L. G. 971. Observations on water content changes in stratified sediments during pit recharge. *Ground Water* **9**(3):29–40.

Zirshky, J. and Harris, D. 1982. Controlling productivity at a hazardous waste site. *Civil Engineering* **52**(9):70–74.

8

Soil Pore-Liquid Monitoring

Lorne G. Everett, Ph.D.
Chief Scientist
Metcalf & Eddy
Santa Barbara, CA

The sampling of soil pore-liquid was reported in the literature in the early 1900s, when Briggs and McCall (1904) described a porous ceramic cup which they termed an "artificial root." The sampling of soil pore-liquid (Everett, 1988), has received increasing attention in more recent years, as concern over the migration of pollutants in soil has increased (Everett et al., 1988). As shown in Figure 8–1, different soils are capable of yielding different levels of water. The unsaturated zone, as described in Chapter 7, is the layer of soil between the land surface and the groundwater table. At saturation, the volumetric water content is equivalent to the soil porosity (see Figure 8–1). In contrast, the unsaturated zone is usually found to have a soil moisture content of less than saturation. For example, the specific retention curve in Figure 8–1 depicts the percentage of water retained in previously saturated soils of varying texture after gravity drainage has occurred. Suction-cup lysimeters are used to sample pore-liquids in unsaturated media because pore-liquid will not readily enter an open cavity at pressures less than atmospheric (the Richard's outflow principle).

Suction-cup lysimeters are made up of a body tube and a porous cup. When placed in the soil, the pores in these cups become an extension of the pore space of the soil. Consequently, the water content of the soil and cup become equilibrated at the existing soil-water pressure. By applying a vacuum to the interior of the cup such that the pressure is slightly less inside the cup than in the soil solution, flow occurs into the cup. The sample is pumped to the surface, permitting laboratory determination of the quality of the soil pore-liquids.

Although a number of techniques are available for indirectly monitoring the movement of pollutants beneath waste disposal facilities, soil core sampling and suction-cup lysimeters remain the principal methods for directly sampling pore-liquids in unsaturated media. The main disadvantages of soil core sampling are that it is a destructive technique (i.e., the same sample location cannot be used again) and it may miss fast-moving constituents. Lysimeters have been used for many years by agriculturists for monitoring the flux of solutes beneath irrigated fields (Biggar and Nielsen, 1976). Similarly, they have been used to detect the deep movement of

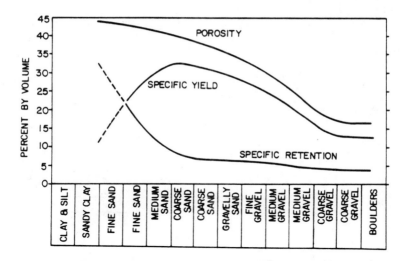

Figure 8-1.   Variation of porosity, specific yield, and specific retention with grain size.

pollutants beneath land treatment units (Parizek and Lane, 1970). This section will discuss soil moisture/tension relationships, soil pore-liquid sampling equipment, installation and operation of the available devices, and sample collection, preservation, storage, and shipping.

SOIL MOISTURE/TENSION RELATIONSHIPS

Unlike water in a bucket, free, unlimited access to water does not exist in the soil. Soil water or, as it is frequently called, "soil moisture," is stored in the small "capillary" spaces between the soil particles and on the surfaces of the soil particles. The water is attracted to the soil particles, and tends to adhere to the soil. The smaller the capillary spaces between the particles, the greater the sticking force. For this reason, it is harder to get moisture out of fine clay soils than it is from the larger pores in sandy soils, even if the percent of moisture in the soil, by weight, is the same.

Figure 8-2 shows the results of careful research work done with special extractors. As described by the Soilmoisture Equipment Corporation (1983), the graph shows the relationship of the percent of moisture in a soil to the pressure required to remove the moisture from the soil. These are called moisture retention curves. The pressure is measured in bars,[1] which is a unit of pressure in the metric system. Figure 8-2 clearly points out that two factors are involved in determining ease of water sampling: 1) moisture content, and 2) soil type.

Moisture in unsaturated soil is always held at suctions or pressures below atmospheric pressure (Stannard, 1988). To remove the moisture, one must be able to develop a *negative* pressure or vacuum to pull the moisture away from around the soil particles. For this reason we speak of "soil suction." In wet soils the soil suction

---

[1]By definition, a bar is a unit of pressure equal to 10^6 dyne/cm². It is equivalent to 100 kPa (Kilopascals), or 14.5 psi, or approximately 1 atmosphere, or 750 mm of mercury, or 29.6 in. of mercury, or 1020 cm of water, or 33.5 ft of water.

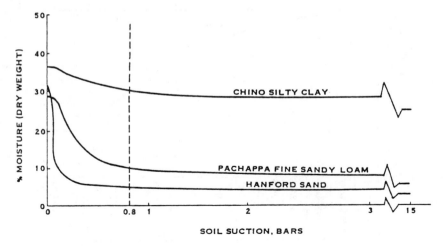

Figure 8–2. Moisture retention curves—three soil types.

is low, and the soil moisture can be removed rather easily. In dry soils the soil suction is high, and it is difficult to remove the soil moisture.

Given two soils (one clay and one sand) with identical moisture contents, it will be more difficult to extract water from the finer soil (clay) because water is held more strongly in very small capillary spaces in clays.

Another fact, brought out by the graphs on Figure 8–2, is that silty clay soil with 30 percent moisture, if placed in contact with a sandy soil with only 10 percent moisture, will actually suck moisture out of the sandy soil until the moisture content in the sandy soil is only 5 percent, as soil suctions are now equal. This is due to the greater soil tension in the fine clay texture.

PORE-LIQUID SAMPLING EQUIPMENT

Well and open cavities cannot be used to collect soil solution flowing in the unsaturated zone under suction (negative pressures). The sampling devices for such unsaturated media are thus called suction samplers or lysimeters. Everett et al. (1983, 1988) provides an in-depth evaluation of the majority of unsaturated zone monitoring equipment. Law Engineering and Testing Company (1982) provides a description of some of the available suction lysimeters. Three types of suction lysimeters are 1) ceramic-type samplers, 2) hollow fiber samplers, and 3) membrane filter samplers.

In areas of macropore flow, pan lysimetry should be employed for soil-pore liquid monitoring, in addition to suction lysimetry. While pan lysimeters (e.g., glass block samplers) are not at present commercially available, it is relatively easy to construct the instrument (R. R. Parizek, personal communication, 1984). However, installation will require more skill and effort than suction lysimeters (K. Shaffer, personal communication, 1984.)

Ceramic-Type Samplers

Two types of samplers are constructed from ceramic material: the suction cup and the filter candle. Both operate in the same manner. Basically, ceramic-type samplers comprise the same type of ceramic cups used in tensiometers. When placed in the soil, the pores in these cups become an extension of the pore space of the soil. Although cups have limitations, at the present time they appear to be the best tool available for sampling unsaturated media, particularly in the field. The use of Teflon for the body tube parts and the porous segment (instead of a porous ceramic) may reduce the operating range of the lysimeter to below a useful level (Everett et al., 1988).

Suction cups may be subdivided into three categories: 1) vacuum operated soil-water samplers, 2) vacuum-pressure samplers, and 3) vacuum-pressure samplers with check valves. Soil-water samplers generally consist of a ceramic cup mounted on the end of a small-diameter PVC tube, similar to a tensiometer (see Figure 8-3). The upper end of the PVC tubing projects above the soil surface. A rubber stopper and outlet tubing are inserted into the upper end. Vacuum is applied to the system and soil water moves into the cup. To extract a sample, a small-diameter tube is inserted within the outlet tubing and extended to the base of the cup. The small-diameter tubing is connected to a sample-collection flask. A vacuum is applied via a hand vacuum-pressure pump and the sample is sucked into the collection flask. These units are generally used to sample to depths up to 6 ft from the land surface. Consequently, they are used primarily to monitor the near-surface movement of pollutants from land disposal facilities or from irrigation return flow.

To extract samples from depths greater than the suction lift of water (about 25 ft), a second type of unit is available, the so-called vacuum-pressure lysimeter. These units were developed by Parizek and Lane (1970) for sampling the deep movement of pollutants from a land disposal project in Pennsylvania. The design of the Parizek and Lane sampler is shown in Figure 8-4. The body tube of the unit is about 2 ft long, holding about 1 l of sample. Two copper lines are forced through a two-hole rubber stopper sealed into a body tube. One copper line extends to the base of

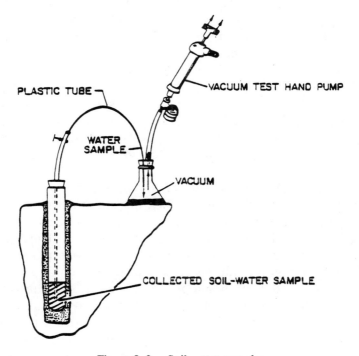

Figure 8-3. Soil-water sampler.

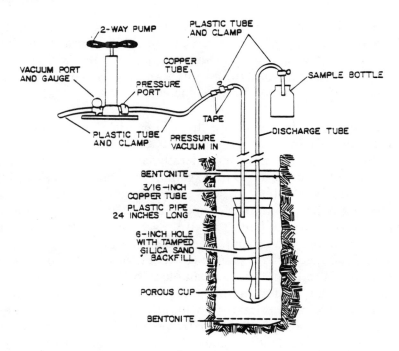

Figure 8–4. Vacuum-pressure sampler.

the ceramic cup as shown and the other terminates a short distance below the rubber stopper. The longer line connects to a sample bottle and the shorter line connects to a vacuum-pressure pump. All lines and connections are sealed. At hazardous waste sites, however, polyethylene or Teflon tubing is recommended.

In operation, a vacuum is applied to the system (the longer tube to the sample bottle is clamped shut at this time). When sufficient time has been allowed for the unit to fill with solution, the vacuum is released and the clamp on the outlet line is opened. Air pressure is then applied to the system, forcing the sample into the collection flask. A basic problem with this unit is that when air pressure is applied, some of the solution in the cup may be forced back through the cup into the surrounding pore-water system. Consequently, this type of pressure-vacuum system is recommended for depths only up to about 50 ft below land surface. In addition to the monitoring effort of Parizek and Lane, these units were used by Apgar and Langmuir (1971) to sample leachate movement in the vadose zone underlying a sanitary landfill.

Morrison and Tsai (1981) proposed a modified lysimeter design with the porous material located midway up the sampling chamber, instead of at the bottom (see Figure 8–5, Morrison and Tsai, 1981). This mitigated the basic problem of sample solution being forced back through the cup when air pressure is applied. The dead space below the porous section, however, will result in potential cross-contamination.

Wood (1973) reported on a modified version of the design of Parizek and Lane. Wood's design is the third suction sampler discussed in this subsection. Wood's design overcomes the main problem of the simple pressure-vacuum system; namely, that solution is forced out of the cup during the application of pressure. A sketch of the sampler is shown in Figure 8–6. The cup ensemble is divided into lower and

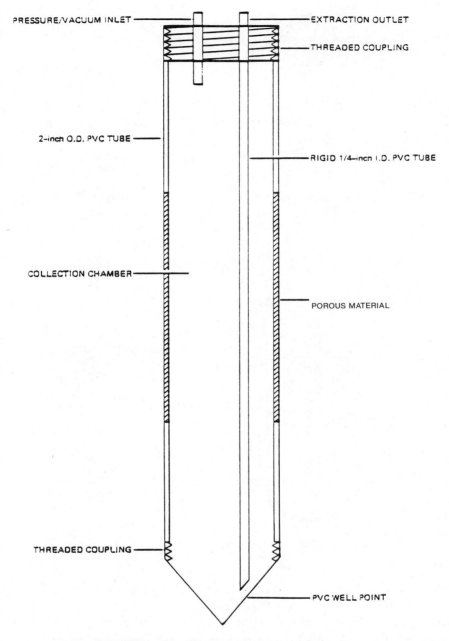

Figure 8-5. Modified pressure-vacuum lysimeter.

upper chambers. The two chambers are isolated except for a connecting tube with a check valve. A sample delivery tube extends from the base of the upper chamber to the surface. This tube also contains a check valve. A second, shorter tube, terminating at the top of the sampler, is used to deliver vacuum or pressure. In operation, when a vacuum is applied to the system, it extends to the cup through the open one-way check valve. The second check valve in the delivery tube is shut. The sample is delivered into the upper chamber, which is about 1 l (0.26 gal) in capacity. To deliver

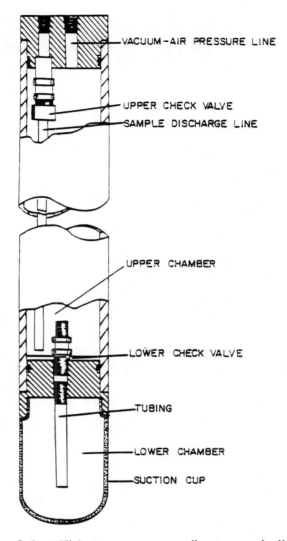

Figure 8–6. "High-pressure-vacuum soil-water sampler."

the sample to the surface, the vacuum is released and pressure (generally of nitrogen gas) is applied to the shorter tube. The one-way valve to the cup is shut and the one-way valve in the delivery tube is opened. The sample is then forced to the surface. High pressures can be applied with this unit without danger of damaging the cup. Consequently, this sampler can be used to depths of about 150 ft below land surface (Soilmoisture Equipment Corporation, 1978). Wood and Signor (1975) used this sampler to examine geochemical changes in water during flow in the vadose zone underlying recharge basins in Texas.

A sampling unit employing a filter candle is described by Duke and Haise (1973). The unit, described as a "vacuum extractor," is installed below plant roots. Figure 8–7 shows an illustrative installation. The unit consists of a galvanized sheet metal

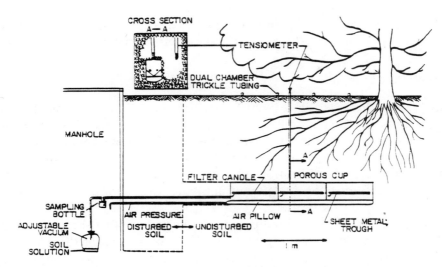

Figure 8-7. Facilities for sampling irrigation return flow via filter candles, for research project at Tacna, Arizona.

trough, open at the top. A porous ceramic candle (12 in. long and 1.27 in. in diameter) is placed into the base of the trough. A plastic pipe sealed into one end of the candle is connected to a sample bottle located in a nearby manhole or trench. A small-diameter tube attached to the other end of the candle is used to rewet the candle as necessary. The trough is filled with soil and placed within a horizontal cavity of the same dimensions as the trough. The trough and enclosed filter candle are pressed up against the soil via an air pillow or mechanical jack. In operation, vacuum is applied to the system to induce soil-water flow into the trough and candle at the same rate as in the surrounding soil. The amount of vacuum is determined from tensiometers. Hoffman et al. (1978) used this type of sampler to collect samples of irrigation water leaching beneath the roots of orange trees during return flow studies at Tacna, Arizona.

Cellulose-Acetate Hollow Fiber Samplers

Jackson et al. (1976) described a suction sampler constructed of cellulose-acetate hollow fibers. These semipermeable fibers have been used for the dialysis of aqueous solutions, functioning as molecular sieves. Soil column studies using a bundle of fibers to extract soil solution showed that the fibers were sufficiently permeable to permit rapid extraction of solution for analysis. Soil solution was extracted at soil-water contents ranging from 50 to 20 percent.

Levin and Jackson (1977) compare ceramic cup samplers and hollow fiber samplers for collecting soil solution samples from intact soil cores. Their conclusion is that: " . . . porous cup lysimeters and hollow fibers are viable extraction devices for obtaining soil solution samples for determining EC, Ca, Mg, and PO_4-P. Their suitability for NO_3-N is questionable." They also conclude that hollow fiber samplers are more suited to laboratory studies, where ceramic samplers are more useful for field sampling. Because of the high potential to alter sample quality, further research is required on these types of samplers before they can be recommended.

Membrane Filter Samplers

Stevenson (1978) presents the design of a suction sampler using a membrane filter and a glass fiber prefilter mounted in a "Swinnex"-type filter holder. Figure 8–8 shows the construction of the unit. The membrane filters are composed of polycarbonate or cellulose-acetate. The "Swinnex" filter holders are manufactured by the Millipore Corporation for filtration of fluids delivered by syringe. A flexible tube is attached to the filter holder to permit applying a vacuum to the system and for delivering the sample to a bottle.

The sampler is placed in a hole dug to a selected depth. Sheets of glass fiber "collectors" are placed in the bottom of the hole. Next, two or three smaller glass fiber "wick" discs that fit within the filter holder are placed in the hole. Subsequently, the filter holder is placed in the hole with the glass fiber prefilter in the holder contacting the "wick" discs. The hole is then backfilled.

In operation, soil water is drawn into the collector system by capillarity. Subsequently, water flows in the collector sheets toward the glass fiber wicks as a result of the suction applied to the filter holder assembly. The glass fiber prefilter minimizes clogging of the membrane filter by fine material in the soil solution.

During field tests with the sampler, it was observed that sampling rates decreased with decreasing soil-water content. The "wick and collector" system provided contact with a relatively large area of the soil and a favorable sampling rate was maintained even when the "collector" became blocked with fine soil. The basic sampling unit can be used to depths of 4 m.

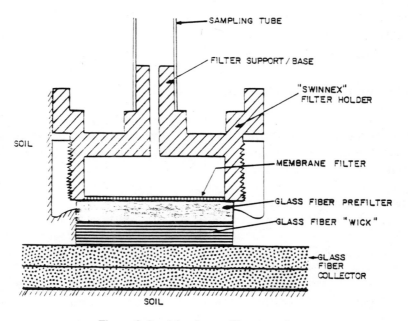

Figure 8–8. Membrane filter sampler.

Pan Lysimeters

The presence of macropore or fracture flow in highly structured soils should be determined. It is important to acknowledge the occurrence of macropore flow under certain soil conditions and its significant potential to contaminate groundwater. The pan lysimeter, which is a free-drainage type lysimeter, is suited for sampling macropore or fracture flow.

There are a number of designs for pan-type lysimeters. Parizek and Lane (1970) constructed a 12 × 15 in. pan lysimeter (Figure 8–9) from 16-gauge sheet metal. Barbee (1983) employed a perforated 12 × 12 in. glass brick, the kind used in masonry construction, as a pan lysimeter (Figure 8–10). Shaffer et al. (1979) devised a 20 cm diameter pan lysimeter with a tension plate capable of pulling 6 centibars of tension. A pan lysimeter can be constructed of any nonporous material, provided a leachate-pan interaction will not jeopardize the validity of the monitoring objectives. The pan itself may be thought of as a shallow-draft funnel. Water draining freely through the macropores will collect in the soil just above the pan cavity. When the tension in the collecting water reaches zero, dripping will initiate and the pan will funnel the leachate into a sampling bottle. The use of a tension plate or a fine sand packing reduces the extent of capillary perching at the cavity face and promotes free water flow into the pan.

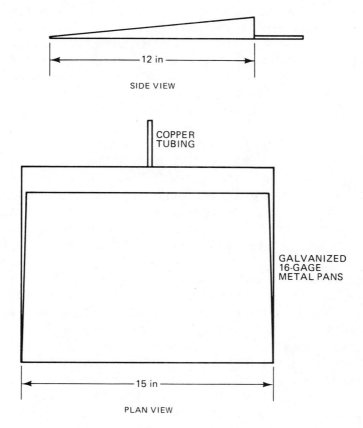

Figure 8–9.   Example of a pan lysimeter.

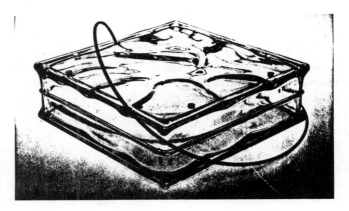

Figure 8-10. Free drainage glass block sampler.

CRITERIA FOR SELECTING SOIL PORE-LIQUID SAMPLERS

In selecting soil-pore liquid sampling equipment, the following criteria should be considered: cost, commercial availability, installation requirements, hazardous waste interaction, vacuum requirements, soil moisture content, soil characteristics and moisture regimes, durability, sample volume, and sampling depth (Everett et al., 1982). Fritted glass samplers, for example, are too fragile for field application. Plastic lysimeters require a continuous vacuum and high soil moisture levels. The vacuum extractor is expensive, and requires intensive installation procedures and a continuous vacuum. The "Swinnex" sampler has difficult installation procedures and produces too small a sample. Some samplers are not commercially available. All Teflon samplers are more expensive than PVC body parts and ceramic cups. The high-pressure vacuum samplers are not required for shallow sampling depths. The simple vacuum lysimeter cannot be used in situ with the sampler totally covered by soil.

In most case, the lysimeters of choice will be pressure-vacuum ceramic lysimeters. Teflon models have certain limitations that preclude their use below field capacity (Everett et al., 1986). Most pressure-vacuum lysimeters are reasonably priced, commercially available, and easy to install. In addition, a constant vacuum apparatus is not required. They can be used in situ at depths well within the requirements of most hazardous waste sites and can produce a large sample volume. Body tubes of various lengths are available to complement the volume and sample depth requirements.

Free drainage lysimeters can only sample and thus monitor the movement of gravitational water when precipitation is equal to or greater than field capacity requirements, or when there is a large water input into the soil (Parizek and Lane, 1970; Tadros and McGarity, 1976; Fenn et al., 1977). However, in the unsaturated zone of soils, most water movement is in the wet moisture range (0 to -50 kPa soil moisture tensions, Reeve and Doering, 1965), and in well-structured soils through macropores (Shaffer et al., 1979), which accounts for the vast majority of the water and chemical constituents that can be lost from the soil by leaching. Free-drainage samplers have the following characteristics:

1. It is a continuously sampling "collection" system without the need for externally applied vacuum.
2. Because vacuum is only used to pull the sample to the surface, there is less potential for losing volatile compounds in the sample obtained.
3. Its defined surface area may allow quantitative estimates of leachate.
4. The method of installation allows for monitoring the natural percolation of liquids through the unsaturated zone without alteration of flow.
5. If made of chemically inert materials (i.e., glass), it has less potential for altering the chemical composition of a sample obtained by it.
6. Since the inside of the glass block type is uneven, the potential exists for cross-contamination from residual samples.
7. If the glass blocks are not installed perfectly level, a sump or collection area can result in dead space where the sample cannot be removed.
8. Pan lysimeters may require trenching to be installed.

Preparation of the Samplers

A decision must be made on the size of pressure-vacuum lysimeters to be installed at the site, and the composition of the pressure-vacuum tubing. According to data by Silkworth and Grigal (1981), the larger commercially available units with a 4.8-cm diameter are more reliable than the 2.2-cm diameter units, influence water quality less, and yield a larger volume of sample for analysis. Although various materials have been used for conducting tubing (e.g., polypropylene and copper tubing), it is advisable to select Teflon® tubing to minimize contamination and interference with the sample.

In order to avoid interferences from chemical substances attached to porous sampling points, it is recommended advisable to prepare each unit using the following procedure, described by Wood (1973). Clean the cups by letting approximately 1 l of 8N HCl seep through them, and rinse thoroughly by allowing 15 to 20 l of distilled water to seep through. This cleaning process can be accelerated by placing the distilled water inside the lysimeter and developing 20–30 psi of pressure to drive the water through the porous material. The cups are adequately rinsed when there is less than a 2 percent difference between the specific conductance of the distilled water input and the output from the cup.

Prior to taking the suction lysimeters in the field, each lysimeter should be checked for its bubbling pressure and for leaks. Complete procedures for testing for leaks and air entry values are given in Everett et al. (1986).

Surveying in the Locations of Sites and Site Designations

The exact location of each sample should be designated on a detailed map. Subsequently, a surveying crew should be sent into the field to precisely locate the coordinates of the sites in reference to a permanent marker. This step is important to facilitate future recovery of any failed samplers.

For convenience, each sampler location should be given a descriptive designation to facilitate all future activities at the site. For example, this designation should be posted at the sampling station and should be marked on all collection flasks to facilitate differentiating between samples.

INSTALLATION PROCEDURES
FOR VACUUM-PRESSURE PORE-LIQUID SAMPLERS

Installing Access lines

The approximate length of the two lines in each sampler should be determined by measuring the distance between the installation point and the above-ground access point (e.g., shelter). The lines should be cut to this length plus an allowance for the distance that the tubes will extend into the sampler. Some excess should be retained at the above-ground access point. The tubes should be installed into a PVC or metal manifold consisting of small-diameter conduit. Although the conduit does provide some structural protection from compression, the main function of the conduit is to discourage rodents, etc., from physically damaging the leads. A convenient method for leading the tubes through the conduit is to first run a cord through the tube, attach the cord to the two lines, and then pull the lines through the conduit. One method for installing the cord is to attach one end to a rubber cork at slightly smaller diameter than the inside diameter of the conduit, then blowing the cork and cord through the conduit using compressed air. Both the sampler and the conduit are then lowered down an open hole or through the center of a hollow-stem auger for deeper installations.

The procedure for installing access tubes (Soilmoisture Equipment Corp., 1983) into the sampler, before placing the unit in a borehole, is as follows:

When installing the tubes, one tube should be pushed through the neoprene plug (see Figure 8–11) so that the end of the tubing reaches down to the bottom of the porous ceramic cup. This "discharge" access tube should be marked at the other end in some fashion to identify it. The other "pressure-vacuum" access tube should be inserted into the neoprene plug so that it extends through the plug perhaps one inch.

After the tubes are installed (see Figure 8–11), tighten the ring clamp with a nail or similar object inserted through the holes provided in the clamp's ring. Tighten only until it meets the body tube.

Step-by-Step Procedures for Installing Vacuum-Pressure Pore-Liquid Samplers

The procedures included in this section are adapted from the operating procedure for a commercially available vacuum-pressure type sampler. (These procedures are generally applicable to similar types of commercially available units, and the ensuing discussion does not constitute an endorsement of this particular sampler.) The procedures are grouped into (a) procedures for preparing the hole, and (b) alternative methods for installing the samplers.

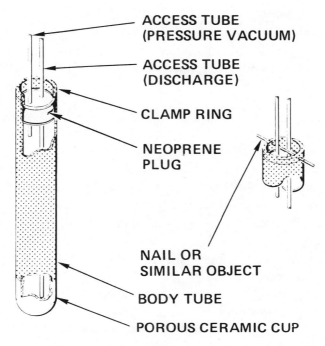

ACCESS TUBE
(PRESSURE VACUUM)

ACCESS TUBE
(DISCHARGE)

CLAMP RING

NEOPRENE
PLUG

NAIL OR
SIMILAR OBJECT

BODY TUBE

POROUS CERAMIC CUP

Figure 8–11. Installation of access tubes in a pressure-vacuum pore-liquid sampler.

Constructing the Hole. In rock-free uniform soils for shallow depths, use a 15.24 cm (6 in.) screw or bucket auger for coring the hole. For deeper installations, a 15.24 cm (6 in.) I.D. hollow-stem auger should be used.

The soil used to backfill around the bottom of the sampler should then be sifted through a $\frac{1}{4}$ in. mesh screen to remove pebbles and rocks. This will provide a reasonably uniform backfill soil for filling in around the soil water sampler.

Sampler Installation Procedure. The goals of a careful installation procedure are: (1) to ensure good contact between the suction cup portion of the sampler and the surrounding soil, and (2) to minimize side leakage of liquid along the sampler wall. Although numerous installation procedures have been used in the past, the bentonite clay method is recommended as the best choice for achieving both of these goals. This method includes a silica sand layer that ensures good contact with the suction cup and a clay plug that prevents leakage down the core hole and along the sampler wall.

Prior to installation, the lysimeters should be checked for leaks and flushed with distilled water. To check for leaks, the lysimeters are totally immersed in a tank of water. It is preferable to use a glass aquarium so that the location of the leaks (bubbles) can be easily identified. One of the tubes going into the suction lysimeter is clamped shut. A pressure line is attached to the second tube. Slowing increase the pressure within the suction lysimeter to 15 psi. On Teflon lysimeters, it is important to check for leaks at all screw fittings. In addition, the Teflon cups may bubble at pressures less than 2 psi. Ceramic units, on the other hand, should not bubble from any location until at least 15 psi. All leaks on Teflon lysimeters should be corrected using Teflon tape. All leaks on ceramic units should be corrected by increasing the pressure at each of the fittings by screwing the pressure couplings down. At this point, it is also assumed that the cups have been prepared and the Teflon access tubes have been installed in the sampler. The cups should be installed while they are wet.

Bentonite Clay Method.   The following is a step-by-step description of the bentonite clay installation method:

1. Core hole to desired depth.
2. Pore in 7.6 cm to 12.7 cm (3 in. to 5 in.) of wet bentonite clay to isolate the sampler from the soil below (see Figure 8–12).
3. Pour in a small quantity of 200 mesh silica-sand slurry and insert soil water sampler. (Slurry contains 1 lb of silica per 150 ml of water.)
4. Pour another layer of 200 mesh silica sand at least 6 in. deep around the cup of the soil water sampler.
5. Backfill with native soil to a level just above the soil water sampler and again add 7.6 cm to 12.7 cm (3 in. to 5 in.) of bentonite as a plug, to further isolate the soil water sampler and guard against possible channeling of water down the hole.
6. Backfill the remainder of the hole slowly, tamping continuously with a long metal rod. Again backfill should be of native soil free of pebbles and rocks.

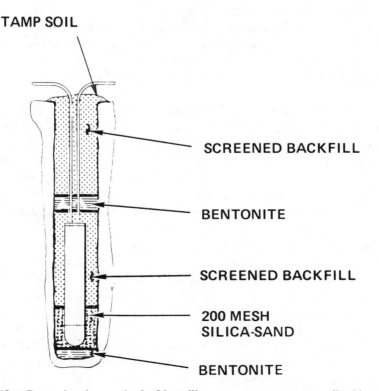

Figure 8–12.   Bentonite clay method of installing vacuum-pressure pore-liquid samplers.

Backfilling and Final Survey.   Upon installation of the sampler in the hole, as described above, and the access tubes, it is time to backfill around the sampling point. First, however, it is advisable to survey-in the exact location of the sampler to facilitate recovery of the unit at some future time. An initial vacuum should be applied to each unit before backfilling to check for leaks and to remove water applied to the slurry. Backfilling should be conducted in stages, using a mechanical tamper to ensure good packing of each layer. It is preferable to backfill on the same day that the installation is made. Delays of 1–2 days can result in a lost of soil moisture in the excavated material and, consequently, problems may occur with packing the soils, i.e., heavy clays. Although in time, the backfill and shafts will return to a natural bulk density, it is preferable to tamp the backfilled material to at least the original bulk density, or preferably higher. If the bulk density is not maintained, the access hole may begin to fill with water. In cases where the bulk density is difficult to maintain, a 25 percent mixture of bentonite and soil should be used. This mixture will preclude any buildup of pooled water in the backfilled borehole.

OPERATION OF VACUUM-PRESSURE SAMPLING UNITS

It is advisable to select a permanent team of two individuals for the sampling program, with one individual being responsible for the operation and the second individual being a helper. A permanent team ensures uniformity in sample collection and chain-of-custody procedures.

Prior to obtaining a sample for analysis, good quality control procedures require that the samplers be evacuated 2–3 days ahead of the actual sampling time. By totally removing any fluid that could have accumulated in the suction lysimeter over time, the field technician is subsequently able to obtain a fresh sample from the unsaturated zone. The procedures required to initially evacuate the sampler are identical to the operational procedures identified below.

The stages in operating a vacuum-pressure sampler are as follows: 1) apply a vacuum to the interior of the sampler, via the vacuum-pressure line, 2) maintain the vacuum for a sufficient period of time to collect a sample in the sampler, 3) release the vacuum, and 4) apply pressure to the vacuum-pressure line and blow the sample through the sample line into a collection flask. Details on each step are included in this discussion.

Two alternatives are available for applying the vacuum and pressure during each collection cycle. The simplest method is to use a vacuum-pressure hand pump, with a vacuum dial. This method is suitable for collecting samples from individual units. In cases in which the access lines from several units are brought together into a common shelter, it may be more convenient to use separate vacuum and pressure bottles connected to a common manifold with outlets to the individual access lines.

The procedure described in the following paragraphs was adopted from the operating instructions for a commercially available sampler. Use of this procedure does not constitute an endorsement of this sampler.

1. Close the pinch clamp on the discharge access tube (see Figure 8–13). All pinch clamps should be tightened with pliers to eliminate the problem of not sealing. Finger-tight pinching of the clamps is not sufficient.

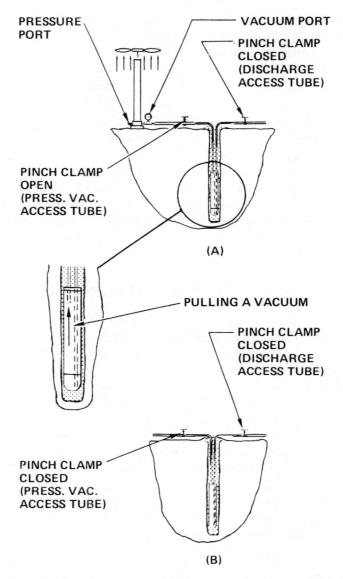

Figure 8–13. Stages in the collection of a pore-liquid sample using a vacuum-pressure sampler.

2. Apply a vacuum to the pressure-vacuum line either by means of a hand pump or by attaching a vacuum bottle. The applied vacuum should be about 60 centibars (18 in. of mercury).

3. When a steady vacuum is obtained, attach a pinch clamp to the vacuum-pressure line. Alternatively, when a vacuum bottle is used, it may be possible to omit using a pinch clamp in an effort to sustain the requisite vacuum.

4. After a period of time that is deemed sufficient to collect a sample (a minimum of 24 hours in some cases), attach sample bottles to the discharge line from each unit.

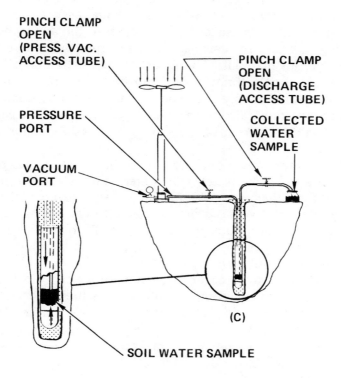

Figure 8-13. *Continued.*

5. Release the vacuum by opening the pinch clamp or removing the vacuum bottle.
6. Apply 1 to 2 atmospheres of air pressure to the pressure-vacuum lines, either by using a hand pump or by installing a container of compressed air, and blow the liquid sample from the sampler into the collection flasks (see Figure 8-13).
7. Remove and seal the flasks.

The volume of sample required is dependent upon the number and kind of analysis to be performed. It may be found during a sampling cycle that the volume of sample obtained from a particular unit or units is not great enough to permit analysis. Alternatively, no sample at all may be obtained. For these cases, it will be necessary to repeat each step using a greater vacuum and longer sampling interval.

Porous Segments in Lysimeters

The vadose (unsaturated) zone consists of a mixture of soil particles, water that is held on the surface of the particles and in small capillary spaces between the particles, and interconnecting air passages that are open to the atmosphere at the soil surface. Removing moisture for chemical analysis from the vadose zone requires the use of special porous materials. Simply exerting a suction on an open tube inserted into the vadose zone will not remove moisture since the interconnecting air passages in the soil will result only in the flow of air into the evacuated tube. However, by using a porous cup sealed to the end of the tube, samples can be removed by suction, providing the diameter of the individual pores in the porous cup do not exceed a critical value.

If the porous cup is fabricated from a hydrophilic material, such as ceramic, water will fill the pores of the cup completely. The water bonds to the porous ceramic and cannot be removed from the pores unless the air pressure differential across the wall of the cup reaches a critical value, which is related to the pore size. If the porous cup is fabricated from a hydrophobic material such as Teflon or PTFE (polytetrafluoroethylene), water will fill the pores of the cup, but the bonding of the water to the hydrophobic material will be less.

The air pressure required to force air through a porous cup which has been thoroughly wetted with water is called the "bubbling pressure" or "air entry value." The smaller the pores in the cup, the higher this pressure will be. The relation of the pore size to the bubbling pressure or air entry value is defined by the equation $D = 30Y/P$, where D is the pore diameter measured in microns, P is the bubbling pressure or air entry value measured in millimeters of mercury, and Y is the surface tension of water measured in dynes/cm.

In order to build a soil water sampling device that can be used successfully in the vadose zone to withdraw moisture from the soil, the device must incorporate a porous cup that has pores so small that the air in the soil, under atmospheric pressure, cannot enter even though a full vacuum is created within the sampler. Under these conditions, water from the capillary spaces in the soil will flow through the pores in the porous cup and into the sampler, but air will not enter.

With respect to the above equation, the maximum size of the pores that will permit this action are as follows: At 20°C, the surface tension of water is 72 dynes/cm. The maximum air pressure is 1 atmosphere, or 14.7 psi, or 760 mm of mercury. In accordance with the equation, the maximum pore size in the porous cup would be $D = (30)(72)/(760) = 2.8$ microns. The pore size of ceramic cups is between 2 and 3 microns. If the pores of the wetted sampler cup do not exceed 2.8 microns in diameter, then a full vacuum can be maintained within the sampler and the water films in the pores of the porous cup will not break down. If the pore size of the cup is twice this amount, namely 5.7 microns, then the maximum vacuum that can be pulled within the sampler is 380 millimeters of mercury, or 50 percent of an atmosphere. Likewise, if the pore size is twice again as large, namely 11.4 microns in diameter, then the maximum vacuum that can be pulled without the cup leaking air is 190 millimeters of mercury, or 25 percent of an atmosphere. Since the majority of the pores used in Teflon or PTFE suction lysimeters is between 70 and 300 microns, the bubbling pressure is only 4 percent of an atmosphere.

Where porous materials are being used in air-water systems, such as in suction lysimeters, the most direct method of evaluating the pore size of the material is through the use of air pressure. By thoroughly wetting the porous material and then exposing one side of it to increasingly higher air pressure values, with the other side under water, one can readily observe when the air pressure becomes high enough to enter the pores and cause bubbling on the opposite side. The specific air pressure at which this bubbling occurs is a direct measurement of the pore size, as defined by the above formula, and indicates directly the effectiveness of the porous materials to withstand air pressure differentials when in use. Evaluating pore size distribution by the mercury intrusion method or other means does not give direct information as to how the porous material will perform in the air-water system in which it is used. PTFE pores are generally round and symmetrical, while ceramic pores are of various ragged shapes. The strength of the water meniscuses in the individual pores are a function of pore shape as well as overall size, and for this reason an accurate measurement of the pressure at which the meniscus will break down and allow air to enter can only be made accurately by direct measurement of the bubbling pressure or air entry value of the wetted porous material.

As shown in Figure 8–14A, the pores within a suction lysimeter will not hold a vacuum in a dry condition. Air can move freely from the soil or silica flour surrounding the lysimeter through the pores into the interior part. Thus, suction lysimeters should be installed in a wetted condition and silica flour should be added as a slurry (1 lb to 150 ml of water). One should recognize that Figure 8–14 is highly diagrammatic and that each pore is representative of a tortuous route through the cup wall. In reality, millions of these tortuous pore routes are located throughout the cups. As shown in Figure 8–14B, the pores become completely filled when the cup has been placed in a wet environment and the pressure on both sides of the cup wall is one atmosphere. As shown in Figure 8–14C, the surface tension of the wetted pore begins to change as a suction is developed within the cup. As noted in Figure 8–14D, the radius of curvature of the surface tension decreases as the suction within the cup increases. The ability of water molecules to withstand these pressure gradients is the reason that air will not enter the cup even though the interior of the cup has been evacuated. Water, on the other hand, will freely move through the wetted pores under the gradient induced by the negative pressure within the cup. As developed under the previous paragraphs, the surface tension is greatly increased by the reduction in pore size. As demonstrated in Figure 8–14E, the surface tension in the pore can be broken by increasing the gradient across the cup wall to greater than the bubbling pressure of the cup. The bubbling pressure of low-flow ceramic cups is between 35 and 45 psi, which translates to a suction of 233 centibars. The bubbling pressure of PTFE cups is between 0.75 and 1 psi, which translates to a very narrow operating range of 7 centibars (Everett et al., 1986).

Once a sample has been obtained in the suction lysimeter, as evidenced by a reduction in the suction gauge, pressure must be applied to the lysimeter interior to push the sample to the surface. When a pressure is exerted on the porous segment, the

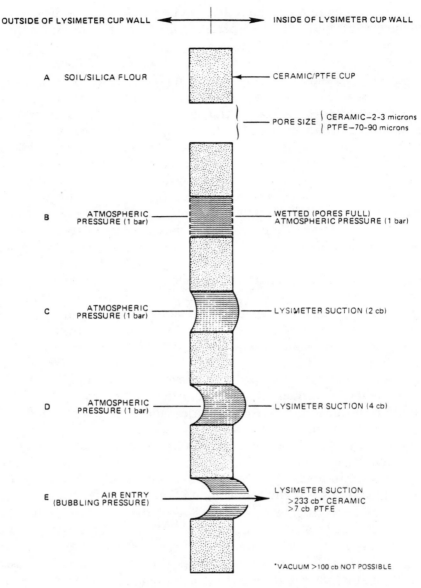

OUTSIDE OF LYSIMETER CUP WALL ◄───────► INSIDE OF LYSIMETER CUP WALL

A   SOIL/SILICA FLOUR ────────── CERAMIC/PTFE CUP

PORE SIZE { CERAMIC–2-3 microns
{ PTFE–70-90 microns

B   ATMOSPHERIC ──────── WETTED (PORES FULL)
    PRESSURE (1 bar)   ATMOSPHERIC PRESSURE (1 bar)

C   ATMOSPHERIC ──────── LYSIMETER SUCTION (2 cb)
    PRESSURE (1 bar)

D   ATMOSPHERIC ──────── LYSIMETER SUCTION (4 cb)
    PRESSURE (1 bar)

E   AIR ENTRY ─────────► LYSIMETER SUCTION
    (BUBBLING PRESSURE)   >233 cb* CERAMIC
                          >7 cb PTFE

*VACUUM >100 cb NOT POSSIBLE

Figure 8-14. Diagrammatic view of lysimeter cup wall.

meniscus will extend away from the center of the suction lysimeter. If too much pressure is applied within the suction lysimeter, the sample may be expelled through the pores. The meniscus behavior, therefore, is a function of the vacuum, pressure, pore size, porous material, soil moisture, and soil texture.

Dead Space in Lysimeters

As a part of the experiments dealing with reduction in flow rate as a function of increased soil suction, Everett et al. (1986) determined that "dead" spaces may exist within suction lysimeters. As shown in Figure 8–15, the ceramic cups are glued to

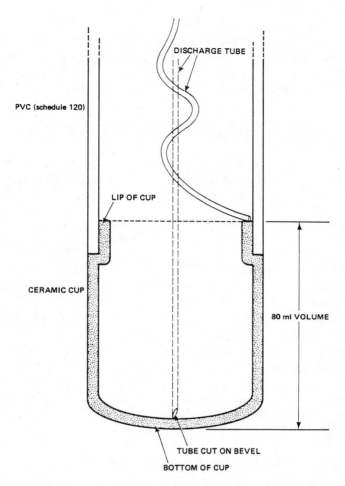

DISCHARGE TUBE

PVC (schedule 120)

LIP OF CUP

CERAMIC CUP

80 ml VOLUME

TUBE CUT ON BEVEL

BOTTOM OF CUP

Figure 8–15. Location of potential dead space in suction lysimeters.

the inner wall of a Schedule 20 PVC body tube. This results in a projection, or lip, on the inside of the suction lysimeter. As polyethylene or Teflon tubes are pushed down or twisted through the two-holed stopper at the top of the lysimeter, they develop a characteristic twist in their length. (The tubing, in most cases, is delivered on a spool and tends to retain a residual bend.) The polyethylene tube may catch on the inside lip of the cup and the operator may conclude that the tube has reached the bottom of the ceramic cup. Since the bottom of the cup cannot be seen, it is difficult to determine whether the tube has actually reached the bottom of the cup. Even measuring with a tape rule may result in the tape rule hanging up on the edge of the cup, giving the impression that the depth to the bottom of the cup has been determined. As a result, an 80-ml error can occur in any rate determinations. This 80 ml of fluid accumulates in the cup and cannot be extracted through the discharge line.

In all PTFE suction lysimeters, the discharge line is a rigid PTFE tube extending to the bottom of the PTFE cup. This design results in zero dead space in the all-PTFE lysimeters. However, the PTFE cups with a PVC body tube have been designed with a rigid interior tube that does not extend to the bottom of the PTFE cup. Since it is impossible to extend this rigidly fixed interior tube, PTFE/PVC units have a constant dead space of 34 ml. The authors are aware of numerous lysimeter investigations where rates of intake and volumes of samples have been reported. To date, most operators, including the manufacturers, have not been aware of the potential dead space within their lysimeters (Everett et al., 1986).

SPECIAL PROBLEMS AND SAFETY PRECAUTIONS

The successful operation of pore-liquid samplers may be restricted by any or all of the following factors: 1) hydraulic factors, 2) soil physical properties, 3) cup-wastewater interactions, and 4) climatic factors.

Hydraulic Factors

The most severe constraint on the operation of pore-liquid samplers involves the soil around the porous segment of a sampler becoming so dry that air bubbles enter the cup and further movement of soil water into the unit is restricted.

If the soil is not excessively dry, a usable sample may still be obtained if suction is applied to the cup for a sufficiently long period of time. Nevertheless, because the yield of suction samplers is greatly reduced under very dry conditions, there may be a situations in which the time required to obtain a sufficiently large sample exceeds the maximum holding time for analysis. Similarly, there may be cases where the soil is so dry that the units simply will not yield a sample. This may be particularly true in arid regions where rainfall is not great enough to wet up the soil profile.

Physical Properties: Soil Texture and Soil Structure

Soil texture refers to the relative proportion of the various soil preparates (particles <2 mm) in a soil (USEPA, 1983). Examples of soil texture classes include silt loam, silty clay, and sand. The successful operation of suction samplers requires a continuity between pore sequences in the porous segment of the sampler and those in the surrounding soils. When soils are very coarse-textured, a good contact between the porous segment of a sampler and the fine pore sequences may be difficult to maintain, and the flow continuum may be destroyed. Unlike the problem of sampling in very dry soils, the problem of poor soil contact is mainly an operational problem, which can be circumvented by using the recommended method of cup installation (see Figure 8–12). In this method, the porous segment of the sampler is placed in close contact with the silica sand, which in turn contacts a larger area of the surrounding soil. This method helps to maintain a continuity in the flow paths that soil water follow in moving from the soil through the silica sand and porous segment into the interior of the sampler.

Soil structure refers to the aggregation of the textural units into blocks. A well-structured soil has two distinct flow regions for liquids applied at the land surface: 1) through the cracks between blocks, or interpedal flow, and 2) through the finer pore sequences inside the blocks, or intrapedal flow. Liquids move more rapidly through the cracks than through the fine pores. Because of the rapid flushing of pollutants through larger interconnected soil openings, the movement of liquid-borne pollutants into the finer pores of the soil blocks may be limited. Inasmuch as suction cup samplers collect water from these finer-pore sequences, the resultant samples will not be representative of the bulk flow.

A primary goal of soil pore-liquid sampling is to detect the presence of fast-moving hazardous constituents. This goal may not be realized if samplers are placed in highly structured soils leading to a flow system such as that described in the last paragraph. The structure of a soil profile is best examined by constructing trenches near the proposed monitoring sites to a depth corresponding to the maximum depth at which the sampling segments will be installed. The extent of large interpedal cracks should be documented at each profile. If such cracks appear to be widespread, alternative sites or monitoring techniques (e.g., pan lysimeters) should be examined. However, it should be borne in mind that even large cracks frequently diminish in width in deeper reaches of the profile. If it is found that structural cracks "pinch out" at the monitoring depth, suction samplers could be installed. As mentioned previously, the extent of macropore flow should be examined to determine the appropriate monitoring approach, for instance, suction or pan lysimetry.

Cup-Wastewater Interactions

For simplicity, the interactions between pore-liquid samplers and wastewater can be grouped into 1) those affecting the operation of the porous segment, principally by plugging, and 2) those that change the composition of pollutants moving through the porous segment.

Plugging.   A basic concern in the use of porous type samples to detect the movement of hazardous waste substances in soils is that the porous segment may become plugged either by particulate matter (e.g., fine silt and clay) moving with the liquid, chemical interactions, or bacterial buildup. The problem of clogging by particulate matter is not as severe as was once thought (Everett et al., 1986). Apparently, soils have the capacity to filter out the fine material before reaching the porous segments. Several studies have been reported involving the use of suction-type samplers for monitoring pollutant movement at land treatment units. Generally, it appears that the sampling units operated favorably without clogging by particulate matter. Examples of such studies include those by 1) Smith and McWhorter (1977), in which ceramic candles were used to sample pollutant movement in soil during the injection of liquid organic wastes; 2) Grier et al. (1977) involving the use of depth-wise suction samplers on fields used for the disposal of animal wastes; and 3) Smith et al. (1977), in which depth-wise suction samples were installed in fields irrigated with wastes from potato processing plants.

Chemical reactions at the surface of a suction sampler may clog the porous network. One type of chemical reaction is precipitation (e.g., of ferric compounds). However, considering the wide variety of chemical wastes, other effects are also possible, leading to the inactivation of suction samplers.

Even though suction samplers may fail because of clogging, the problem may still be an operational difficulty that can be overcome. For example, installing silica sand around the cup may filter out particulate matter. Unless this filter becomes clogged, the samplers should continue to operate. However, this approach may not be sufficient to prevent clogging by chemical interactions.

Change in the Composition of Hazardous Constituents During Movement Through Pore-Liquid Samplers.   It is fairly-well established that the porous segments of suction samplers filter out bacteria, but not viruses. Similarly, a reduction may occur in the metal content of liquids moving into samplers because of interactions within the porous segment. This problem can be reduced by acid leaching of the cups before they are installed in the field.

Because a major concern at polluted areas is the fate of hazardous organic constituents, the amount of organic-cup interactions should be estimated before field-installing sampling units. Change in the composition of hazardous constituents during liquid moving through suction samplers can be demonstrated by laboratory studies. Basically, during such studies, suction samplers are placed in liquids of known composition contained in beakers. Samples are drawn into the cups and extracted for analysis. The change in composition is then easily calculated. In preparing these tests, it is essential that each cup be preconditioned in accordance with recommended practice, for instance, flushing with 8N HC1, followed by rinsing with distilled water.

Climatic Factors

A major factor limiting the operation of suction samplers in very cold climates is that the soil water may become frozen near the cups. This means that a sample cannot be obtained during freezing conditions. Another undocumented problem which conceivably could occur is freezing of samples within the cups and lines, so that the samples cannot be brought to the surface. Since the samplers are located at depth, it is unlikely that freezing would occur. Prior to winter setting in, the lines should, however, be flushed.

Another effect of freezing temperatures is that some soils tend to heave during freezing and thawing. Consequently, suction samplers may be displaced in the soil profile, resulting in a break in contact. In addition, if the cups are full of liquid when frozen, the cups may be fractured as a result of expansion of the frozen liquid. The extent of these problems, however, has not been determined.

Safety Precautions

Worker safety is of paramount importance when installing systems of pore-liquid samplers and during sample handling. In some cases, all contact with the waste and liquid samples should be avoided, and toxic fumes should not be inhaled. Similarly, certain wastes are highly flammable, and precautions should be taken to avoid the creation of sparks. No smoking should be allowed. The degree of precaution that should be exercised, including the type of protective clothing, must be decided on a case-by-case basis. Further safety precautions are discussed in Chapter 7.

Lysimeter Failure Confirmation

In the event that a sample cannot be retrieved from an installed suction lysimeter under conditions where the operator knows that the soil suction levels should be high enough to obtain a sample, specific procedures should be followed. Adjacent to a suction lysimeter that appears to have failed, a soil suction determination can be made to determine whether the available soil moisture is high enough to obtain a sample. Soil suctions are determined using tensiometers. Tensiometers are commercially available and are produced with various designs and lengths.

A tensiometer consists of a tube with a porous ceramic tip on the bottom, a vacuum gauge near the top, and a ceiling cap. When it is filled with water and inserted into the soil, water can move into and out of the tensiometer through the connecting pores in the tip. As the soil dries and water moves out of the tensiometer, it creates a vacuum inside the tensiometer, which is indicated on the gauge. When the vacuum created equals the "soil suction," water stops flowing out of the tensiometer. The dial gauge reading is then a direct measure of the force required to move the water from the soil. If the soil dries further, additional water moves out until a higher vacuum level is reached. When moisture is added to the soil, the reverse process takes place. Moisture from the soil moves back into the tensiometer through the porous tip, until the vacuum level is reduced to equal the lower soil suction value, then water movement stops. If enough water is added to the soil so that it is completely saturated, the gauge reading on the tensiometer will drop to zero. Because water can move back and forth through the pores in the porous ceramic tip, the gauge reading is always in balance with the soil suction.

The effective operational range for suction lysimeters is between saturation and 60 centibars of suction as determined by the tensiometer. Above 60 centibars of suction, a ceramic lysimeter will operate (Everett et al., 1986). However, the flow rates will be so low that effectively one cannot get a sample. If the tensiometer readings are between 0 and 60 centibars of suction, the suction lysimeter should obtain a sample. If no sample is obtained under these soil suction ranges, the suction lysimeter will be deemed to have failed and should be excavated or abandoned.

Tensiometers can be installed in the soil adjacent to the suspect suction lysimeter by using conventional soil-drilling tools. The body tube and porous sensing tip of tensiometers are $\frac{7}{8}$ in. (2.2 cm) in diameter. Installation must be made so that the porous ceramic sensing tip is in tight contact with the soil. Commercially available insertion tools can be used in rock-free soils. Standard $\frac{1}{2}$ in. (U.S.) steel pipe can also be used to drive a hole into the soil to accept the tensiometer. In rocky soils, a soil auger can be used to bore a larger hole, and then the soil is sifted and packed around the porous ceramic tip to make good contact before the hole is backfilled. The surface soil is tightly tamped around the body tube to seal surface water from entering. The tensiometers should be installed at the same depth so that they will be reading soil suction conditions at the depth of the installed suction lysimeter. Tensiometers require 2–3 hours to come into balance with the ambient soil suction. As such, the tensiometers should be read 3 hours after their initial installation. Tensiometers can be left in place in the field to determine if the soil suction is high enough for the suction lysimeters to operate and to obtain a sample.

PAN LYSIMETER INSTALLATION AND OPERATION

As mentioned above, pan lysimeters are more effective in soils in which macropore flow dominates. The two pan lysimeters that appear to have the most application are the trench lysimeter and the free-drainage glass-block sampler. Parizek (Parizek and Lane, 1970) is responsible for the majority of the available information on trench lysimeters, while Barbee (1983) is the principal author of the research on glass-block samplers. Other devices are being developed (i.e., drum lysimeters) that can be considered as part of a monitoring system.

Trench Lysimeters

Trench lysimeters are lysimeters made of galvanized, 16-gauge metal, with dimensions of 0.305 × 0.45 m (12 × 15 in.) that are installed in a trench. Parizek and Lane (1970) developed an installation technique for these devices (see Figure 8–16). Their approach includes installing the trench lysimeter in the sidewall of a trench shelter. Copper tubing is soldered to a raised end of the pan to allow soil water to drain into a sample container located inside the sampling pit. The trench shelter is covered with a sloping roof and a ladder is placed at one end of the shelter to allow access for sampling.

The design by Parizek and Lane (1970), however, introduces a sampling bias problem. If the trench lysimeters are installed close to the side of the trench shelter, as represented in Figure 8–16, the collected sample will be biased. This bias results from the fact that the trench shelters, which project above the land surface, will cause waste application equipment to avoid the actual sampling area to prevent

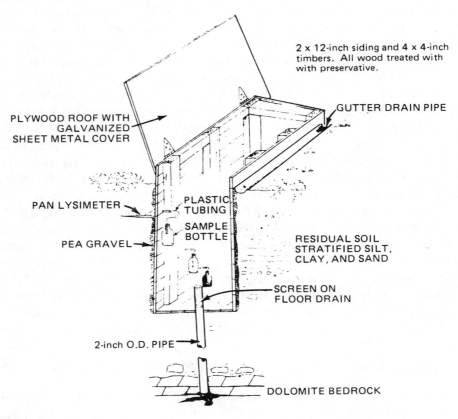

Figure 8–16. Trench lysimeters installed in trench shelter.

damage to the shelter. To alleviate this problem, a slightly modified installation approach is recommended below.

Figure 8–17 illustrates the recommended installation approach for pan lysimeters, including trench lysimeters and glass-block samplers. At a randomly selected site, dig a 1.22 m (4 ft) wide, 3.66 m (12 ft) long trench, excavated to a depth of 2.44 m (8 ft). The trench sidewalls should be temporarily supported with timbers and siding to reduce the risk of cave-ins. The entire seepage face should be inclined 1 to 5 degrees from the vertical.

The trench lysimeter is installed into the sidewall of the trench at a specified level, but a significant distance above the trench floor (see Figure 8–17). A discharge line is installed from the trench lysimeter to a discharge point at the surface. The distance between the lysimeter and the discharge point should be at least 10 m (30 ft) to preclude any sampling bias above the lysimeter. When a sample is required, a vacuum is placed on the discharge line and a sample is retrieved.

After the sampling lines are installed, the lysimeter installation trench is backfilled according to the same procedures described below for glass-block samplers.

Free-Drainage Glass-Block Samplers

One technique for measuring gravitational water in the unsaturated zone was developed by Barbee (1983). The hollow glass-block free-drainage sampler was developed as a technique for improving the capability for monitoring fluid movement in the unsaturated zone.

The free-drainage sampler is made from a hollow glass block (obtained from the PPG Company, Houston, TX 77020) 30 cm × 30 cm × 10 cm deep, with a capacity of 5.5 l (Figure 8–10). A rim, approximately 0.158 cm high around the edge of the upper and lower surfaces, enhances the collecting effectiveness of the blocks. To collect a sample, nine 0.47 cm diameter holes are drilled near the edge around the upper surface of the block. The block is then thoroughly washed with distilled water. A 0.47 cm O.D. nylon tube is then inserted into the block and coiled on the bottom so that all of the accumulated liquid can be removed. A sheet of 0.158-cm-thick

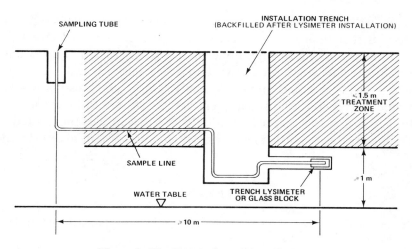

Figure 8–17. Pan lysimeter installation.

fiberglass is cut to fit over the upper surface, including the holes, without overhanging the edge. This sheet enhances contact with the overlying soil and also prevents soil from contaminating the sample and plugging the holes.

The sampler is installed by digging a 2.44 m (8 ft) deep trench with a backhoe at a randomly selected site. A tunnel of about 45 cm is then excavated into the side of the trench, but a significant distance above the trench floor (see Figure 8–17). The tunnel is correctly sized by using a wood model slightly larger than the glass block. Extreme care is taken to keep the ceiling of the tunnel level and smooth to ensure water will not run off the block .and also to have a smooth surface against which the block can be pressed. Jordan (1968) noted that, unless the edges of the free-drainage sampler are in firm contact with the soil for the entire perimeter of the sampler, water will tend to run out through spaces between the sampler and soil, particularly if the ceiling of the tunnel has many irregularities. In clay soil, it is necessary to use a small knife to lightly score the ceiling of the tunnel because of smearing and compaction of the surface during excavation. One glass block is then carefully placed in the tunnel and then pressed firmly against the ceiling, being held in place by soil packed tightly beneath and to the sides of it.

The sampling lines must be carefully installed to prevent sampling bias. The nylon sampling tube is run underground in a trench to a sampling location (see Figure 8–17). The trench for the sampling tubes usually need only be approximately 3 ft (1 m) deep to prevent damage from operating equipment. The nylon sampling tube is then run to the soil surface and a sample drawn by applying a vacuum.

After installation, the trenches for sampling line and lysimeter installation are backfilled. Prior to backfilling the lysimeter installation trench, aluminum foil, 46 cm wide, should be pressed against the side of the trench into which the lysimeter was installed. The aluminum foil prevents lateral movement of liquid from the back-filled soil into the undisturbed soil above the glass-block lysimeter. Any temporary sidewall support structures may be removed prior to backfilling the trench. Careful attention should be paid to properly tamp the soil in the trench after backfilling.

PAN LYSIMETER LIMITATIONS

Pan lysimeters will only function when the soil moisture is greater than field capacity. This implies that their use must coincide with a continuously wetted soil, with most of the flow occurring through macropores (i.e., cracks). This situation could exist at certain sites at which highly structured soils are present.

Pan lysimetry will, as noted previously, only sample gravitational water. Because the pan lysimeter is a continuous sampler, the device should be emptied after each precipitation event in order to prevent sample loss. Because of the limited experience with pan lysimetry, there is little knowledge of clogging potential or effective operating life.

REFERENCES

Apgar, M. A. and Langmuir, D. 1971. Ground-water pollution potential of a landfill above the water table. *Ground Water* 9(6): 76–93.

Barbee, G. C. 1983. A Comparison of Methods for Obtaining "Unsaturated Zone" Soil Solution Samples. M. S. Thesis. Texas A&M University, College Station, TX.

Bigger, J. W. and Nielsen, D. R. 1976. Spatial variability of the leaching characteristics of a field soil. *Water Resour. Res.* **12**(1): 78–84.

Briggs, L. J. and McCall, A. G. 1904. An artificial root for inducing capillary movement of soil moisture. *Science* **20**: 566–569.

Duke, H. R. and Haise, H. R. 1973. Vacuum extractors to assess deep percolation losses and chemical constituents of soil water. *Soil Sci. Soc. Am. Proc.* **37**:963–964.

Everett, L. G. 1988. Moderator Session IV, Vadose Zone Investigations. In *Proceedings Conference on Southwestern Groundwater Issues,* March 23–25, 1988. Albuquerque, New Mexico: Association of Groundwater Scientists and Engineers.

Everett, L. G., McMillion, L. G., and Eccles, L. A., 1988. Suction lysimeter operation at hazardous waste sites. In *Groundwater Contamination-Field Methods,* Collins/Johnson (Eds.), ASTM, STP-963. Philadelphia, PA: ASTM.

Everett, L. G., McMillion, L. G., and Eccles, L. A. 1986. *Suction Lysimeter Operation at Hazardous Waste Sites.* Cocoa Beach, FL: ASTM.

Everett, L. G., Hoylman, E. W., and Wilson, L. G. 1983. *Vadose Zone Monitoring for Hazardous Waste Sites.* Washington, DC: U.S. Environmental Protection Agency.

Everett, L. G., Wilson, L. G., and McMillion, L. G. 1982. Vadose zone monitoring concepts for hazardous waste sites. *Groundwater* **20**(3): 76–87.

Fenn, D., Cocozza, E., Isbister, J., Briads, O., Yare, B., and Roux, P. 1977. Procedures Manual for Ground Water Monitoring at Solid Waste Disposal Facilities. EPA/530/SW 611. Washington, D.C.: U.S. Environmental Protection Agency.

Grier, H. E., Burton, W., and Tiwari, C. 1977. Overland cycling of animal waste. In *Land as a Waste Management Alternative,* pp. 693–702. Ann Arbor, MI: Ann Arbor Science.

Hoffman, G. J., Dirksen, C., Ingvalson, R. D., Maas, E. V., Oster, J. D., Rawling, S. L., Rhoades, J. D., and van Schilfgaarde, J. 1978. Minimizing salt in drain water by irrigation management. In *Agricultural Water Management,* Vol. 1, pp. 233–252. Amsterdam: Elsevier Scientific Publishing Company.

Jackson, D. R., Brinkley, F. S., and Bondietti, E. A. 1976. Extraction of soil water using cellulose-acetate hollow fibers, *Soil Sci. Soc. Am. J.***40**: 327–329.

Jordan, C. F. 1968. A simple, tension-free lysimeter. *Soil Sci.* **105**:81–86.

Law Engineering and Testing Company. 1982. Lysimeter Evaluation, a report to the American Petroleum Institute.

Levin, M. J. and Jackson, D. R. 1977. A comparison of in-situ extractors for sampling soil water. *Soil Sci. Soc. Am. J.* **41**: 535–536.

Morrison, R. D. and Tsai, T. C. 1981. *Modified Vacuum-Pressure Lysimeter for Vadose Zone Sampling.* Huntington Beach, CA: Calscience Research, Inc.

Parizek, R. R. and Lane, B. E. 1970. Soil-water sampling using pan and deep pressure-vacuum lysimeters. *J. of Hydrol.* **11**:1–21.

Parizek, R. R. 1984. Personal communication. Penn. State U., University Park, PN.

Reeve, R. C. and Doering, E. J. 1965. Sampling the soil solution for salinity appraisal. *Soil Sci.* **99**:299–344.

Scott, V. H., and Scalmanini, J. C. 1978. Water wells and pumps: Their design, construction, operation and maintenance. Bulletin 1889, Div. of Agr. Sci., The Univ. of Calif.

Shaffer, K. A. 1984. Personal communication. Central County Planning Commission, Bellefonte, PA.

Shaffer, K. A., Fritton, D. D., and Baker, D. E. 1979. Drainage water sampling in a wet, dual-pore soil system. *J. Environ. Qual.* **8**:241–246.

Silkworth, D. R. and Grigal, D. F. 1981. Field comparison of soil solution samples, *Soil Sci. Soc. Am. J.,* **45**:440–442.

Smith, J. L. and McWhorter, D. M. 1977. Continuous subsurface injection of liquid organic wastes. In *Land as a Waste Management Alternative,* pp. 646–656. Ann Arbor, MI: Ann Arbor Science.

Smith, J. H., Robbins, C. W., Bondurant, J. A., and Hayden, C. W. 1977. Treatment of Potato Processing Wastewater on Agricultural Land: Water & Organic Loading, and the Fate of Applied Plant Nutrients, in Land as a Waste Management Alternative, Ann Arbor Science, pp. 769–781.

Soilmoisture Equipment Corp. 1978. Operating instructions for the *model 1900* soil water sampler. Santa Barbara, CA: Soilmoisture Equipment Corp.

Soilmoisture Equipment Corp. 1983. Internal memo on soil tension. Santa Barbara, CA: Soilmoisture Equipment Corp.

Stannard, D. I. 1988. Tensiometers—Theory, Construction, and Use. Denver, CO: U.S. Geological Survey Professional Paper.

Stevenson, C. D. 1978. Simple apparatus for monitoring land disposal systems by sampling percolating soil waters. *Environ. Sci. and Tech.* **12**:329–331.

Tadros, V. T. and McGarity, J. W. 1976. A method for collecting soil percolate and soil solution in the field. *Plant Soil* **44**:655–667.

U.S. Environmental Protection Agency. 1983. RCRA Guidance Document: Land Treatment Units. U.S. Environmental Protection Agency (draft).

Wood, W. W. and Signor, D.C. 1975. Geochemical factors affecting artificial recharge in the unsaturated zone. *Trans. Am. Soc. of Agr. Eng.* **18**(4): 677–683.

Wood, W. W. 1973. A technique using porous cups for water sampling at any depth in the unsaturated zone. *Water Resour. Res.* **9**:486–488.

9

Control of Subsurface Migration

by James C. S. Lu

Calscience Engineering and Laboratories, Inc.
Cypress, California

INTRODUCTION

Overview

Control technologies for subsurface contamination can be applied at several points in the contamination process. Disposal to land can be avoided by preventing the generation of the waste, or by destroying it in some treatment process. Waste that must be landfilled (or that was previously landfilled) can be modified so that it does not generate contaminated groundwater. If landfilling of dangerous materials is inevitable, it can be done at a site where containment measures will prevent infiltration and escape of water. Finally, for wastes already in place, or for groundwater and soils already contaminated, the site can be isolated so that no further migration takes place. Isolation may be followed by extraction and treatment.

Selection of Control Alternatives

The most appropriate control technology for a subsurface contaminated site can be selected only after a thorough evaluation of the problem, waste and site characteristics, and the available control options (Figure 9–1). The selection process is described in the functional steps shown in Figure 9–2 and discussed below.

1. Collection of existing data. Some data may be readily available for identifying waste and site characteristics and the nature and extent of contamination. Such information may be in records of the site owner and operators, related government agencies (United States Geological Survey, regional water quality control agency), historical topographic data, and surveys of known and suspected waste generators and types of wastes.
2. Site-specific investigations. Five types of site-specific investigations are important for evaluating the nature and extent of contamination and developing control alternatives. These five types are: geophysical, waste and soil, hydro-

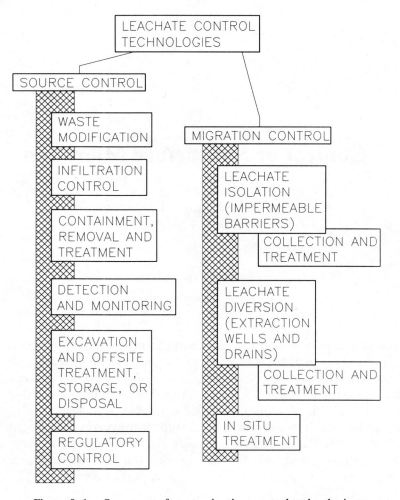

Figure 9–1.   Summary of contamination control technologies.

geologic, surface water and sediment, and biota investigations. The objective of the geophysical investigation is to qualitatively or semiquantitatively determine spatial relationships of waste, groundwater, and geologic units. This investigation allows a more cost-effective focusing of subsequent site investigation tasks.

Examples include seismic reflection or refraction techniques to identify the depth and shape of the water table and bedrock, vertical electrical sounding or electromagnetic surveys to estimate the areal and vertical extent of plumes, and magnetometry or ground-penetrating radar to determine the location and distribution of buried drums. Waste and soil analyses are necessary to determine the type, concentration, and extent of contaminants and soil characteristics. Hydrogeologic investigation determines subsurface lithography, the extent of the leachate plume, and groundwater depth, flow rate, and direction. Aquifer characteristics, such as permeability, thickness of subsurface formations, and sediment type will be determined. Water and sediment studies assess the possibility of surface contamination and potential pathways of contami-

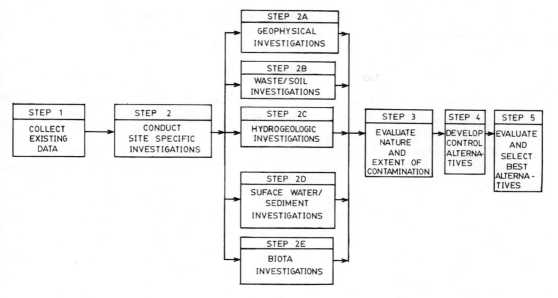

Figure 9-2. Flow chart for selection of contamination control alternatives.

nant transport to surface waters. Biological investigations assess any impacts to aquatic or terrestrial organisms at downstream locations and on the site to provide guidance for remediation.

3. Nature and extent of contamination. Prior to the selection of any control technology for a specific site, it is necessary to describe the nature and extent of contamination. Data obtained in Step 2, described above, can be used for such evaluation.

4. Development of control alternatives. The next step is to develop control alternatives in detail, as they apply to the specific site. Two limiting data sets, restrictive site characteristics and waste characteristics, are addressed in the development of each alternative. Other considerations include the level of technology development, comparable applications, performance records, availability of components, operational complexity, and construction needs.

5. Evaluation and selection of best alternative(s). Control technologies developed under Step (4) undergo an initial screening before a detailed analysis. This screening has three parts: cost, acceptable engineering practices, and effectiveness. After the cost estimates have been made, the alternatives are compared with a cost-benefit analysis, and those whose costs are an order of magnitude higher than the others are eliminated, unless they offer substantially greater benefits or are the only applicable technology. The technical feasibility and effectiveness is assessed for the location and conditions of the site. For example, data received from the hydrogeologic study may indicate that a groundwater barrier wall cannot be used because of the lack of a confining layer below the site to which the barrier wall can be sealed.

After screening, all of the remaining alternatives are subjected to a more vigorous and detailed evaluation. Five kinds of data are evaluated: technical (performance, reliability, constructability, implementation time, and safety), public health, environmental, institutional, and costs.

Contaminant Detection and Monitoring

Leachate detection and monitoring are vital to leachate control technology. They provide early warning, so that steps can be taken to reduce or avoid leachate generation. Once control measures are in place, monitoring determines their effectiveness or detects breakdowns. The information also determines the extent of contamination and identifies leachate migration rates and directions, so that proper remedial action can be developed.

Vadose zone monitoring and groundwater monitoring are both widely practiced for these objectives (Chapters 6, 7, and 8). For storage facilities such as underground tanks, in-tank and beneath-tank detection and monitoring technologies are also available.

Regulatory Control

Government regulation is the basic means for enforcing the reduction or elimination of subsurface contamination problems. While it is not possible here to describe existing regulations in any detail, some general characteristics of regulation can be recognized.

Regulation can control existing, planned, or abandoned hazardous waste sites. It also includes hazardous materials processing, treatment, transport, storage, and disposal facilities. There are three types of regulatory controls: siting or land use restrictions, construction standards, and operational requirements.

Siting criteria can be selected for new hazardous waste facilities to exclude unsuitable sites and to select the best potential sites. Environmentally sensitive areas such as wetlands, high population-density areas, prime farm land, and major resource areas may be excluded. Factors such as geologic setting, flooding, seismic risk, landslides, subsidence, settlement, and tidal wave effects must be evaluated.

Construction standards and operational requirements can be enforced to ensure the integrity of the hazardous waste sites. Specific structural requirements, such as materials used for construction, types of liners, barriers, and leachate monitoring control can be specified for the sites. Other operational requirements such as permits, inspection, preparedness and prevention, contingency plans and emergency procedures, manifesting, record keeping, and reporting requirements can also be provided for the site owner-operators, for the prevention of subsurface contamination.

WASTE MODIFICATION

Technologies are available for the modification of wastes, both before disposal and after they are in land disposal facilities. The treatments may eliminate the hazards of the waste itself or reduce the leachate quantity or hazardousness. The most important advantage of waste modification over other methods is that it affords destruction of hazardous or undesirable species *before* groundwater is contaminated.

In most cases, this technology can be considered a permanent solution. However, many of the waste modification technologies are not fully developed, and applications and reliability are not well demonstrated. Use of these technologies may require considerable bench or pilot tests. Waste modification may be pretreatment, *in-situ* treatment, sorbent addition, and fixation.

Pretreatment

Pretreatment technologies change waste characteristics before disposal. A wide variety of thermal, physical, chemical and biological treatment processes is available for pretreatment purposes depending on waste characteristics and the level of treatment required (Figure 9–3). Thermal treatment can be used to destroy wastes by converting them to harmless products like carbon dioxide. Heat may glassify or calcinate some inorganic wastes, leaving inert solids. It can be considered as an energy recovery process if a significant net heat energy is generated. It can also be considered as a volume reduction process, in that only the noncombustible inorganic materials remain after treatment.

There are many types of thermal treatment technologies which may prove adequate for destruction of hazardous species, though some are in the development stage. Incinerators of several different designs have been proven either by pilot tests

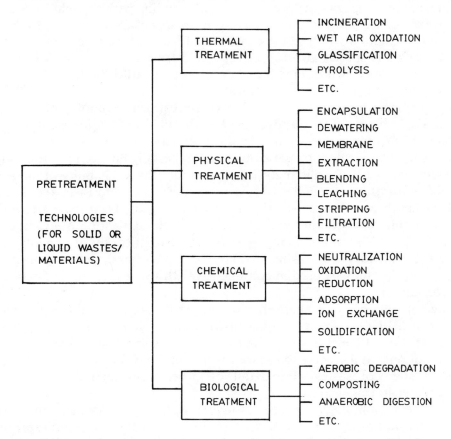

Figure 9-3.   Examples of treatment technologies for hazardous waste and contaminated groundwater.

or field scale units to be effective for certain hazardous species (USEPA, 1978; USEPA, 1975). Other thermal treatment methods, such as the molten salt process, wet air oxidation, pyrolysis, glassification, plasma arc and supercritical water destruction may also be applicable in the near future (Hanchak, 1984).

Residues produced by thermal treatment units have characteristics significantly different from those of the wastes. Leachate quality, or both quality and quantity, generated from the disposal of the pretreated wastes can be greatly altered. Concentrations of toxic organic species can usually be eliminated or reduced. Inorganic species, such as heavy metals, can also be reduced in the leachates if glassification or calcination are used.

Physical pretreatment processes can be used to encapsulate, separate, or remove hazardous or undesirable species from the wastes before disposal, and thus eliminate or reduce these species from the future leachate. Encapsulation techniques have been tested for the purpose of isolating hazardous wastes by mixing or covering them with relatively inert material (Pojask, 1979; USEPA, 1977). Thermoplastic techniques (using paraffin, polyethylene, or bitumen materials), organic polymer processes (using urea-formaldehyde, vinyl ester-styrene, or polyester resin), and surface encapsulation techniques (using a binder and polyethylene) are major methods employed for physical encapsulation.

Dewatering technologies, such as centrifugation, clarification, coagulation, filtration, thickening, evaporation, distillation, and blending with dry materials are used to remove or reduce moisture in the wastes before disposal (Berkowitz, 1978). These techniques substantially reduce the chances of future leachate generation. Hazardous or undesirable species also can be separated from the wastes by various phase separation processes such as membrane processes (ultrafiltration, reverse osmosis), extraction, leaching, stripping, and filtration.

Many chemical processes are applicable as pretreatment technologies for detoxifying hazardous wastes before disposal (Berkowitz, 1978). pH adjustment, oxidation, reduction, adsorption, ion exchange, and solidification are proven technologies for pretreatment. Pretreatment by pH adjustment, especially lime or limestone addition for the increase of pH of solid materials, can reduce the leachability of metals from solids in the disposal sites (Fuller and Artiola, 1978; Chen and Eichenberger, 1981). Oxidation has been used to detoxify reduced forms of hazardous species in the wastes. Examples are sodium hypochlorite for the detoxification of cyanide wastes (Tolman et al., 1978), and permanganate for the detoxification of phenols, formaldehyde, and pesticides (Berkowitz, 1978). On the other hand, reducing agents, such as ferrous sulfate, may be used in conjunction with hydroxides to precipitate hexavalent chromium.

Mixing sorbents or ion exchangers with wastes before disposal has gained attention (Lu et al., 1984). These will not reduce leachate volume, but can greatly reduce the leachate contaminant concentrations. The use of solidification techniques is also growing, not only as a scheme to reduce leachability, but also as a method to reduce moisture contents of wastes. More details of sorbent addition and solidification are presented later in this section.

Pretreatment by biological processes involves placing a waste stream in contact with a suspension of microorganisms, so that the hazardous or undesirable organic compounds in the waste stream are decomposed before disposal. Most of conventional secondary waste water treatment processes (activated sludge, aerated lagoons, trickling filters, anaerobic digestion, and waste stabilization ponds) are applicable as pretreatment processes for liquid hazardous waste streams. Composting and anaerobic digestion processes may be used for solid or semisolid hazardous waste streams.

In Situ Treatment

In situ treatment technologies are those which are applied to soil and waste while they are in place, without excavation. These technologies are most likely to be feasible for sites where wastes are well defined, and shallow, and the extent of contamination is small. Such limitations suggest applicability to relatively shallow landfills, surface impoundments, and contaminated soil. Four major *in situ* treatment technologies have been practiced—solution mining, neutralization and detoxification, stripping and aeration, and microbial degradation.

Solution mining, also called soil flushing or washing, is the process of flooding the disposed wastes or contaminated soils with a solvent and collecting the emerging solution with a series of shallow wells. Five types of solvent may be used, depending on the hazardous species to be removed and waste characteristics (USEPA, 1982):

1. Water, applicable for wastes containing readily water soluble species.
2. Acids, such as sulfuric, hydrochloric, nitric, phosphoric, and carbonic acids, applicable for extracting basic metal salts.
3. Alkalis, such as sodium hydroxide, applicable for extracting aluminum, lead, tin, and zinc metals, as well as certain organic sulfur compounds and phenols.
4. Complexing and chelating agents, such as EDTA, citric acid, and ammonium salts, applicable for heavy metals.
5. Surfactants, applicable for oil and grease.

The advantage of solution mining is its relatively low cost for implementation. However, its effectiveness is difficult to determine, and the solvent may become a pollutant itself, if the system has been poorly designed.

In situ neutralization and detoxification can be achieved by applying or injecting a substance into the contaminated area to immobilize or destroy a pollutant. Again, a well system can be used for the injection of chemical agents. In contrast to solution mining, *in situ* neutralization and detoxification techniques do not inherently incorporate seepage collection systems. Four types of chemical reactions may be applicable for *in situ* modification of hazardous wastes or contaminated soils: neutralization, precipitation, chemical oxidation, and chemical reduction. Examples are numerous. Alkalis can be used to neutralize acidic wastes, and vice versa. Heavy metals may be precipitated by alkalis or sulfides. Free fluorides may be precipitated by calcium salt. Cyanides can be detoxified (oxidized) by sodium hypochlorites, permanganates, or hydrogen peroxide. Hexavalent chromium can be precipitated (reduced) by ferrous sulfate.

Other restrictions and problems exist for *in situ* treatment. It is limited to processes whose byproducts are either less toxic or nontoxic, or relatively insoluble. It must be applied to wastes which are readily soluble or can be reacted easily, and seepage collection systems may be required if the byproducts are still toxic. It is difficult to determine the thoroughness of this technology.

Air stripping or aeration technologies have been suggested, tested, and practiced for the removal of volatile organics from contaminated sites. Aeration of contaminated soil by turning and mixing a thin soil layer has been used for trichloroethylene (degreaser) removal and tested for perchloroethylene (cleaning fluid) removal. Stripping by well systems is also applicable for the decontamination of volatile organics from contaminated sites.

Microbial degradation may be applicable for *in situ* treatment of biodegradable wastes (Chapter 4). An aerobic process is usually used, especially for complex organics. Specialized strains of bacteria have been developed for the breakdown of certain chemicals and may be ordered in dry bulk quantities (Polybac Corporation, 1978). *In situ* microbial degradation could be the least-cost alternative, if it is feasible. However, it is a relatively slow process. Situations where it can be applied best are those where complete mixing and aeration can be achieved, such as a surface impoundment, a landfarming site, or a chemical spill area.

Sorbent Addition

Laboratory studies by Liskowitz et al. (1976) and Fuller and Artiola (1978) have demonstrated the effectiveness of various natural and synthetic materials in removing contaminants from water. Selected materials included bottom ash, limestone, fly ash, vermiculite, illite, ottowa sand, activated carbon, kaolinite, natural zeolites, activated alumina, and cullite. No single sorbent can reduce the concentration of all constituents in leachate to acceptable levels (Liskowitz et al., 1976). The test results, however, suggested that combinations of different sorbents can be used to reduce most leachate contaminants to acceptable levels. For leachate source control, the sorbent can be either placed in a waste disposal site as a bottom layer or mixed with waste before land disposal.

Fixation

Fixation involves a number of techniques designed to seal or solidify the wastes in a hard, stable mass (USEPA, 1979). The major objective of fixation processes is to reduce or eliminate the leachability of hazardous or undesirable species. Because of land disposal restrictions of wastes containing free flowing liquid, fixation has also been used as a dewatering process for slurry or sludge land disposal.

Physical fixation and glassification processes were discussed previously in this section. Other chemical fixation processes can be classified into two categories: cement-based or lime-based. Cement-based solidification involves mixing the wastes with Portland cement, or cement and one of the following materials: sodium silicate, fly ash, fluorosilicate, and lime (Pojask, 1979; USEPA, 1977). Hazardous wastes can be incorporated into the rigid matrix of the hardened concrete. This method is especially effective for metallic wastes and certain acids and strong oxidizers (e.g., chlorates and nitrates). However, there are disadvantages. It requires large volumes of solidifiers and is affected by salts of zinc, copper, magnesium, tin, and lead, and sodium salts of arsenate, borate, phosphates, iodates and sulfide. Its long-term effectiveness is unknown and it is not effective for organic species.

Lime-based solidification involves mixing the wastes with lime and one of the following materials: fly ash, clay, Portland cement, plaster-of-paris, or sodium silicate. To be compatible with lime-based solidification, wastes should be stable at high pH. The most common use of this technique is to solidify flue gas cleaning sludge. Its advantages and disadvantages are similar to those of cement-based solidification. In recent years, many proprietary chemical solidifiers have become available. Most of these solidifiers are based on the pretreatment principle: a small amount of chemical compound(s) is added into cement or lime-based agents to pretreat (e.g. precipitate, oxidize, reduce, or other reactions) the wastes so hard-to-fix wastes can be handled.

EXCAVATION

Excavation is commonly used for remediation of contaminated sites. It is effective for materials such as solid hazardous wastes and contaminated soils, sediments, and slurries. Its obvious disadvantage is that it does not alter the wastes: their hazards are only moved to another site. Often workers, neighbors, and even others on the freeway, are exposed to the dangers during the operation. It may be a logical alternative, however, if the contaminated materials are moved to a secure landfill or to an effective treatment facility. Liquid hazardous materials, such as liquids stored in surface impoundments, also can be disposed in this manner, using vacuum trucks.

Pre-Excavation Activities

Excavation requires disturbance of the buried materials. Direct physical contact with the hazardous waste and spills may occur. In order to ensure the success of the cleanup project, the project manager should:

1. Gather existing relevant data, such as type and effects of hazardous species involved, and geohydrologic conditions of the site.
2. Conduct site investigations, such as the six types of investigations discussed previously (geophysical, geochemical, hydrogeologic, air, surface water and sediment, and biota).
3. Evaluate data, such as preparation of a profile of the materials, identification of potential problems, selections of treatment, storage, and disposal facilities (TSD) alternatives.
4. Prepare cleanup plans, such as:

   • excavation and disposal plan;
   • air monitoring plan;
   • health and safety plan;
   • odor and toxic vapor control plan;
   • water pollution (e.g., leachate, runoff) control plan;

- noise control plan;
- traffic and circulation control plan;
- trespassing control plan; and
- aesthetic and dust-generating control plan.

5. Prepare compliance procedures and obtain regulatory approval.

Excavation and Disposal Plan

Depending on site conditions and waste material, work zones must be established at the site for the excavation operation. A typical example of the site work zones is shown in Figure 9–4. The innermost zone is called the exclusion zone, where contamination does or could occur. All people entering this zone must wear prescribed levels of protection, which may include gloves, suits, and respirators. They must also have had appropriate training.

The outermost zone, the support zone, is considered to be an uncontaminated area where all support equipment and material can be supplied. Decontamination stations are established in the transition zone, or contamination reduction zone. If a contamination site is less hazardous, a less stringent site control may be utilized.

The major equipment used for excavation should be chosen based on the type of material, site conditions, and cost-effectiveness analysis. Typical choices are:

1. Solid wastes, soils—dragline, backhoe, or clamshell.
2. Sediments, slurries—dewatering equipment and excavation equipment as above.

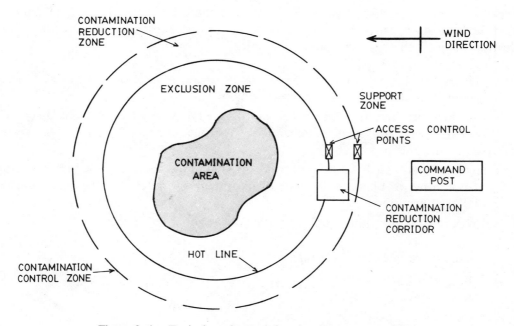

Figure 9–4. Typical work zones for site cleanup operation.

3. Sediment, sludges—hydraulic dredging equipment, such as centrifugal pumping systems or portable hydraulic pipeline dredges.

Air Monitoring Plan

Air monitoring during excavation is necessary to ensure the safety of workers and the general public. Hazardous atmospheres may be explosive, oxygen-deficient, radioactive, or toxic. Suitable monitoring equipment and methods must be used. In general, air monitoring can be either *in situ* monitoring or air sampling followed by laboratory analysis.

Health and Safety Plan

A site health and safety plan must be prepared for routine (but hazardous) response activities and for unexpected site emergencies. Levels of protection must be specified for the site or certain portions of the site. Criteria for the selection of type of levels can refer to EPA "Standard Operating Safety Guides" (USEPA, 1984). Site entry procedures, decontamination methods, and emergency response procedures must also be specified. Personnel involved in the cleanup activities must be trained for all health and safety requirements.

Other Mitigation Measures

During excavation and transportation activities, control plans may be necessary for the mitigation of impacts of odor and toxic vapor, surface runoff, subsurface leachate, noise, traffic, trespassing, and dust-generation. Various materials such as masking agents, absorbents, neutralizing agents, and sprayed-on liners may be used to reduce odor and toxic vapor problems.

Drainage and water diversions, and leachate containment and removal, can be used for runoff and leachate control. Noise is generally not a major problem, but it can be reduced by choosing quieter equipment or controlled through better scheduling. Waste transportation routes and alternative routes should be carefully selected to avoid traffic problems. A fence or other barrier must be constructed to control trespassing. Dust-generation problems may be reduced by well-controlled watering, wind barriers, and better operating practices.

CONTAMINATED SITE CONTROL

The approaches to cleanup of contaminated soils and groundwater share a common characteristic: they include steps to gain control of the site. In some cases, for mild contamination of dry soils, it may be sufficient to provide a cover to prevent the entry of water or the spreading of dust, and leave the contaminants in place.

There is much greater difficulty when flowing groundwater is contaminated. Here, control must include hydrologic management. In many cases, operation of a collection system is necessary to extract the contaminated water. In turn, the extracted water must be treated to meet discharge requirements.

Cleanup efforts have commonly utilized five devices for leachate and groundwater migration control:

1. Impermeable subsurface barriers can prevent the migration of groundwater, which may either stop clean water from entering the contaminated site, or stop contaminated water from leaving.
2. In some cases, the barrier is a permeable treatment bed. These are relatively porous materials which allow the water to pass through, but which adsorb or neutralize contaminants.
3. Drains may be used to collect leachate and groundwater. These can be installed either as the waste is disposed, in a controlled landfill; or afterwards, for an uncontrolled site.
4. Extraction systems will collect and remove contaminated groundwater and exert control on groundwater movement. Injection systems may add clean water to the groundwater system, in order to gain control of flow velocity and direction.
5. Water treatment systems, installed at the surface, will remove the contaminants from the groundwater before disposal or reinjection.

These elements have been used in a wide variety of combinations, providing the flexibility to meet specific site conditions (Figures 9-5 through 9-8). Barriers and drains can be combined with injection and extraction wells to avoid accumulation of contaminated water underground. Permeable treatment beds are sometimes used alone, if site conditions are appropriate. Under appropriate conditions, the contaminant control and cleanup system may consist only of appropriately placed injection and extraction wells, with treatment of the removed water (Figure 3-57).

The selection of migration-control technologies is site-specific, and the success of a specific technology relies heavily on the subsurface geohydrologic conditions. This makes site investigations critically important. Thorough subsurface surveys and testing are necessary to identify geological formations, rates, gradients and directions of groundwater migration, the vertical and areal extent of contamination, and effects of subsurface water isolation and well field pumping.

INFILTRATION CONTROL

Groundwater contamination has often occurred when rainwater or other precipitation has passed through solid wastes or polluted soils to the water table. Infiltration controls are approaches to reduce or prevent the entry of water to hazardous waste disposal sites or contaminated soils. These controls minimize or eliminate hazardous leachate generation and reduce erosive transport of cover materials and exposed buried wastes. They include drainage and water diversion, surface sealing, and revegetation with irrigation scheduling (Figure 9-9).

Drainage and Water Diversion

Drainage and water diversion techniques modify the natural topography and runoff characteristics of waste disposal sites and other subsurface contamination sources, to control infiltration and erosion. The best systems are those installed before waste

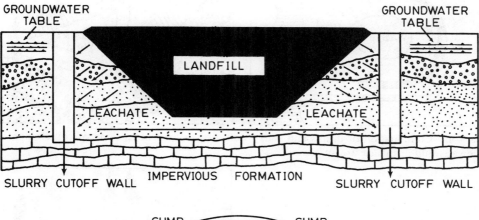

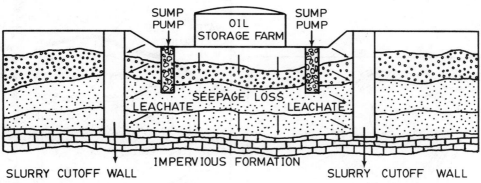

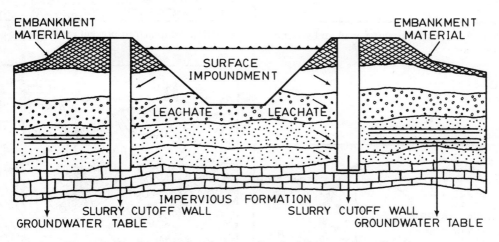

Figure 9-5. Cross-sectional views of typical slurry wall applications.

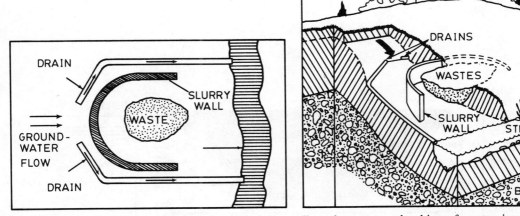

Figure 9–6. Combination of drains and slurry wall used to prevent leaching of contaminants into groundwater and a surface stream.

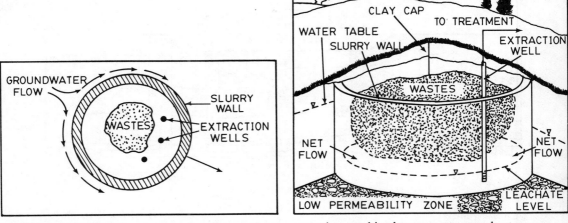

Figure 9–7. Circular slurry wall and groundwater extraction combined to prevent groundwater contamination.

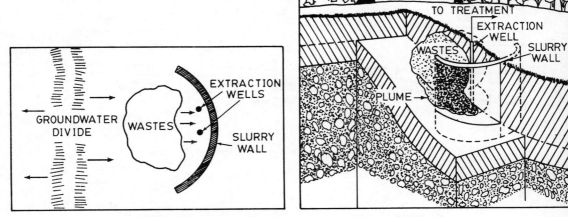

Figure 9–8. Slurry wall and extraction wells used to collect leachate.

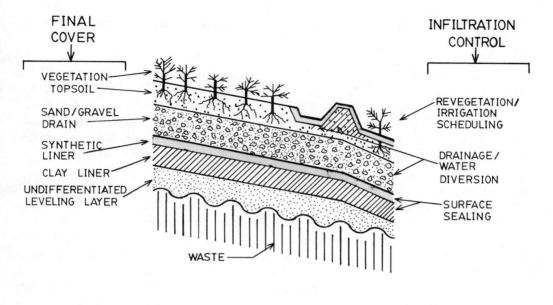

FINAL COVER

VEGETATION
TOPSOIL
SAND/GRAVEL DRAIN
SYNTHETIC LINER
CLAY LINER
UNDIFFERENTIATED LEVELING LAYER
WASTE

INFILTRATION CONTROL

REVEGETATION/ IRRIGATION SCHEDULING
DRAINAGE/ WATER DIVERSION
SURFACE SEALING

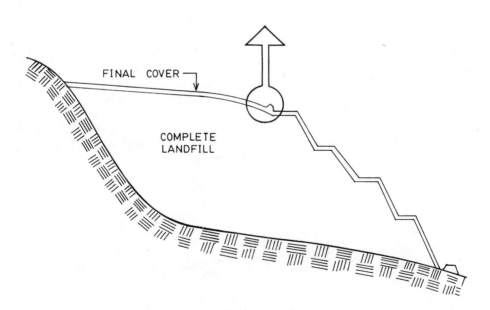

FINAL COVER
COMPLETE LANDFILL

Figure 9-9. Infiltration control technologies.

is disposed of, in a well-designed landfill. It is also possible, however, to install some at uncontrolled sites.

Grading. Compacting and grading the surface of contamination sites is a simple and economical procedure. A profile with a maximum slope of 12 percent and a minimum of 6 percent has been reported to minimize infiltration and prevent erosion (Tolman et al., 1978; Lutton, 1978; and Lutton et al., 1979). For equal surface areas of land, doubling the slope length will increase soil losses by 1.5 times (USEPA, 1976). In order to prevent erosion while reducing infiltration, some water diversion techniques (benches, terraces, dikes, berms, ditches, sedimentation basins) can be used, as discussed later in this section.

Sand or Gravel Drain.   A sand or gravel layer is usually installed above an impermeable liner (to be discussed in the following subsection, "Surface Sealing") to collect the infiltrated water and move it away from the site. Poorly graded sands, gravels, or sand or gravel mixtures can be used for the drainage layer. The thickness of the layer can be designed based on permeability, slope, and distance between drains, and the amount of liquid that reaches the drain (Figure 9–10).

Water Diversion.   Several well-established water diversion techniques are available for diverting surface water flow in critical areas of hazardous waste sites. Those methods most applicable are dikes, berms, ditches, bench terraces, chutes, downpipes, seepage basins, and sedimentation basins. Dikes and berms are well-compacted earth ridges or ledges constructed immediately upslope from or along the perimeter of hazardous waste sites. They can be used to intercept storm runoff, divert the flow to natural or manmade drainage ways, stabilize outlets, or trap sediments. Ditches are excavated drainageways used on the upslope or downslope perimeters of hazardous waste sites or landfills to intercept and divert runoff. They

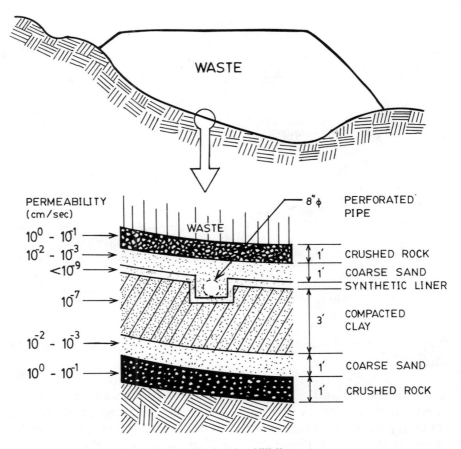

Figure 9–10.   Typical landfill liner system.

also can be placed along the contour of graded slopes and in conjunction with a supporting dike or berm constructed along the downhill edge of the drainageway.

Bench terraces are relatively flat areas constructed along the contour of very long or very steep slopes to slow down runoff and divert it into ditches. Chutes or open channels and downpipes can be used to carry surface runoff from one level to a lower level without erosive damage. Seepage and sedimentation basins are designed to intercept runoff and suspended solids entrained in the runoff. Seepage basins are designed to recharge collected runoff downgradient from the site. Sedimentation basins are usually the final step in control of diverted runoff prior to discharge into a receiving water body. At any given hazardous waste facility, the most effective method for managing infiltration may be a combination of two or more drainage and water diversion techniques discussed above. The selection of individual techniques will depend on the size and topography of the site, local climate, and hydrology, and the soil characteristics.

Surface Sealing

Surface sealing, or capping, is the technique of using impermeable materials to seal the surface of a contaminated area to prevent infiltration, control erosion, and isolate hazardous chemicals. A variety of impermeable cover materials and sealing techniques is available for such purposes. Surface sealing is essentially a cover system applied on completed hazardous waste sites, like landfills. It is also used, however, for sealing the upper surfaces of uncontrolled dump or spill sites (sometimes as a temporary measure). In order to protect the integrity of the impermeable liner, a leveling layer and a sand or gravel drainage layer are usually placed below and above the liner (Figure 9–11). The choice of sealing material and method of application are dictated by site-specific factors such as the desired functions, liner availability, costs, nature of wastes, topography, hydrology, and meteorology. There are five general types of sealing material: compacted soils and clays, admixes, flexible polymeric membranes, sprayed-on linings, and soil sealants.

Compacted Soils and Clays.   Fine-grained soils such as clays and silty clays have low permeabilities, and are therefore best suited for sealing purposes. The permeability of compacted clays is in the range of 10^{-9} to 10^{-6} cm/sec.

Admixes.   Admixed lining materials, such as hydraulic asphalt concrete (HAC), soil cement, soil asphalt, soil lime, soil fly ash, and bentonite-soil mixtures can create stronger and less permeable sealing materials. HAC has been used for centuries as a water-resistant material. HAC is resistant to inorganic leachates, but is not applicable for sealing waters with organic solvents and hydrocarbons. Permeability as low as 10^{-9} cm/sec can be achieved with HAC. For a soil-cement cover, approximately 8 percent (by weight) dry cement is blended into the in-place soils with a rotary hoe or tiller as water is added. Permeability of soil cement is usually in the range of 10^{-7} to 10^{-5} cm/sec, depending on soils and soil/cement ratios used. Soil cements are good materials to resist weathering and chemicals such as alkali, organic matter, and inorganic salts. But soil cement has a tendency to crack and shrink on drying.

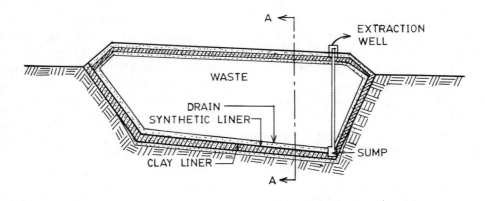

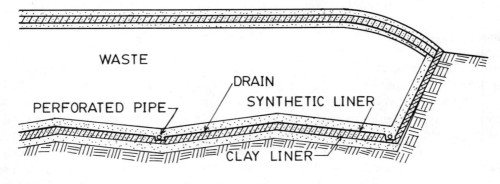

SECTION A-A

Figure 9-11. Schematic of hazardous waste landfill leachate collection system.

Soil asphalt is a mixture of on-site soil and liquid asphalt. Permeabilities as low as 10^{-8} to 10^{-7} cm/sec can be obtained with soil asphalt. Increases in asphalt/soil ratio will decrease permeability.

Cover soils may also be treated with lime, fly ash, and bentonite clay. Lime applied as 2–8 percent (by weight) calcium oxide or hydroxide is suitable for cementing clayey soils. Soils also can be treated to reduce permeability and shrink and swell behavior by lime or fly ash mixture. The addition of lime is recommended for neutralizing acidic cover soils, thereby reducing the leaching potential of heavy metals.

Flexible Polymeric Membranes. Flexible polymeric membranes (synthetic liners) include a wide range of rubbers and plastics. Examples can be listed:

Rubbers (elastomers)
    Butyl (polyisobutylene/isoprene)
    EPDM (ethylene propylene diene monomer)
    Neoprene

Thermoplastics (uncured)
    PE (polyethylene)
    PVC (polyvinyl chloride)

CPE (chlorinated PE)
Elasticized polyolefin (3110)

Thermoplastic elastomers
Hypalon (Chlorosulfonated polyethylene)
EPDM (ethylene diene propylene monomer 4060)

Synthetic liners are generally more expensive and labor-intensive sealing materials that require special field installation methods. The sheets that form the liner must be individually laid in place, then sealed together. Care must be taken to ensure that the liner is not ruptured during subsequent construction activity.

Analysis and evaluation of the leachate or waste characteristics, site conditions, environmental factors, and performance data of the liner should be carefully conducted before final selection of a liner (Geswein, 1975; Geswein et al., 1978; Haxo, 1976a; Haxo, 1976b; Haxo, 1976c; Matrecon, 1980; Ware and Jackson, 1978; and Cope et al., 1984). Incompatible liner materials may be weakened or dissolved by waste solvents, allowing the site to leak.

Sprayed-On Linings. Sprayed-on liners are formed by spraying a liquid that solidifies to form a continuous membrane onto a prepared surface. Examples are air-blown asphalt, emulsified asphalt, urethane modified asphalt, and rubber and plastic latexes.

Soil Sealants. Application of certain chemicals or latexes to soils can change the characteristics of soils, such as an increasing soil density, decreasing soil permeability, or facilitating soil compaction. Examples of soil sealants include sodium chloride, tetrasodium pyrophosphate, sodium polyphosphate, sodium carbonate, and organic polymers.

Revegetation and Irrigation Scheduling. Revegetation provides an inexpensive and cost-effective means of stabilizing the landfill cover, reducing infiltration, and minimizing erosion by wind and water. Revegetation can also reduce water runoff, so that more enters the soil. Increased infiltration is offset at least partly by transpiration from vegetation. A subsidiary benefit is that revegetation enhances the appearance of the landfill or contaminated site. In order to prevent increased infiltration due to runoff reduction or irrigation water requirements, water balance models can be used, or moisture monitoring in the cover material can be conducted. Based on modeling or monitoring results, irrigation can be scheduled to reduce or eliminate infiltration. Other water balance factors (surface slope, soil types, vegetation types, etc.) also can be considered.

MIGRATION BARRIERS

Leachate Containment

Containment of leachate or contaminated groundwater within the waste disposal site can be achieved by a drain or liner system situated beneath the site to intercept the leachate before it migrates into the surrounding environment. Use of a subsurface drain or liner system at existing sites is limited by the logistics of placing the system beneath the wastes. Although prefabricated drain or liner systems have widespread use as leachate containment systems for new sites, it is not technically feasible to use prefabricated systems for existing disposal sites. However, slurries and grouts may be injected to form a bottom seal under certain limited conditions.

A mixture of sand and gravel layer, or a combination of crushed rock and coarse sand layers can be used as the drainage layer (Figure 9–10). Leachate collection trenches with perforated pipes are situated at low areas of the drainage layer. Beneath the drainage layer, a synthetic liner and a compacted clay layer are installed. If groundwater intrusions or subsurface springs occur beneath the bottom of the liners, a combination of a coarse sand and rock layer or a sand and gravel layer will also be provided to prevent liner damage due to buildup of the hydraulic pressure underneath.

The types of liners applicable for leachate containment were discussed under "Surface Sealing." Selection of an appropriate liner should be based on the cost-effectiveness of the liner. The following factors should be considered for evaluation of liner effectiveness:

1. Waste compatibility: including the effects of types, compositions and strengths of the wastes on candidate liners.
2. Performance: including durability (service life), permeability, strength (capability of preventing cracking and puncturing), and resistance (capability of resisting the effects of ozone, ultraviolet light, microorganisms, burrowing animals, and plant roots).
3. Site conditions: including characteristics of soil, occurrence and depth of groundwater, settlement or subsidence, and seismic conditions.
4. Environmental conditions: including temperature extremes, weather, sunlight, humidity, pH, and redox conditions.

Once a liner is selected that has the required effectiveness, costs can be considered. The design engineering should consider the preliminary design criteria and any site-specific factors which would influence material and installation costs. Contingency planning should be included to account for unusual site and environmental conditions.

Slurry Walls

Objective. A slurry wall is an underground water barrier which prevents the horizontal subsurface movement of groundwater. It is constructed by the excavation of a trench of the desired configuration and filling it with materials of very low permeability. Because the trenches are commonly deep, it is often necessary to use a bentonite and water slurry to support the sides during excavation.

The slurry wall may isolate landfill leachate by providing a complete seal extending to an impermeable liner or may divert groundwater flow. The slurry wall may also be utilized as a down-gradient barrier from the landfill (Figures 9–5 through 9–8). In this case, leachate impounded by the barrier may be collected for recycling or treatment (Tolman et al., 1978; Lu et al., 1984; and Spooner et al., 1984).

Historical Review. The slurry wall was used as a subsurface water barrier for the improvement of oil well stability as early as 1914 by petroleum engineers (Nash, 1978; Xanthakos, 1979; Ryan, 1980). The next 20 to 30 years involved investigations of slurry properties, such as thixotrophy and viscosity control. In 1929, bentonite clays were first used in drilling operations to stabilize deep wells in unconsolidated materials and to bring cuttings to the surface. The first slurry trench cutoff wall was built at Terminal Island near Long Beach, California in 1948. The 1950s marked a period of continuous development and improvement of the slurry trench technique. By the mid-1960s, slurry walls had become an established method for use in earth dam construction. For pollution control, slurry walls have been constructed to control sewage, mine wastes, landfill leachates, and subsurface migration from contaminated sites.

Function and Principles. Bentonite slurry has an unusual "supporting" capability to prevent the collapse of the cutoff wall during excavation. This supporting action was credited for the stability of trench excavations and made it possible to reach considerable depths. This support arises in three steps:

1. Penetration (Filtration) of slurry. Bentonite particles (in colloidal sizes) can penetrate the surface of the excavated soils to a thickness of a few inches to several feet.
2. Formation of filter cake. The penetrating bentonite particles and soil particles will form a tightly packed zone of gelled material (filter cake) within a few minutes to hours. The filter cake has a plastering effect on the interface so that individual soil particles and grains are held in the earth structure.
3. Formation of a protective film. Within a few seconds, the cake is covered by a thin layer of bentonite particles that further stabilizes the surface of the cutoff wall. In essence, the trench becomes completely lined with a layer of filter cake and a protective film of extremely low permeability.

When the trench has been excavated to the desired depth, backfilling is begun. Often one section of trench is backfilled while a new section is being excavated. Backfilling is sometimes done with only the excavated soil material, but most often other materials are used.

Excavation and Preparation of Slurry. Backhoes, draglines, buckets, or clamshells can be used for slurry wall excavation. The backhoe is most commonly used because of its low cost and high speed. Backhoes can excavate to 30 and 60 ft for standard and extended equipment, respectively. Dragline buckets (6 to 15 tons) can routinely excavate to 60–80 ft, and will reach 150 ft or more in ideal conditions. Hydraulic clamshells can also excavate to 150 ft or more. If the excavated depth is beyond 60 ft, it is best to combine the backhoe and clamshell. The bentonite slurry can be prepared by slowly adding bentonite to the water and mixing with high-shear recirculating pumps until the clay is hydrated. The slurry may be discharged directly to the trench, or to a holding tank. The density of the slurry is usually from 1.04 to 1.15 gm/ml (65–72 lbs/ft^3). The ratio of bentonite to water is usually 0.04–0.06 to 1.

Types of Slurry Wall. Slurry walls can be classified according to the type of backfill materials, the degree of stiffness, and the construction methods:

1. Earth cutoffs. Earth cutoffs are slurry walls backfilled with clay materials and bentonite. They are flexible slurry walls.
2. Clay-cement cutoffs. These are slurry walls backfilled by mixtures of clay and cement, sometimes called solidified walls.
3. Cement-bentonite cutoffs. A particular type of clay-cement cutoffs, these are classified separately because of their wide application. Cement-bentonite slurries normally contain about 6 percent (by weight) bentonite, 18 percent ordinary Portland cement, and 76 percent water.
4. Plastic-concrete cutoffs. Plastic concrete has an ultimate strength of 1,400 psi or less. It usually consists of gravel, sand, cement, clay, and bentonite mixed with water. In general, the strength of plastic concrete depends on the constituent materials and mainly on the water-cement ratio.
5. Rigid-concrete cutoffs. These are plain or reinforced concrete slurry walls.
6. High-resistance noncorrosive cutoffs. These are slurry walls resistant to extremely low or high pH groundwater or groundwater containing corrosive salts such as sulfates. Examples of such noncorrosive cutoff materials are sulfate-resisting cement (ASTM Type V), fly-ash concrete, and bituminous mixes (generally 70 percent aggregate, 10 percent lime filler, and 19 percent bitumen, etc.).

Applications. Placement of the slurry wall depends on the location of the waste and the direction and gradient of groundwater flow. If the slurry wall is placed on the upgradient side of a waste site, groundwater can be forced to flow around the wastes. Walls can be either "keyed" into a low permeability formation below the aquifer, or placed to only intercept the upper portion of the aquifer (Figures 9–5 to 9–8).

Advantages and Disadvantages. Slurry walls have been installed to retard the movement of groundwater, contaminants, and leachate at many waste sites with various site conditions. Slurry walls are relatively simple to construct, contaminant-resistant bentonites are available, and maintenance costs are low. They are not, however, an answer to all waste-site problems. Problems associated with slurry walls include the cost of shipping bentonite, overexcavation in rocky ground because of boulders, and the possible deterioration of certain bentonites when exposed to high ionic-strength leachates. In a few cases, there have been serious and costly failures. Slurry walls are most effective when groundwater is shallow, an impermeable soil or bedrock strata is available for contact with the barrier, and hydraulic gradients across the wall are small (Tolman et al., 1978).

Grout Curtains

Objective. Like a slurry wall, a grout curtain provides a vertical barrier to either divert groundwater flow around a hazardous waste site or intercept migrating leachate. A grout curtain is created by injecting solutions under pressure into soils and underlying earth materials. The grout fills the voids, preventing the passage of liquids.

Grouting Techniques.   Three types of grouting techniques are practiced in the construction industry: area grouting, high-pressure grouting and contact grouting. Area grouting is a relatively low-pressure blanket used to seal and consolidate soils near the ground surface. High-pressure grouting is performed at depth to seal fissures or small void spaces. Contact grouting is conducted at the outer surface of an excavation to seal water flow. High-pressure grouting is most commonly used for leachate migration control.

Grouting Equipment.   Basic grouting requires a mixer, an agitator, a pump, circulation lines, and control fittings. The mixer and agitator mix the ingredients and keep the grout stirred, ready for pumping. The pump drives the grout through the circulation line, past the grout hole and back to the agitator. Control fittings are used on the grout hole to control injection rates and pressures.

Types of Grout Curtain.   There are two groups of grouting materials:

Common grouts
    Cement
    Bentonite
    Chemical grouts
    Sodium silicate formulations
    Acrylamide grouts
    Lignosulfonate grouts (chromelignins)
    Phenoplast grouts
    Aminoplast grouts.

Specialized grouts
    Epoxy resins
    Silicone rubbers
    Lime
    Fly ash
    Bituminous compounds.

Portland cement is widely used. For grouting, a water-cement ratio of 0.6 to 1 or less is effective (Bowen, 1975). Portland cement is often used with a variety of additives that modify its behavior. Among these are clays, sands, fly ash, and chemical grouts. Bentonite is ideal for grouting because of its injectability and low permeability. Bentonite grouts can be injected into materials with lower permeabilities than can other suspension grouts. Chemical grouts are a more recent development than suspension grouts (cement or bentonite grouts). The oldest and still the most commonly used chemical grouts are silicate-based, which account for over 75 percent of the grouting performed for waterproofing. Acrylamide grouts are composed of an acrylamide monomer, a cross-linking agent (i.e., methylene-bis-acrylamide), an accelerator (usually ammonia persulfate), and a catalyst (such as dimethylaminopro-prionitrite).

Chromelignins consist of lignosulfonate or lignosulfite reacted with a hexavalent chromium compound. The former two materials are byproducts of the wood pulp processing industry. Phenoplasts are formed by the polycondensation of a phenol and an aldehyde. Aminoplast grouts are still in the development stage and seldom used.

Applications. The use of a particular grout material is a function of soil characteristics encountered at the facility. American Cyanamid Co. (1975) has categorized the use of chemical grouts with various soil properties (Figure 9–12). As with slurry walls, placing a grout curtain upgradient from a waste site can redirect the flows through the wastes. Grout curtains also can be constructed surrounding the entire waste site and keying into the bedrock or impermeable formation. A semicircular grout curtain can be placed around the upgradient end of a landfill (Figure 9–13; Tolman et al., 1978).

Advantages and Disadvantages. Grouting technology is well developed and is applicable to a wide range of soil properties. However, grouting is usually considered effective only in soils with permeabilities of 10^{-5} cm/sec or greater. Grouting offers a useful approach in consolidated or cohesionless soils, although costs may be prohibitive for large landfills. The use of grout curtains for diverting groundwater around landfills or intercepting leachate is undocumented; however, the principle should be applicable for landfilling practices.

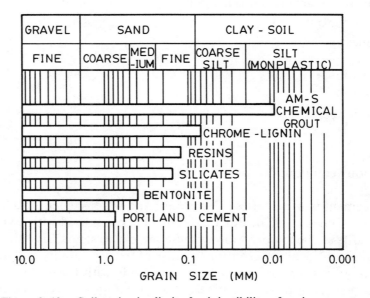

Figure 9–12. Soil grain size limits for injectibility of various grouts.

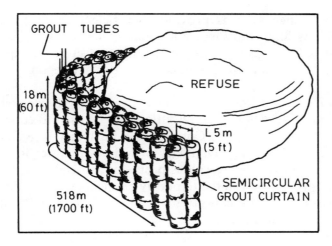

Figure 9–13. Grout curtain used to prevent groundwater entry to waste mass.

Sheet Piling Cutoff Wall

Cutoff walls made of flat steel sheets may be used to retard the flow of water under and through a landfill. The walls consist of lengths of steel sheet piling permanently driven into the ground. Sheet piling is readily available in various shapes and weights, the selection of which is based on costs and ease of installation. The sheet piling is assembled before being driven into the ground. The sheet piling wall is not watertight because of imperfections in the interlocking edges.

The approach offers several distinct advantages over slurry trenches and grout curtains. Among these are ease of construction (no excavation is necessary), availability of contractors to perform the work, the maintenance-free nature of the wall, and the relatively low construction costs. Limitations of the sheet piling cutoff walls include a potential for corrosion, problems of driving the sheet through rocky ground, and the slight initial leakiness of the wall.

GROUNDWATER DIVERSION

Subsurface Drains

Description. Subsurface drains are underground gravel-filled trenches, generally lined with perforated pipes or tiles. The purpose of drains is to intercept up-gradient groundwater that can be redirected from a waste disposal site to lower the water table (Figure 9–6). They can also be used to collect groundwater and leachate down-gradient of a waste disposal site. Drains are constructed by excavating a trench to a desired depth, partially backfilling the trench with highly permeable sand or gravel, placing a plastic or ceramic drain tile in the sand and gravel bed, and completing the backfilling. The drainage can be collected in an impermeable sump and directed or pumped back to the waste disposal site or the treatment facility.

Applications. Drainage systems provide an efficient control system where the groundwater table is high. Subsurface leachate collection systems have been proposed or constructed at several existing waste disposal sites. Examples include Love Canal (Glaubinger, 1979) and the Rossman's Landfill in Oregon City (Anon., 1979).

Advantages and Disadvantages. Subsurface drains are relatively cheap in installation and operation (Figure 9-14). Gravity flow can be used to collect leachates and expensive liners are not necessary. Considerable flexibility is available in terms of spacing and configuration. Disadvantages include difficulties in trenching operations for coarse or rocky soils, as well as problems associated with deep trenching. Subsurface drains are not well suited to poorly permeable soils and require adequate monitoring to assure leachate collection. In most cases, it will not be feasible to situate underdrains beneath the site.

Drainage Ditches

Description. Drainage ditches or surface drains are similar to subsurface drains except that no collection pipes or tiles and backfills are used (Figure 9-15). Drainage ditches can be an integral part of a leachate collection system or as interceptor drains.

Applications. Drainage ditches may be used for flat or gently rolling disposal sites underlain by impermeable soils where subsurface drainage is impractical or uneconomical. A system used by Union Carbide is a typical example of the drainage ditch system (Slover, 1975). This system is used around the periphery of a impoundment to carry off impoundment surface waters and peripheral waters that may contribute to ponding and leachate formation.

Advantages and Disadvantages. In general, drainage ditches are simpler than subsurface drains to install, as long as soils are firm so that the integrity of the ditches can be maintained. Drainage ditches can be used for collecting leachates in poorly

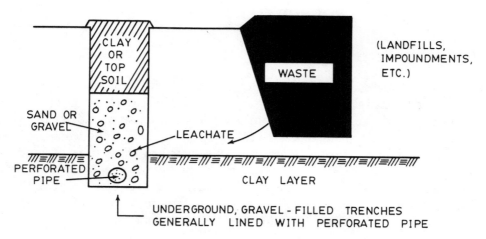

Figure 9-14. Subsurface drain for collection of leachate.

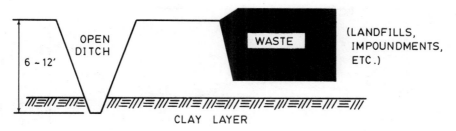

Figure 9-15. Drainage ditch for collection of leachate.

permeable soils where subsurface drains cannot be used. Drains are also applicable for the diversion of disposal site side seepage and runoff. Installation and operation costs for drainage ditches are lower than for subsurface drains. However, maintenance costs for drainage ditches are usually higher. Because of the stability concerns, drainage ditches are not suitable for deep disposal sites. Safety and security measures are needed for drainage ditches to prevent accidents.

REMOVAL OF CONTAMINATED GROUNDWATER

Leachate Removal

Leachate removal from hazardous waste units can be achieved by the combination of leachate extraction wells, bottom drain/liner systems, and horizontal leachate collection pipes (Figure 9-11). A sump, basin, or wetwell is also provided to facilitate leachate collection. The major design problems for leachate removal are estimation of leachate generation rates, determination of the optimum spacing and size of the drainage pipe, and selection of the depth, slope, and orientation of the drainage layer. Determination of these specifications is usually based on practical experience, experimental data, and calculations using water balance and drainage equations.

Leachate generation or design flow estimates can be made using water balance models such as the EPA HELP Model or those described in Lu et al. (1984) "Production and management of leachate from municipal landfills—Summary and assessment."

Drain spacing can be determined from Hooghoudt's formula (Baver et al., 1972).

Well Point Systems

Well point systems can be placed up-gradient from a disposal site to prevent groundwater flow through the fill (Figure 9-16). Well points are also used for leachate collection down-gradient of the disposal site, although care must be exercised to prevent the system from accelerating the rate of groundwater flow through the contaminated zone. Well points are short lengths of plastic or Teflon well screen which can be placed in the saturated zone. The well points are connected to a pump which evacuates the air. The vacuum forces the groundwater to flow into the wells. Well point systems are usually effective for dewatering to about 30 feet, and may provide drawdowns of about 15 feet within the zone of influence (Nobel, 1963). The advan-

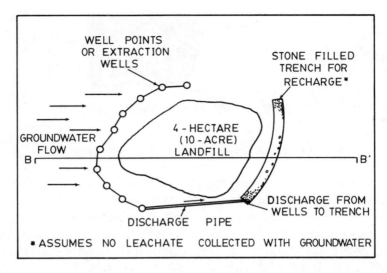

Figure 9–16. Well points or extraction wells for preventing groundwater flow through landfill.

tages of a well point system are moderate installation costs and relatively simple construction methods. The disadvantage of this system is that the power costs are high, continued maintenance is necessary, and a long-term financial commitment is required.

Deep Well Systems

Deep well withdrawal refers to an installation in which a greater vertical depth of water can be removed than can be handled by a vacuum-extracted well point system. Deep wells are capable of lowering the groundwater table as much as 40 ft in uniform sand (Lu et al., 1984). Each well is usually equipped with its own submersible pump and is capable of dewatering large areas.

TREATMENT OF CONTAMINATED GROUNDWATER

Groundwater Treatment

Contaminated groundwater can be treated by a wide variety of physical, chemical and biological processes (Figure 9–3). Treatment techniques demonstrated in recent years are comprised of the traditional secondary or tertiary treatment processes common to wastewater systems. The following treatment processes have proven applicable:

  Biological treatment
    Activated sludge process
    Trickling filters
    Rotating biological discs

Stabilization ponds/aerated lagoons
Anaerobic filter
Anaerobic digesters

Physical treatment
Air stripping
Ultrafiltration
Reverse osmosis

Chemical treatment
Coagulation/precipitation
Carbon adsorption
Ion exchange
Chemical oxidation

Figures 9–17 and 9–18 show the effectiveness of various treatment processes for the removal of organic and inorganic species from contaminate groundwater (Lu et al., 1984). Because landfill leachates can have high concentrations of both organic and inorganic contaminants, treatment of leachates at the field scale requires integration of the basic treatment processes into a systematic approach. The design of such a system should account not only for the volume and quality of leachate to be treated, but also on the changes in leachate quality over time.

In Situ Treatment

Leachate or contaminated groundwater may be treated *in situ,* without disturbing the soil, by a variety of chemical, physical or biological technologies. Examples are: permeable treatment beds; *in situ* chemical treatment; and *in situ* biological treatment. These methods may have considerable potential for reducing the quantities of contaminants. However, most of them are still in the development stage.

Permeable treatment beds can remove contaminants physically and chemically from leachate or groundwater (Figure 9–19). Relatively few materials can be used: activated carbon; limestone or crushed shell; glauconitic greensand or zeolite; and synthetic ion exchange resins.

An activated carbon bed is most suitable for leachate or groundwater contaminated with organic compounds and certain heavy metals (copper, lead, selenium, zinc, cobalt, arsenic, and chromium). A limestone or crushed shell bed may be applicable where neutralization of acidic groundwater or removal of heavy metals is needed. Zeolite and ion exchange resins are very effective in removing ionic species such as heavy metals. A drawback is that permeable treatment beds may become saturated or plugged over time, and may need periodic replacement.

Injection of chemicals (such as neutralization agents, lime slurry, adsorbents, and oxidants) into the subsurface environment to neutralize, immobilize or destroy contaminants in leachates and groundwater has been considered in the past. Although

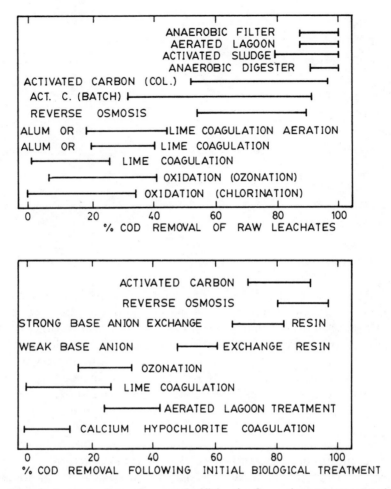

Figure 9–17.   Chemical oxygen demand removal efficiencies for various treatment technologies.

the effectiveness of this technology is not well documented, it is theoretically a feasible *in situ* treatment method. Difficulties may come from the uneven distribution of chemicals underground, unknown treatment conditions and nonhomogeneous nature of the subsurface environments.

In situ biological treatment is also used. Most schemes involve modifying the subsurface environment to promote the activity of indigenous microorganisms for biodegradation of readily decomposible organics. Below the water table, injection and extraction wells are first installed to gain hydrologic control. Injection water is enriched with oxygen and nutrients such as nitrate and phosphate. As the injection water moves through the contaminated area, biodegradation rates are increased.

In the vadose zone, *in situ* biodegradation is commonly inhibited because the soils are too dry. Treatment may consist of installation of infiltration galleries, which consist of horizontal trenches or perforated pipes. These are filled periodically with nutrient-enriched water, which migrates outward by unsaturated flow. Again, the objective is to provide the indigenous microorganisms with an appropriate environment for rapid biodegradation.

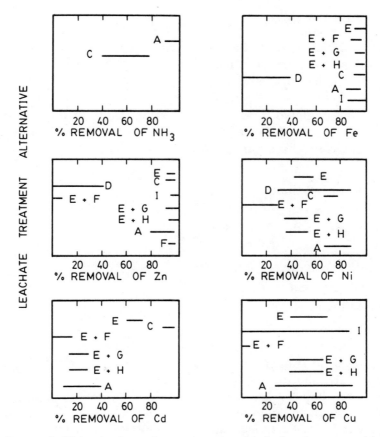

Figure 9-18. Removal efficiencies for various treatment technologies. A = anaerobic filter; B = air stripping; C = activated sludge; D = activated carbon; E = lime coagulation; F = sodium hypochlorite coagulation; G = alum coagulation; H = sodium hydroxide coagulation; I = aerated lagoons.

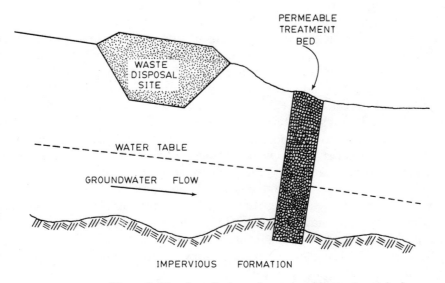

Figure 9-19.   Installation of a permeable treatment bed.

On-Site or Off-Site Treatment

Control systems which involve the interception and collection of contaminated groundwater will also require treatment of the collected wastewater prior to ultimate disposal. Treatment has been the subject of considerable research over the past decade.

For slightly contaminated groundwater (contaminants in the low ppm or ppb ranges) extraction may be followed by several systems, depending on the type of contaminants, environmental conditions and cost-effectiveness:

Spraying in an open pond;
Stripping in an air-stripping tower;
Stripping in a stream-stripping tower;
Carbon adsorption;
Stripping and carbon adsorption;
Other physical-chemical treatment; and
Biological treatment.

The stripping technologies discussed are suitable mainly for volatile organic compounds. They result in transfer of the contaminants to the air, however, and may be prevented by air pollution regulations in some cases.

The methods can be used for both volatiles and nonvolatiles. In general, undiluted leachates are more contaminated than polluted groundwater, and require greater treatment effort. Fresh leachates from most waste disposal sites are characterized by very high BOD and COD concentrations (in the range of 10,000–50,000 mg/l or higher). These concentrations, however, are likely to decrease with continued leaching and biostabilization. High concentrations of major cations, anions, and suspended solids are also present near the onset of leaching, and similarly tend to decrease as percolation through the solid waste continues.

The presence of significant concentrations of metals, including copper, cadmium, lead, mercury, and zinc, are of particular environmental and public health concern. Unlike the organic and major ionic concentrations of leachate, heavy metal concentrations are generally unpredictable as leaching continues. In addition to long-term changes in the composition of leachates, short-term variability of specific components is observed. These short-term changes are produced by seasonal effects such as increased infiltration or evapotranspiration, or through hydrologic events. The combined effects of short and long-term leachate variability makes the design of a treatment system particularly difficult. This information indicates that leachate can be quite different from contaminated groundwater and domestic wastewaters, and thus may not be treatable through conventional municipal wastewater treatment systems. Leachates may have a much higher organic content than wastewater, as well as generally higher concentrations of toxic substances such as metals. Also, whereas a particular domestic wastewater may exhibit stable concentrations over time, leachates can be highly variable.

Treatment Methods

The suggested general approaches to groundwater and leachate treatment are biological treatment with aerobic and anaerobic biostabilization systems, and physical and chemical treatment utilizing precipitation, adsorption, coagulation, chemical oxidation, and reverse osmosis.

After an extensive literature survey of treatment methods for landfill leachates, Chian and DeWalle (1976) concluded that newly formed landfill leachate is best treated by biological treatment processes. Leachate from stabilized landfills, with a low organic content, are best treated with physical and chemical methods.

Biological Treatment of Leachate

Various researchers have investigated the potential of aerobic and anaerobic processes for the stabilization of leachate. The primary goal of these approaches is to reduce high BOD and COD concentrations in leachate. A basic problem in biological treatment is that the leachate metals and other contaminants may poison the biological treatment culture.

Aerobic Treatment.   Laboratory-scale research has demonstrated that activated sludge processes can effectively remove organic matter and metals from leachate (Boyle and Ham, 1974; Cook and Foree, 1974; Uloth and Mavinec, 1977; Palit and Qasim, 1977). Removal of 90 to 99 percent of the leachate BOD and COD and some metals removal were commonly reported. The mixed liquor volatile suspended solids concentrations (MLVSS, a measure of the density of the microorganism culture) in these lab-scale reactors were higher than those typically observed in wastewater treatment: 5,000 to 10,000 mg/l for leachates compared to 1,000 to 2,000 mg/l for typical municipal wastewater processes. Because of the high suspended solids concentrations, the food to microorganism ratio was low (0.02–0.06).

Necessary treatment times are substantially higher for leachates than for municipal wastewater (1 to 10 days for leachate compared to several hours for wastewater). Solids detention times were 30 to 60 days for leachate compared to 5 to 15 days for municipal wastewater. These operational characteristics indicate that the large amount of organic matter in leachate is not readily oxidized, and requires extensive biological activity for stabilization. The long treatment times indicate that extensive aeration energy requirements will be required for aerobic treatment of leachate.

Various other operational problems were observed during aerobic treatment, including foaming, nutrient deficiencies, and toxic inhibition. Uloth and Mavinec (1977) indicated that excessive aeration in conjunction with high concentrations of metals contributed to foaming, and that mechanical mixing independent of aeration and anti-foaming admixtures, could help obviate this problem. Palit and Qasim (1977) indicated leachate stabilization could be hampered by nutrient deficiencies, and that the addition of nutrients may be necessary in some cases. The toxic effects of metals and other constituents appear to inhibit biological removal of oxygen demanding material, as indicated by the increased time required for biostabilization.

Metals removal from leachate during aerobic treatment was reported by Uloth and Mavinic (1977). Following a biological detention time of 10 days, activated sludge digester effluents showed less than 10 mg/l of iron, dropping from an original concentration in raw leachate of 240 mg/l. More that 95 percent of the mixed liquor aluminum, cadmium, calcium, chromium, manganese, and zinc were also removed by the settling biological floc. An average of 85 percent of the lead and 76 percent of the nickel was associated with the sludge solids. Between 49 percent and 69 percent of the mixed liquor magnesium was removed by settling. The authors noted that the percentage removal for all metals is generally higher than that observed by other researchers for activated sludge processes.

Anaerobic Treatment. Anaerobic treatment of leachates has been effective in reducing organic loads. Continuous culture lab-scale reactors have demonstrated 90 to 99 percent organic removal, which is comparable to that achieved by aerobic treatment (Pohland, 1975). Similar removal efficiency is reported by Chian and DeWalle (1976) using a lab-scale anaerobic filter.

Anaerobic treatment offers several advantages over aerobic treatment. First, the build-up of biological solids in the reactor is low, implying a reduction in sludge disposal requirements. Second, treatment times demonstrated for anaerobic treatment of leachate (about 10 to 15 days) are comparable to the time required for anaerobic digestion of municipal wastewater. Finally, the absence of aeration requirements allows a saving of power costs, though some energy may be required to maintain reactor temperatures at the necessary 30–35°C. A major product of anaerobic digestion is methane, which can be used as an energy source for maintaining reactor temperatures, or for other purposes.

Physical and Chemical Treatment of Leachate

Various physical and chemical treatment processes have been applied to leachate treatment. These are particularly useful in treating leachates from older landfills whose organic content is negligible, or as a polishing step for leachates previously treated by biological methods.

Precipitation and Coagulation. Chemical precipitation and coagulation experiments have been fairly successful in removing iron, color, and suspended solids in leachates. Ho et al. (1974) and Thornton and Blanc (1973) have demonstrated that precipitation with lime is effective in the removal of iron and other multivalent ions, color, suspended solids, and COD. BOD concentrations, however, are apparently unaffected. Coagulation with alum or ferric chloride is reported by each of the aforementioned research projects. Although some removal of iron and color was demonstrated, both coagulants were found to be of limited value in removing COD, chloride, hardness and total solids. Results were highly pH dependent, and chemical dosages were high, producing large amounts of waste solids.

Ion Exchange. Leachate treatment by ion exchange is reported by Pohland (1975) for leachates which had been previously treated biologically. Pohland reported good results using a combination of cationic and anionic exchange resins, removing many ionic species as well as dissolved solids and nutrients. Very little residual organic removal was reported; however, the use of mixed resins appears promising as a treatment approach for the nonorganic fraction of leachate.

Carbon Adsorption.   The removal of leachate organics by carbon adsorption was studied by Pohland (1975), for leachates previously treated by ion exchange. Though moderately successful, activated carbon was found to release solids which adversely affected the total solids concentration of the leachate. Pohland suggested that if leachate were to be treated with both carbon and mixed resins, the carbon treatment should precede the mixed resins. Activated carbon column tests performed by Ho et al. (1974) demonstrated complete color and odor removals at a detention time of 20 minutes, along with 55 percent COD removal.

Reverse Osmosis.   Chian and DeWalle (1976) examined reverse osmosis treatment of leachates and found effective removal of COD and dissolved solids. However, the potential for membrane fouling by suspended solids, colloidal material, and iron hydroxides, was noted. Thus, reverse osmosis is perhaps most effective as a post-biological treatment step for removal of residual COD and dissolved solids. Chian and DeWalle also note that membrane efficiency is sensitive to pH.

Chemical Oxidation.   Chemical oxidation was shown to be reasonably effective in removal of COD, iron and color (Ho et al., 1974), though only at high concentrations of the oxidizing agent. Oxidizing agents included chlorine, calcium hypochlorite, potassium permanganate, and ozone. The various treatment processes reported in the literature for leachate can be summarized (Figures 9–17 and 9–18).

Leachate Treatment Systems

Because leachates can have high concentrations of both organic and inorganic contaminants, treatment of leachates at the field scale requires integration of the basic treatment methods into a systematic approach. The design of such a system should account not only for the volume and quality of leachate to be treated, but also for the changes in leachate quality over time. Several leachate treatment system options are: complete treatment of leachate, pretreatment or combined treatment with municipal wastewater, land application of leachate, aerated lagoons, and leachate recirculation through the landfill.

Combined with hydrologic control and appropriate site containment, treatment systems offer an attractive alternative for ending subsurface migration of hazardous wastes.

REFERENCES

American Cyanamid Co. 1975. *All About Cyanamid A5-9, Chemical Grout.* Wayne, NJ: American Cyanamid Co. 95 pp.

Anonymous. 1979. Oregon landfill seeks to control leachate under difficult conditions. *Solid Waste Management,* **9**:21.

Baver, L. D., et al. 1972. *Soil Physics.* New York: John Wiley and Sons, Inc.

Berkowitz, J. B. 1978. Unit Operations for Treatment of Hazardous Industrial Wastes. Noyes Data Corp.

Bowen, R. N. C. 1975. *Grouting in Engineering Practice.* New York: Halstead Press.

Boyle, W. C. and Ham, R. K. 1974. Biological treatability of landfill leachate. *J Water Pollution Control Federation,* **46**(5):860–872.

Chen, K. Y. and Eichenberger, B. A. 1981. Evaluation of Costs For Disposal of Solid Wastes From High-BTU Gasification. Argonne, IL: Argonne National Laboratory.

Chian, E. S. and DeWalle, F. B. 1976. Sanitary landfill leachates and their treatments. *J. Env. Eng. Div., ASCE,* **102**(EE2):215–239.

Cook, E. N. and Foree, E. G. 1974. Aerobic biostabilization of leachate. *J. Water Pollution Control Federation,* **46**:380–392.

Cope, F., Karpinski, G., Pacey, J., and Steiner, L. 1984. Use of liners for containment at hazardous waste landfills. *Pollution Engineering:* **16**(3):22–32.

Fuller, W. H. and Artiola, J. 1978. Use of limestone to limit contaminant movement from landfills. In *Land Disposal of Hazardous Wastes.* U.S. Environmental Protection Agency EPA-600/1-9-78-016.

Geswein, A. J., Landreth, R. E., and Haxo, H. E. 1978. Use of Liner Materials for Land Disposal Facilities. U.S. Environmental Protection Agency WW-562.

Geswein, A. J. 1975. Liners for Land Disposal Sites: An Assessment. U.S. Environmental Protection Agency EPA/530/SW-137.

Glaubinger, R. S. 1979. *Love Canal aftermath. Chemical Engineering,* **86**(23), 86–92.

Hanchak, M. J. 1984. Future technologies for management of hazardous wastes. *Waste Age* **15**(10):24–26.

Haxo, Jr., H. E. 1976a. Assessing synthetic and admixed materials for lining landfills: Formation, collection and treatment. In *Gas and Leachate From Landfills.* U.S. Environmental Protection Agency EPA-600/9-76-004.

Haxo, Jr., H. E. 1976b. Evaluation of selected liners when exposed to hazardous wastes. In *Proceedings Hazardous Waste Research Symposium,* Tucson, Arizona. EPA-600/9-76-015.

Haxo, Jr., H. E. 1976c. Evaluation of liner materials exposed to leachate: Second interim report. U.S. Environmental Protection Agency EPA-600/2-76-255.

Ho, S., Boyle, W. C., and Ham, R. K. 1974. Chemical treatment of leachates from sanitary landfills. *J. Water Pollution Control Federation* **46**(7):1776–1791.

Liskowitz, J. W., Chan, P. C., and Trattner, R. B. 1976. Evaluation of selected sorbents for the removal of contaminants in leachate from industrial sludges. In *Residual Management by Land Disposal.* U.S. Environmental Protection Agency EPA 600/9-76-015.

Lu, James C. S., Eichenberger, Bert, and Stearns, Robert J. 1984. Production and Management of Leachate From Municipal Landfills: Summary and Assessment. U.S. Environmental Protection Agency EPA-600/S2-84-092.

Lutton, R. 1978. Selection of cover for solid waste in land disposal of hazardous wastes. In *Proceedings Fourth Annual Research Symposium,* D. Shultz (Ed.). U.S. Environmental Protection Agency EPA-600/9-78-016.

Lutton, R., Regan, G., and Jones, L. 1979. Design and Construction of Covers for Solid Waste Landfills. U.S. Environmental Protection Agency EPA-600/2-79-165.

Matrecon, Inc. 1980. Lining of Waste Impoundment and Disposal Facilities. U.S. Environmental Protection Agency EPA SW-870.

Nash, J. K. T. L. 1978. Slurry trench walls, pile walls, trench bracing. In *Sixth European Conference on Soil Mechanics and Foundation Engineering,* Vienna, Austria.

Nobel, D. G. 1963. Well points for dewatering groundwater. *Ground Water* **1**(3):21–36.

Palit, T. and Qasim, S. R. 1977. Biological Treatment Kinetics of Landfill Leachate, *J. Env. Eng. Div., ASCE,* **103**(EE2):353–366.

Pohland, F. G. 1975. *Sanitary Landfill Stabilization with Leachate Recycle and Residual Treatment.* Cincinnati, OH: U.S. Environmental Protection Agency EPA-600/2-75-043, 105 pp.

Pojask, R. B. (Ed.). 1979. *Toxic and Hazardous Waste Disposal, Volume I: Processes for Stabilization/Solidification.* Ann Arbor, MI: Ann Arbor Sci.

Polybac Corporation. 1978. Technical Data Sheets, Allentown, PA.

Ryan, C. R. 1980. Slurry cut-off walls, methods and applications. Presented at: GEO-Tec, Chicago, IL.

Slover, E. 1975. A Case History: Implementing a Chemical Waste Landfill. In *Proceedings Fourth National Congress, Waste Management Technology and Resource and Energy Recovery,* Atlanta, GA.

Spooner, P. A., Wetzel, R. S., Spooner, C. E. Furman, C. A., Tokarski, E. F., and Hunt, G. E. 1984. Slurry Trench Construction for Pollution Migration Control. Cincinnati, OH: U.S. Environmental Protection Agency.

Thornton, R. J. and Blanc, F. C. 1973. Leachate treatment by coagulation and precipitation, *J. Env. Eng. Div., ASCE,* **99**(EE4):535–544.

Tolman, A. L., Ballestero, Jr., A. P., Beck, Jr., W. W., and Emrich, G. H. 1978. Guidance Manual for Minimizing Pollution from Waste Disposal Sites. Cincinnati, OH: U.S. Environmental Protection Agency EPA-600 12-78-142.

Uloth, V. C. and Mavinec, D. S. 1977. Aerobic treatment of a high strength leachate. *J. Env. Eng. Div., ASCE,* **103**(EE4):647–745.

United States Environmental Protection Agency. 1978. Burning Waste Chlorinated Hydrocarbons in a Cement Kiln. SW-147C, PB-280-118.

United States Environmental Protection Agency. 1977. Development of a Polymeric Cementing and Encapsulating Process for Managing Hazardous Wastes. EPA-600/2-77-045.

United States Environmental Protection Agency. 1976. Erosion and Sediment Control, Surface Mining in the Eastern U.S. EPA-625/3-75-006.

United States Environmental Protection Agency. 1982. Handbook for Remedial Action at Waste Disposal Sites. EPA-625/6-82-006.

United States Environmental Protection Agency. 1975. Incineration in Hazardous Waste Management. SW-141, PB-261-049.

United States Environmental Protection Agency. 1984. Standard Operating Safety Guides. Office of Emergency and Remedial Response.

United States Environmental Protection Agency. 1979. Survey of Solidification/Stabilization Technology for Hazardous Industrial Wastes. EPA-600/2-79-056.

Ware, S. A. and Jackson, G. S. 1978. Liners for Sanitary Landfills and Chemical and Hazardous Waste Disposal Sites. EPA-600/9-78-005.

Xanthakos, P. P. 1979. *Slurry Walls,* New York: McGraw-Hill Book Company.

Index

Index

Acclimation period, definition, 126
Acetone, 214
Acetone rinse, 225
Acidity of wastes, 26
Adsorption, 216, 251, 252, 259
  and desorption, 157
  and microbial growth, 131
  and microorganisms, 137
  and soil type, 160
  by chemical bond formation, 151
  by hydrophobic bonding, 152
  by ion exchange, 152
  by multiple mechanisms, 153
  by Van der Waals forces, 153
  capacity, 156
  isotherms, 154
  kinetics, 156
  Langmuir or Freundlich models for, 154
  on soils, 151
Advection, influence on contaminants, 77
Aeration, 231, 235
Aerobic chemoautotrophy, 120
Aerobic conditions and redox potential, 27
Aerobic respiration, 119
Agriculture, as contaminant source, 10
Air compressor, 183
Air-lift sampler, 243. *See also* gas-lift method
Air lift sampling methods, 226. *See also* gas-lift
    method
Air quality, 258
Air rotary drilling, 97, 182–183
  advantages, 175
  disadvantages, 175
Air stripping, 344, 368
Alconox, 225
Alkalinity of wastes, 26
All-purpose tricone bit, 181
Ammonia, microbial oxidation of, 121
Anaerobic conditions and redox potential, 27
Anaerobic decomposition, in landfills, 135
Anaerobic respiration, 122
  definition, 119
Analytes, 260, 261
Annular space, 181

Aquifer, general characteristics, 199
Aquifer restoration, 263
Aquifer stabilization, 210
Aquifers,
  artesian description, 45
  artesian recharge, 48
  confined cone of depression, 73
  confined description 45, 46
  confined discharge, 71
  confined formula, volume of water released, 56–57
  confined storativity, 55
  confined transmissivity, 65
  definition, 44, 51
  effect on contaminant plumes, 78, 81
  flow diagrams, used in, 60–62
  groundwater occurrence, 45
  measuring water transmitted, 63–66
  monitoring well, 88–89
  multiple, 88, 91
  perched definition and description, 45
  pumping effects, 72–76
  recharge and discharge areas, 47
  rock composition, 50
  semiconfined description, 46
  tests, 98
  unconfined description, 45
  unconfined formula, volume of water released, 56–
    57
  unconfined recharge, 71
  unconfined transmissivity, 65
  unconfined water-level, 53–55
  water-table cone of depression, 73
  water-table flow patterns, 70
  water-table recharge, 48
Artificial filter pack, 204
  average specific gravity, 199
  impurities, 204
  settling, 204
  washed, 205
Artificial gravel pack, 199
Artificial recharge sources, 48–49
Assessment program,
  field work, 97–98
  health and safety plan, 96

Assessment program (*cont.*)
    key steps, 92–93
    quality assurance plan 96, 100
    record search, 93–95
    regulatory agencies, 95
    sampling plan, 96–97
    work plan, 95–97
ASTM specifications, 188
ASTM standards, 188
Atmospheric invasion, 216
Audits, 261, 262
Auger drilling, 177, 192
Auger drill rig, 180

Backwashing with air, 211
Bacterial growth, 252
Bailers, 184, 216, 224, 226, 248
    advantages, 227, 232
    disadvantages, 227, 233
    photo, 232
    precautions, 232
    PVC, 225
    sampling procedures, 231
    Teflon, 225
    well purging, 232
Bailing, 212
Bennett pump. *See* double-acting piston pump
Bentonite,
    advantages, 206
    disadvantages, 206
    grout, 205
    pellets, photos, 207
    seal placement, 206
Biodegradation,
    in situ, 138
    on-site, 137
Biological alterations, 259
Biological oxygen demand: definition, 28
Biological treatment, 364
Bit types, 181
Borehole,
    diameter, 180, 184, 187
    logging, 188
    wall, 180, 181, 184
Boring log, 174
Breakthrough, definition, 157
Bridge-slot screen. *See* manufactured louvre-type
        screen
Bridging, 205, 206, 210
Bucket auger drilling, 177

Cable tool drilling, 184–185, 211
    advantages, 176
    disadvantages, 176
    suitable medium, 185
Calcium hypochlorite, 214
Calibration, standard solutions, 216
Calibration standards, 224
Carbohydrate: microbial reduction of, 124
Carbon adsorption, 371

Carbon dioxide,
    degassing, 226
    microbial reduction of, 124
Casing driver, 185
Casing,
    driving of, 187
    material cost, 196
Casing pipe,
    filter packs, 204
    fittings, 203
    flush threads, 204
    threads and couplings, 204
Caving, 180
Cellulose acetate filters, 250
Cement grout, 205
    advantages, 206
    disadvantages, 206
Centrifugal pump,
    advantages, 228
    diagram, 240
    disadvantages, 228, 238, 235
Chain-of-custody, 258
    example, 257
    form, 256
Chemical additives, 259
Chemical alterations, 233, 235, 259
Chemical analyses, 192
Chemical oxidation, 371
Chemical oxygen demand, definition, 28
Chemical rating curves, 214
Chemical treatment, 365
Chemoautotrophy, definition, 119
Churn method. *See* cable tool drilling
Circulating fluid, 184
CLP, 252. *See also* contract laboratory program
Column length, and dissolution, 147
Cometabolism, 125
    and hazardous waste, 125
Complexes, 19
    organometallic, 20
Compressors. *See* mud pumps
Conductivity, 212, 213, 221, 224, 248, 258
Conductivity meters, 216
Conductivity probe, 213
Consolidated rocks, characteristics (porosity), 50
Contaminant concentration reduction, 170
Contaminant plume,
    case study, 106–108
    sorption, 83
Contaminant sources, 171–172
Contaminants,
    adsorption, 170
    advection, 170
    characteristics, 170
    decay, 170
    densities, 81
    dispersion, 170
    entering pumping well, 74–75
    inorganics, 251
    ion exchange, 170

migration, 207, 263
organics, 251
plume shape and size, 77–83
sources, 94
transportation, 170
types of hazards, 214
volatilization of, 165
volume, 81, 82–83
Contaminated materials, disposal procedures, 215
Contaminated wells, 250
Contamination, 192, 216, 243
Continuous-flow gas-drive sampling device, 242
Contour maps,
  case study, 106
  example and use, 62–63
Contract laboratory program, 252, 256
Control alternatives, 339
Conventional centrifugal pump modificaton, 239
Conventional core barrel samplers, 192
Conventional pump, 237. See also submersible centrif-
  ugal pump
Core barrel samplers, 188, 192–193
Core sampling, 180
Coring method, 192
Coring purpose, 192
Corrective actions, 261
Corrosion, 196
Cross-contamination, 205, 207, 215, 225, 226, 234,
  248, 250
Cuttings, 178, 181, 183, 184
  definition, 193
  purpose, 193
Cylindrical soil samplers, 275
  continuous sample tube systems, 276, 284–286, 293
  peat samplers, 277, 281, 285, 294
  split-barrel drive samplers, 276, 284–286, 293
  thin-walled volumetric samplers, 275–276, 285, 291,
    293

Darcy's Law,
  as a groundwater model, 101
  definition, flow quantity, 40
  formula, 66
Data analysis, 262
Data evaluation, 263
Data sufficiency, 263
Data validity, 263
Decontamination, 289, 300–301
  field decontamination, 300–301
  laboratory cleanup, 300
  procedures, 215, 225, 250
    for drilling and soil sampling equipment, 214
  regulations, 225
  solutions, 214
    acetone, 224
    alconox, 224
    distilled water rinse, 224
    hexane rinse, 224
    methanol rinse, 224
    methylene chloride, 224

purpose, 225
recommended uses, 214, 225
steam cleaning, 224
trisodium phosphate, 224
Deep well injection, as contaminant source, 11
Deep well systems, 364
Deflocculants, 212
Degassing, 216, 232, 234, 235, 236, 238, 239
Deionized water, 215, 225
Denison sampler, 191
Depressurization, 226, 251
Depth-specific sampling device, 233
Derrick, 182
Desorption, 216, 251, 252
  by flushing, 157
Detergent wash,
  alconox, 214
  trisodium phosphate, 214
Detoxification, 343
Dewatered wells, 224
Diamond core conventional barrel sampler, 188
Diffuser pump, 238
Diffusion, through clay barriers, 162
Direct air rotary drilling, 182. See also air rotary
  drilling
Direct circulation rotary drilling. See mud rotary
  drilling
Direct rotary drilling. See mud rotary drilling
Discharge pipe, 212
Discharges, threat ranking of, 37
Dispersion,
  effects, 69–70
  influence on contaminants, 77
  in gases, 17
  in liquids, 17
  in rock formations, 145
  of contaminants, 145
Dissolution, 259
  by microorganisms, 136
  of contaminants, 146
Dissolved oxygen, 213, 224
Dissolved oxygen meters, 216
Distilled water rinse, 214
Document control, 258
Documentation, 252, 256, 258, 262
  blanks, 258
  duplicates, 258
  splits, 258
Double-acting piston pump, 236, 248
  diagram, 239
  photo, 240
Double check-valve bailer, 231
  diagram, 234
Double-tube core barrel sampler, 192
Down-hole air hammer, 183
Downhole pneumatic hammer. See down-hole air
  hammer
Drag bit, 181
Drain, sand or gravel, 352
Drawdown, 199

Drawworks, 182
Drill bit, 184
Drill collar, 181
Drill line, 184
Drill pipe, 181, 183, 184
Drill rods, 181
Drill stabilizer, 181
Drill stem, 178, 184
Drilling case study, 106
Drilling contaminant, lubricating oil, 183
Drilling equipment, 183
Drilling fluid, 184. *See also* drilling mud
Drilling foam, 183
Drilling jars, 184
Drilling methods. *See* well drilling methods
Drilling mud, 181
Drive point, 186
Drive point well. *See* driven wells
Drive samplers,
    piston samplers, 188
    split-spoon samplers, 188, 190
    thin-walled tube samplers, 188
Drive shoe, 183
Drive weight, 186
Driven wells,
    advantages, 176
    diameter, 186
    disadvantages, 176
    suitable material, 186
Dry ice, 212. *See also* solid carbon dioxide
Dual-wall method, 184. *See also* dual-wall reverse circulation rotary drilling
Dual-wall reverse circulation rotary drilling, 184
    advantages, 176
    disadvantages, 176
    figure, 185
Duplicates, 259

Earth resistivity survey, description, 105–106
Effective porosity, measuring saturated flow, 67
Eh, 213
Electrochemical effects on well material, 194
Endogenous phase,
    and waste treatment, 127
    definition, 126
Energy head formula/equation, 57
Engine, 182
Engineering properties, 263
Environmental protection agency (EPA),
    data requirements, 36
    regional offices, 93
Equipment calibration, 261
Equipment cleaning protocol, 214
Evaporation, 251
Exoenzymes, 129
Exponential phase,
    and waste treatment, 127
    definition, 126

Extractability,
    measured by extraction procedure, 30
    of wastes and soils, 29
Extraction procedure, 30, 147
Extraction systems, 348
Extraction wells, 198, 199

Factory slotted perforated pipe, 200
    advantages, 201
    disadvantages, 201
Faulty grouts, 207
Faulty seals, 207
Fermentation, definition, 119
Ferric iron,
    microbial oxidation of, 121
    microbial reduction of, 124
Field data sheets, 217
Field logbook, 217, 219
Field measurements, 256, 258
Field procedures, 216
Field slotted pipe, 200
    advantages, 201
    disadvantages, 201
Filing systems, 258
Filter pack placement, 205
    radial thickness, 205
Filtration, in soils, 150
Fixation, 344
Flame ionization detector (FID), 219
Flow-through box, 213, 224, 248, 250
Flow regimes, 270
    Darcian flow, 270–271
    macro pore flow, 271–272
Fluorescence microscopy, of soils, 116
Forms,
    boring field log, 178
    chain of custody record, 189
    water level measurement sheet, 218
    water quality field sampling data sheet, 217
    well construction summary, 179
    well development data form, 211
Formulas,
    Darcy's law, 66
    energy head, 57
    groundwater velocity, 68
    porosity, 52
    specific return, 52
    specific storage, 55
    specific yield, 52
    transmissivity, 65
    water level decline, 53–54
    water table decline, 54
Fractionation, for analysis of wastes, 30
Fractured media, 170
Freezing, 259
Freundlich isotherm for soils, 161
Freudlich model, definition, 154

Galvanic effects, 194
Gas-drive devices, 235

Gas-drive sampling device, 240–243
  advantages, 241
  diagram, 241
Gas-driven piston pump, sampling advantages, 236, 237
Gas-driven pumps, 226
Gas-driven single-acting piston pump, 236
  diagram, 238
Gas-lift (air lift) samplers, 235
  advantages, 229–230
  disadvantages, 229–230
Gas-lift method, 243
  advantages, 243
  diagram, 244
Gas-operated bladder pump, 235, 243
  advantages, 230, 244–245
  disadvantages, 230, 245
  sampling technique, 245
  schematic, 246
  variations, 244
Gas-operated double-acting piston pump,
  advantages, 229
  disadvantages, 229
Gas drive devices,
  advantages, 229
  disadvantages, 229
Gasification, by microorganisms, 137
Gear-drive electrical submersible pumps, 231
  advantages, 228
  disadvantages, 228
Genetically engineered organisms, 140
Geologic environment, 170
Geologic formations, characteristics, 49
Geophysical investigation, 338
Geotechnical properties, 192
Grab samples, 189
Grain size analysis, 199
Gravel pack, 180
Grease, 252
Gross reactions, of microorganisms, 117
Groundwater, 194
  acid-base reactions and hydrolysis in, 163
  analysis types, 99
  as drinking water, 41
  availability, 41
  complexes in, 164
  contaminants,
    inorganics, 195
    leaching, 195
    organics, 195
    plasticizers, 195
    solvation, 195
  contamination,
    in situ treatment, 365
    treatment, 364
  control, 347
    by removal, 363
    drainage ditches for, 362
    grout curtains for, 358
    sheet piling for, 361

  slurry walls for, 356
  subsurface drains for, 361
corrosion, 194
definition, 40
dilution in, 144
direction, 170
discharge, 170
  determined by Darcy's Law, 66–67
  effect on contaminant plume, 86–87
  location, 49, 71–72
  pumping, 72–75
  sources, 71–72
flow system, 170
in situ treatment of, 139
logging method, 213
model construction, 102–104
  definition, 100
  misuse, 104
  selection, 101–102
  types, 101
  uses, 105
monitoring network installation, 171
monitoring of contaminants in, 145
monitoring plan, 248
monitoring program case study, 108–114
  data management, 99–100
  design, 170, 264
  elements, 98–99
  investigation, 173
  laboratories, 99, 100
  reconnaissance drilling, 171
  well construction and drilling, 171
monitoring wells,
  diagram, 172
  installation, 171
  location in fractured rock, 91
  location with multiphase contaminants, 90–91
  location with pumping wells, 90
  purpose, 171
occurrence in aquifers, 45
parameters, 213
preservatives,
  nitric acid, 250
  sulfuric acid, 250
  table, 252
quality, 226
  monitoring, 213
  monitoring devices, 226
quality sampling. See groundwater, sampling
recharge, 170
  affecting aquifers, 76
  case study, 106–107
  effect on contaminant plume, 84–85
  example of area of recharge, 63
  location, 71–72
  sources, 70–72
redox reactions in, 164
sample,
  collection sequence, 250
  preservation, 259

Groundwater (*cont.*)
  preservation methods, 250
  preservation precautions, 260
  preservation purpose, 259
  preservation techniques, 259
  pretreatment, 173
  shipping, 259
  storage, 250, 259
  sampling,
    analyte preservation techniques, 252–253
    bottles, 250
    case study, 106
    chemical alteration, 226
    containers, 252–255
    device selection, 226
    objective, 226
    physical alteration, 226
    procedures, 250
    storage time, 252–255
    volume, 252–255
  sampling devices,
    characteristics, 249
    grab devices, 231
    labels, 250
    limitation, 231
    maintenance, 231
    selection of, 248
  sampling methods, 226, 251
    adsorption, 173
    affecting factors, 216
    atmospheric invasion, 173
    degassing, 173
    desorption, 173
    determining types, 173
    sampling error, 173
    volatilization, 173
  sources, 4
  technology, 169
  threat of contamination judged by EPA, 34
  treatment, 364
  velocity, 170
    estimated by Darcy's Law, 67–70
    formulas, 68
Groundwater divide definition, 63
Groundwater flow case study, 110–112
  changes due to pumping, 72–76
  characteristics, 169
  figure, 222
  in fractured rock system, 91
  lines, 58–60, 221
  nets, 60
  patterns, 41, 49
  preparation of diagrams, 60
  rate, 41
  shown by contour map, 62–63
Grout, 180
Grout curtains,
  applications, 360
  techniques, equipment, and types, 359

Grouting, 180
  materials, 205

Hand-operated drilling devices, 278
  barrel augers, 278, 285, 287, 295–297
  Dutch-type augers, 280, 285
  mud augers, 281, 285
  post-hole augers, 280, 285, 295
  regular or general barrel auger, 280–281, 285
  sand auger, 281, 285
  screw-type augers, 278, 285
  tube-type samplers, 281–283, 285, 297–298
Hand-operated equipment, 272, 285–287, 289, 295
  barrel augers, 278, 285, 287, 295–297
  sample collection, 285, 288, 295, 301
  screw-type augers, 278, 285, 295
  tube-type samplers, 281–283, 285, 297–298
  Veihmeyer Tube, 283, 285, 287, 299, 300
Hand-vacuum pumps, 235
Hand augers, 177
Hazardous, classification as, 33
Hazardous waste sources, 263
Heavy metals, 173
Helical rotor submersible pump,
  advantages, 228–229
  disadvantages, 228–229
Hexane, 214
  rinse, 225
Holding times, 260
Hollow-stem auger drilling, 97, 177, 181, 192
  advantages, 174
  disadvantages, 174
Hollow bit, 180
Homogenous, definition, 51
Hydraulic conductivity, 170, 192, 216
  controlling leakage rates, 71
  Darcy's Law, 66
  definition, 40
  description, 64
  determine effects of pumping, 75
  effect on contaminant movement, 78
  use, 63–64
Hydraulic gradient,
  Darcy's Law, 66
  definition, 40–41, 47, 58
  determined by water-table map, 63
  effects of pumping, 75
Hydraulic pressure, 191
Hydraulic rotary drilling. *See* mud rotary drilling
Hydrochloric acid, 212
Hydrofluoric acid, 212
Hydrogen,
  microbial oxidation of, 120
  microbial reduction of, 124
Hydrogeologic characterization, 173
Hydrogeologic cycle, definition of process, 47
Hydrogeologic investigation, 338
Hydrogeologic units, 170
  characteristics, 170, 171

Hydrostatic head due to recharging water, 47
Hydrostatic pressure, 181

Igneous rocks,
    aquifers composed of, 50
    description, 43
Impermeable subsurface barriers, 348
Industrial waste, 259
Inert plastic gloves, 215
Infiltration,
    control, 348
    description, 47
In-line measuring chamber. See flow-through box
Inorganic, 196
Inorganic compounds, 194
In situ treatment, 343, 365
    biological, 366
Installation procedures for pore-liquid samplers, 318
    access line installation, 318
    bentonite clay method, 320
    hole construction, 319
    sampler installation, 319
Instrument calibration, 219, 261, 262
    field instruments, 224
Intermittent pumping, 210
Inventory tracking, 258
Inverse drilling, 184
Ion exchange, 370
Ion probes, 213
Iron, 251

Jet-drive point, figure, 187
Jet percussion drilling, 187
    advantages, 177
    disadvantages, 177
Jet pumps, 245–247
    advantages, 230, 246
    disadvantages, 230, 247
    schematic, 246
Jetted wells, 186–187
    advantages, 177
    disadvantages, 177
    suitable material, 187
Jetting, 210
Joints. See pipe fittings, flush thread

Karst terrains, 145
Kelly, 181
Kelly bar, 177
King tubes, 188
Kynar, 194, 196
    advantages, 195, 197
    disadvantages, 195, 197

Labeling, 258
Laboratory,
    analysis, 224
    cleaning certification, 226
    sampling container preparation, 226
Lag phase, and waste treatment, 126
    definition, 126

Landfills,
    codisposal in, 143
    description, 8
    dilution in, 144
    irrigation of, 355
    microbiology of, 134
Langmuir model, definition, 154
Leach, heavy metals, 196
Leachate,
    composition, 35
    containment, 355–361
    drains, 348
    sources, 6
Leachate treatment, 369
    anaerobic, 370
    biological, 369
    physical and chemical, 370
    systems, 371
Legal evidence, 258
Lift pump, 212
Liner,
    landfill, 352
    sprayed, 355
    synthetic, 355
Litigation, 256, 260
Log death phase, definition, 126
Log phase, definition, 126
Lubricants, 196
Lysimeter, 324
    dead space, 326–327
    failure confirmation, 330–331
    porous segments, 324–325
    special problems, 328–330

Major ions,
    calcium, 194
    chlorine, 194
    magnesium, 194
    sulfate, 194
Manufactured louvre-type screen, 200
    advantages, 201
    description, 202
    disadvantages, 201
Mast, 182, 184
Mechanical vibrator, 186
Membranes, 354
Metals, 195, 243, 240, 251, 252, 259
    filtration, 251
Metamorphic rock,
    aquifers composed of, 50
    description, 43
Methane:
    in landfills, 18, 135
    microbial oxidation of, 120
    transport of contaminants by, 18
Methanol rinse, 214, 225
Methylene chloride, 214
Microbial decomposition, products of, 136
Microbial degradation, in situ, 344
Microbiological contents of wastes, 33

Microbiological decomposition, promotion of, 137
Microcellulose acetate filters, 250
Microorganism communities, and biodegradation, 131
Microorganisms,
   and in situ treatment, 138
   inhibition of, 128
   in soils, 132
Mild steel, 196
   advantages, 195, 197
   disadvantages, 195, 197
Mineral nutrients and microbial growth, 130
Monitoring program. *See* Groundwater, monitoring
   program case study
Monitoring well, 210
   above- and below-grade wells, 208
   capping, 208
   casing and screen materials, 196–197
   casings material selection, 194
   Christy-type box, 208
   completion, 207
   construction material's groundwater corrosiveness,
     195
   construction material's stress, 195
   construction material's temperature, 195
   corrosion, 210
   depth and diameter, 194
   development objectives, 213
   diagram, 209
   encrustation, 210
   identification label, 209
   labeling, 208
   leaks, 207
   measurement point, 208, 209
   sand pumping, 210
   screens material selection, 194
   security casing, 208
   security devices, 208
   surface protection, 208
   ventilation, 209
Motor, 184
Mud pit, 181
Mud pumps, 183, 184
Mud rotary drilling, 97, 180
   advantages, 175
   disadvantages, 175
   figure, 182
   suitable medium, 181
Multi-purpose drill rigs, 285, 289
   auger drills, 272, 285
   continuous flight-auger drilling and sampling, 273–
     274, 276–277, 285, 290
   hollow-stem auger drilling and sampling, 272–273,
     285
   multi-purpose auger core rotary drill rigs, 272, 285
   sample collection, 285, 288–289
Multiline hoist, 184

National Enforcement Investigations Center Policies
   and Procedures Manual, 258
Natural filter, 204

Neat cement, 206
Negative (suction) displacement methods, 234–235
   centrifugal, 226
   jet pumps, 226
   peristaltic, 226
   suction-lift pumps, 226
Neutralization, 343
   and precipitation, 149
Nitrate, microbial reduction of, 123
Nitric acid, 216, 251
Nitrogen gas,
   microbial oxidation of, 122
   microbial reduction of, 124
Nonplumb wells, 235
Nonvolatile organics, 250, 251, 252
Nylon rope, 231

ODEX bit, 183
Organic chemicals, adsorption on soils, 161
Organic chemicals,
   analysis of, 23
   definition, 22
   gas chromatography/mass spectrometry, 26
   liquid chromatography, 25
   vapor phase analysis of, 23
Organic compounds, 194–195
Organic vapor analyzers, 192
Organics, 194, 196, 252, 258
Oxidation, 196, 251
Oxidizing conditions, 259
Oxygen,
   and in situ treatment, 138
   and microbial growth, 130

Pan lysimeter, 331
   free drainage glass block samplers, 333
   limitation, 334
   trench lysimeters, 332
Partition coefficient,
   and soil characteristics, 160
   definition, 156
Percussion drilling. *See* cable tool drilling
Peristaltic pumps, 235
   advantages, 228
   disadvantages, 228
Permanent files, 258
Permeability, 204, 212
Permeable treatment bed, 348, 365
pH: 212, 213, 221, 224, 226, 248, 258, 259
   and dissolution, 148
   indicator paper, 216
   in landfills, 135
   meters, 216
   of wastes, 26
Photochemical degradation, 165
Photoionization detector (PID), 219, 224
Physical treatment, 365
Pilot borehole, 199
Pipe and screen casing fitting types, description, 203
Pipe-base screen, figure, 202

Pipe fittings,
    flush thread, 203
    plain square ends, 203
    threads and couplings, 203
Piston pumps, 235, 236
Piston sampler, 191
Pitcher sampler, 191
Plan,
    air monitoring, 347
    excavation and disposal, 346
    health and safety, 347
Plug, 178
Plumes, contaminant, 170
Pneumatic hammer, 186
Pneumatic sampling device, 233
Point source bailer. See double check-valve bailer
Polymeric fluids, 206
Polyphosphates, 212
Polypropylene, 194, 196, 226, 231
    advantages, 195, 196–197
    disadvantages, 195, 196
Polyvinyl chloride (PVC), 194, 196, 226, 231
    advantages, 195, 196
    disadvantages, 195, 196
Pore-liquid sampler operation, 323–324
Pore-liquid sampling equipment, 308
    ceramic-type samplers, 308–310
    hollow-fiber samplers, 313–314
    membrane filter samplers, 308, 314
    pan lysimeters, 308, 315, 329, 331, 333–334
    sample location, 317
Pore space, 173
Porosity, 199, 204
    and dissolution, 148
    definition, 40
    formulas, 52
    of wastes and soils, 16
    primary definition and description, 44
    representative ranges, 53
    secondary definition and description, 44
Porous media, 170
Porous unconsolidated formations, 183
Positive displacement methods, 235–236
    bladder pumps, 226
    electrical pumps, 226
    gas-drive devices, 226
    gear-drive, 226
    helical-rotor, 226
    piston pumps, 226
    sampling advantages, 235
Potentiometric maps, 219
Potentiometric surface,
    definition, 46
    in unconfined aquifer, 47
Potentiometric tables, 219
Power-driven earth augers,
    bucket augers, 177, 180
    hollow-stem augers, 177
    solid-stem augers, 177

Precipitation,
    and coagulation, 370
    and groundwater and soil quality, 150
    by microorganisms, 137
Preservatives, 226, 235
Procedure modifications, 216
Production wells, 198, 199
Pumping, 212
    and surging. See intermittent pumping
    effects on aquifers, 72–76
    rate, 224
    well,
        case study, 110–114
        effect on assessment program, 95
Pumps, 182, 216, 224
Purged water storage, 216
Purging, 224
Pushed tube, 191
PVC. See polyvinyl chloride

QA. See quality assurance
QC. See quality control
Quality assurance, 260, 262
    site characterization, 261
Quality control, 260, 262
    field, 261
    laboratory, 261

Radial spray wand, 215
Rawhiding. See intermittent pumping
Reamer, 181
Receptors, 263
Recordkeeping, 252
Redox potential,
    and dissolution, 148
    definition, 27
Redox reactions of microorganisms, 118
Reducing conditions, 259
Refrigeration, 259
Regional analytical quality control coordinators, 260
Regulatory agencies, 225
    requirements and recommendations, 251
Remedial action, 173
Respiration, definition, 119
Respirators, 219
Retardation, and fluid composition, 161
Retardation factor, 158
    and attenuation, 158
    definition, 157
    derivation, 158
    for chlorinated hydrocarbons, 162
    formula for, 159
Reverse circulation rotary drilling, 183–184
    equipment, 184
    suitable medium, 184
Reverse osmosis, 371
Rod pump, 236
    diagram, 237
Roller cone bit, 181
Rotary-percussion drilling. See down-hole air hammer

Rotary-type drill rig, 192
Rotary bucket drilling. *See* bucket auger drilling
Rotary drilling. *See* mud rotary drilling
Runoff, as a contaminant source, 11

Safety,
    equipment, 219
    precautions, 301–302
    precautions for material and tool handling, 215
    requirements, 216
Sample,
    activities, 288–289
    collection, 285, 288
    containers, 216
    contamination, 236
    identification, 258
    label, 256, 260
    possession, 258
    transfer, 234
Sampler preparation, 317
Samplers, 285, 291
    continuous sample tube systems, 284–286, 293
    peat samplers, 277, 281, 285, 294
    piston samplers, 284–285, 292
    split-barrel drive samplers, 276, 284–286, 293
    thin-walled volumetric tube samplers, 275–276, 285,
        291, 293
Sampling,
    containers, 226
    plan, 252
    quality assurance, 261
    tube, 181
Sand, 180
    migration, 210
    pump, 184
Saturated zone,
    characteristics, 170
    definition, 2
    description, 42–43
Screened pipe,
    filter packs, 204
    fittings, 203
    flush threads, 204
    threads and couplings, 204
Seal, 180
Sedimentary rock,
    aquifers composed of, 50
    description, 43
Selecting soil samplers, 283, 285
    compositing, 287–288
    for sample size, 286
    for site accessibility and traffic ability, 286
    for various soil types, 284–286
    obtaining various sample types, 283–285
    personnel, 287
Selenium, adsorption, 163
Semivolatile organics, 251
Septic tanks, as leachate sources, 10
Settling basin. *See* mud pit
Shallow-depth pumps, 235

Shelby tubes, 191
Shipment, 260
Sieve analysis, 205
Single-tube core barrel sampler, 192
Single check-valve bailer, 231
    diagram, 233
Single source bailer. *See* double check-valve bailer
Site, control, 347
Site assessment, by EPA, 12
Slot size, 199
Sloughing, 180
Slurry walls,
    applications, 358
    definition, 356
    function, 357
    history, 357
    preparation, 357
    types, 358
Sodium bicarbonate, 214
Sodium carbonate, 214
Soil,
    and dissolution, 149
    and microbial growth, 128
    biofilms in, 133
    flushing, 343
    formation of, 133
    for surface sealing, 353
    moisture, 192
    moisture relationships, 309
    permeability, and dissolution, 147
    pore-liquid monitoring, 306
    probes, 189
    samplers,
        drive sampler, 189, 190
        hand-operated, 189
        soil probes, 189
        soil tubes, 189
        Veihmeyer tubes, 189
    sampling,
        below water table, 190, 193
        decontamination, 190
        description, 188
        disturbed, 180, 188
        equipment, 188
        equipment decontamination, 188
        handling techniques, 191
        intervals, 190
        labeling, 189, 191
        logging, 188
        preservation, 190
        purpose, 188, 192
        regulations, 192
        shipping, 189, 192
        soil tubes, 188
        storage, 189, 190, 191
        techniques, 187–189
        transportation, 191
        undisturbed, 180, 189, 192
Solid carbon dioxide, 212
Solid-stem auger drilling, 177

Solids,
   adsorption on, 16
   shape, 15
Soluble constituents, 259
Solution mining, 343
Solvent cement, plain square ends fittings, 203
Solvents, and adsorption, 162
Sorbent addition, 344
Sorption,
   effect on contaminant plume, 83
   influence on contaminants, 77
Sorptive/desorptive properties, 194
Specific return formulas, 52
Specific storage formula, 55
Specific yield,
   formulas, 52
   measuring saturated flow, 67
   representative ranges, 53
Spiral flights, 178
Split-barrel samplers. *See* split-spoon samplers
Split-spoon samplers, 190–191
   best suited medium, 190
   blow count, 190
   cohensionless soil samples, 190
   retainers, 190
   ring-lined barrel sampler, 190
   sand traps, 190
   thin-walled sleeves, 190
   thin-walled split shell, 190
   very soft soils, 190
Split-tube samplers. *See* split-spoon samplers
Stainless steel, 196, 231
   advantages, 195, 197
   disadvantages, 195, 197
Standard drilling method. *See* cable tool drilling
Standing water, 235
Steam cleaning, 214
Steel casing, 183
Storage, 260
Storage coefficient,
   determine effects of pumping, 75
   determining response to pumping, 55
Storage drums, 216
   temporary storage areas, 214
Storage,
   incompatible materials, 215
   labels, 215
   purged water, 225
   waste water, 224, 225
Submersible centrifugal pumps, 237–239
Submersible piston pump, 235, 236
Submersible pump, 212, 245, 247
Substrate,
   and microbial growth, 129
   toxicity, 129
Subsurface materials, 187
Suction-lift limit, 238
Suction-lift mechanisms,
   advantages, 227–228
   disadvantages, 227–228

Suction-lift methods, 232, 234–235
   sampling advantages, 235
   sampling limitations, 235
Suction pump, 245
Sulfate, microbial reduction of, 124
Sulfide, microbial oxidation of, 120
Surface impoundments,
   description, 7
   liners, 7
Surface probes, 189
Surge block with air, 211–212
Surging, 210
Swivel socket, 184
Synthetic reactions, of microorganisms, 117
Syringe devices,
   advantages, 227
   disadvantages, 227
Syringes, 226, 233, 235

Table-driven machines, 181
Table. *See* Top head drive
Teflon, 194, 196, 226, 231
   advantages, 195, 197
   coated wire, 231
   disadvantages, 195, 197
Temperature, 213, 221, 224, 248, 258
   and dissolution, 147
   and precipitation, 149
   probes, 216
Temporary casing, 180
Thermal equipment, 214
Thermometers, 216
Thin-walled tube sampler,
   methods, 191
   parts, 191
   types, 191
Top head drive, 181, 182, 184
Total head, definition, 57
Total organic carbon, definition, 28
Trace metals, 233, 194, 235
   adsorption on soils, 163
   analysis of, 21
   definition, 21
   toxicity of, 21
Trace organic constituents, 233
Traffic reports,
   high-hazard, 256
   inorganic, 256
   organic, 256
Training sessions, 216
Transmissivity,
   formula, 65
   use, 65
Transporting devices, sampling equipment, 247–248
Treatment, on-site or off-site, 368
Tremie pipe, 207
   bridging, 205
   used with cement grout, 208
Truck-mounted generator, 248

Tube-type samplers, 281–283, 285
    hand-held power augers, 283, 285
    soil sampling tubes, 281, 285
    Veihmeyer tube, 283, 285, 287, 299–300
Tube samplers, 191
Turbine pump, 238
Turbulence, 226, 238, 246, 251

UD (undisturbed) tubes, 191
Unconsolidated deposits, porosity, 52
Unconsolidated formations, characteristics (porosity),
    51
Under-reamer, 181
Underground storage tanks,
    detection and monitoring, 340
    leakage from, 5, 9
    numbers, 6
Unified Soil Classification System (USCS), 188
Unsaturated zone (vadose zone),
    capillary fringe, 43
    characteristics, 170
    geologic materials, 42
    intermediate zone, 43
    soil zone, 43
USEPA, 225
    procedures, 260

Vadose zone,
    capillary fringe, 269–270
    definition, 2
    description, 267–272
    intermediate unsaturated zone, 268–269
    soil zone, 267–268
    unsaturated zone, 267–272
    zone of aeration, 267–268
Veihmeyer tubes, 188
Velocity stagnation point, 72
Vials, 251
Volatile-gas stripping, 226
Volatile organics, 235, 242–244, 250–252
Volatilization, 165, 216, 232
    and water table, 166
Volute pump, 238, 240

Walking beam, 184
Wash boring. See jetted wells
Wash samples,
    definition, 193
    purpose, 193
Waste,
    disposal facilities,
        assessment of groundwater quality, 92
        discharged materials, 92
    mixing, and precipitation, 149
    modification, 340
    water,
        disposal, 251
        storage, 250
        storage drums, 225–226

Wastes,
    chemical treatment, 342
    cleanup plans, 345
    codisposal of, 143
    dewatering, 342
    dilution of, 142
    elemental composition of, 18
    excavation, 345
    gases, 17
    liquids, 17
    mixing with soil, 143
    physical treatment, 342
    pretreatment, 341
    solidification, 342
    speciation of, 19
    thermal treatment, 341–342
Water,
    and sediment studies, 338
    diversion, 352
    flow, 258
    level decline formulas, 53–54
    production, 210
    quality, 216
        sampling, 215
    table, 222
        decline formula, 54
    treatment systems, 348
Water-level measurement,
    depth to water, 258
    electric sounder, 220
    form, 219
    indicator probe, 219
    measuring point, 219
    measuring tape, 219
    procedure, 219
    purpose, 219
Water-table maps example, 62
Waterloo cohesionless aquifer core barrel sampler, 193
Welding plain square ends fittings, 203
Well,
    casing, 184
    conditions, 220
    construction Summary Form, 174
    depth, 223, 258
    development,
        backwashing with air, 212
        chemicals, 212
        contaminated groundwater, 210
        definition, 210
        form, 214
        hydraulic jetting, 212
        methods for small diameter wells, 212
        surge block, 210
    diameter, 198
    drilling methods,
        air rotary, 174
        auger, 174
        cable tool, 174
        driven wells, 174
        dual-wall reverse circulation rotary, 174

efficiency, 183
jet percussion, 174
jetted wells, 174
medium, 183
mud rotary, 174
performance of, 183
problems, 183
reverse circulation rotary drilling, 174
drilling process, 173–174
efficiency, 199
evacuation. *See* well purging
point systems, 363
pump,
gasoline motor, 239
priming of, 238
pumping rates, 231, 235, 238
purging, 216, 231
bailer, 232
estimate of well water volume, 219, 221
procedure, 223
protocol, 221
pump placement methods, 222, 223
screen,
costs, 199

lengths, 199
purpose, 198
slot size, 199
strength, 199–200
types, 198, 200
water,
casing volume, 224
measurement level, 223
sampling equipment, 216
yield, 198
Wire-wound continuous-slot screen, 200
advantages, 201
description, 202–203
disadvantages, 201
Wire-wound perforated pipe, 200
advantages, 201
description, 202
disadvantages, 201
Wire line core barrel sampler, 188, 192
Wire line piston core sampler,
methods, 193
purpose, 193

Z tubes, 191
Zobell solution, 214